SPRINGER TRACTS IN MODERN PHYSICS

Ergebnisse der exakten Naturwissenschaften

Volume 57

Editor: G. Höhler

Springer-Verlag Berlin Heidelberg GmbH 1971

Manuscripts for publication should be addressed to:

G. Höhler, Institut für Theoretische Kernphysik der Universität, 75 Karlsruhe 1, Postfach 6380

Proofs and all correspondence concerning papers in the process of publication should be addressed to:

E. A. Niekisch, Kernforschungsanlage Jülich, Institut für Technische Physik, 517 Jülich, Postfach 365

ISBN 978-3-662-15592-9 ISBN 978-3-540-36401-6 (eBook)
DOI 10.1007/978-3-540-36401-6

Originally published by Springer-Verlag Berlin Heidelberg New York in 1971
Softcover reprint of the hardcover 1st edition 1971

Library of Congress Catalog Card Number 25-9130.

Strong Interaction Physics

Heidelberg-Karlsruhe International Summer Institute in Theoretical Physics (1970)

Contents

Some Consequences of Unitarity and Crossing Existence and Asymptotic Theorems

D. ATKINSON

Contents

1. Introduction

In these lectures, I want to concentrate on two aspects of S-matrix theory. Wanders has already explained some proven properties of the S-matrix in axiomatic field theory. My first task will be to demonstrate that there exist functions which satisfy these properties. I will show that there exists a large infinity of functions satisfying analyticity, crossing and unitarity, corresponding both to a freedom in choosing an inelasticity function, and to the celebrated CDD ambiguity. Wanders will be explaining how some of this ambiguity can be reduced by physically motivated parametrizations of the low-energy amplitude.

The second aspect of the theory I will consider concerns the high-energy bounds that any such amplitude must satisfy. I will first prove the Froissart bound [1], and the fact that any s-channel dispersion relation only needs two subtractions below the t-channel threshold. Then I will describe some very recent developments by *Yndurain* and *Common* [2], in which a rigorous inequality on cross-sections is derived that works at finite energies, unlike the Froissart bound, which is only asymptotic. Lastly, I will mention certain aspects of the Pomeranchuk theorem [3], and distinguish it carefully from the Froissart bound. We will see that the Pomeranchuk theorem does not follow from the axioms of field theory alone, and that, if the Serpukhov data [4] are correct, and the Pomeranchuk theorem does not hold in some cases, there are a number of escape routes available to the agile theoretician.

Let me now go back to my first topic, the task of proving the existence of crossing-symmetric, unitary functions. Before we become engrossed in the equations, I want to make some general remarks. I am going to demonstrate the existence of functions, candidate pion amplitudes, $F(s, t)$, that satisfy crossing symmetry, elastic unitarity up to the four-pion threshold, and the inelastic unitarity inequalities above it, and that obey the Mandelstam representation. As you are probably aware, it is not possible to prove the Mandelstam representation from axiomatic field theory. In fact, *Martin* [5] has produced counter examples that have the analyticity required by axiomatic field theory, but which have non-Mandelstam singularities. However, any amplitude that satisfies the Mandelstam representation, and the other properties I enumerated, also satisfies the requirements of axiomatic field theory.

First of all, I am going to give the proof in the most trivial case, which is that in which the Mandelstam representation has no subtractions, and where the isospin complications are ignored. This will serve to explain the technique of the proof to you without unnecessary complications. Then I will sketch the proof in its most complete form, where we can have subtractions, with full isospin, and with CDD poles. This matter of CDD poles deserves some comment for physical reasons, and also in order to dispel some ideas some of you might have that CDD poles are undesirable.

It has been known for a long time that the one-dimensional partial-wave N/D equations, with given left-hand cuts, possess the CDD ambiguity. It has been known for several years that, when there are strong forces in a channel that is coupled to the system of interest (in this case the two pion system), but which is not included explicitly in the calculation (except for its contribution to the elasticity factor), then the relevant solution of the partial-wave dispersion relation may require CDD poles. We might call such a CDD pole an "inelastic CDD pole", to distinguish it from an "elementary particle" CDD pole. Even if one wants to rule out the latter kind of CDD pole by an ad hoc assumption, one would not want to exclude inelastic CDD poles. As we shall see, the CDD ambiguity is still there in the exact equations.

2. Existence Proof with No Subtractions

We are going to construct, first of all, an amplitude, $F(s, t)$, that satisfies an unsubtracted Mandelstam representation with the simple crossing symmetry of neutral pions [6]. That is, all three double spectral functions, $\varrho(x, y)$, are the same and

$$\varrho(x, y) = \varrho(y, x). \tag{2.1}$$

Mandelstam [7] showed that the part of the double spectral function determined by elastic unitarity is

$$\varrho^{\mathrm{el}}(s,t)=\iint_4^\infty \mathrm{d}t_1\mathrm{d}t_2 K(s;t,t_1,t_2)\,d^*(s,t_1)\,d(s,t_2)\,, \tag{2.2}$$

where the "unitarity kernel" is

$$K(s;t,t_1,t_2)$$
$$=\frac{4}{\pi}\theta[t_-(s;t,t_1)-t_2]\,\{s(s-4)\,[t_+(s;t,t_1)-t_2]\,[t_-(s;t,t_1)-t_2]\}^{-1/2} \tag{2.3}$$

with

$$t_\pm(s;t,t_1)=t+t_1+\frac{2tt_1}{s-4}\pm 2\left\{tt_1\left(1+\frac{t}{s-4}\right)\left(1+\frac{t_1}{s-4}\right)\right\}^{1/2}, \tag{2.4}$$

and where $d(s,t)$, the t-discontinuity of the complete amplitude, is given by

$$d(s,t)=\frac{1}{\pi}\int_4^\infty \mathrm{d}s'\left[\frac{1}{s'-s}+\frac{1}{s'-u}\right]\varrho(s',t)\,, \tag{2.5}$$

with $u=4-s-t$.

The expression (2.2) is only constrained to be equal to the complete double spectral-function, $\varrho(s,t)$, in the elastic region, $4\leqq s\leqq 16$. On the other hand, a detailed examination of the domain of integration in (2.2) shows that

$$\varrho^{\mathrm{el}}(s,t)\equiv 0\,, \tag{2.6}$$

for $t<16$ and all s. This leads us to the Ansatz

$$\varrho(s,t)=\varrho^{\mathrm{el}}(s,t)+\varrho^{\mathrm{el}}(t,s)+v(s,t)+v(t,s)\,, \tag{2.7}$$

where $v(s,t)$ can be any function that vanishes when $s\leqq 16$ and $t\leqq 16$. For then we satisfy both the symmetry (2.1), and the requirement that $\varrho=\varrho^{\mathrm{el}}$ for $s\leqq 16$, since the last three terms in (2.7) vanish for $s\leqq 16$.

The Eqs. (2.2), (2.5), and (2.7) can be regarded as a nonlinear system that, for a given $v(s,t)$, determines $\varrho^{\mathrm{el}}(s,t)$ in terms of itself:

$$\varrho^{\mathrm{el}}(s,t)=M[\varrho^{\mathrm{el}};s,t]\,. \tag{2.8}$$

I will show that, under certain restrictions on v, we can find a solution of this nonlinear equation. This is done essentially by showing that, for suitable v, the iteration

$$\varrho^{\mathrm{el}}_{n+1}=M[\varrho^{\mathrm{el}}_n]\,, \tag{2.9}$$

starting from $\varrho^{\mathrm{el}}_0(s,t)\equiv 0$, converges to a limit in a complete function space.

One first has to define a complete, normed, linear space of functions that have sufficiently nice properties to ensure that all the integrals, in particular the singular integral in Eq. (2.5), are well-defined. One restricts attention to the class of functions that are Hölder-continuous. A function, $f(x)$, is Hölder-continuous with index $\mu > 0$, if the difference $|f(x+\theta) - f(x)|$ goes to zero at least as quickly as θ^μ as $\theta \to 0$.

In our case we need Hölder-continuity with respect both to s and to t, and it turns out to be satisfactory to define the space of all functions, $F(s,t)$, for which

$$\|F\| = \sup \frac{|F(s_1,t_1) - F(s_2,t_2)|\,[\log\bar{s}\log\bar{t}]^{1+\varepsilon}}{\left|\dfrac{s_1 - s_2}{s_1 s_2 \bar{t}}\right|^\mu + \left|\dfrac{t_1 - t_2}{t_1 t_2 \bar{s}}\right|^\mu} \tag{2.10}$$

and

$$F(s,\infty) = 0 = F(\infty, t)\,. \tag{2.11}$$

Here $\varepsilon > 0$, $0 < \mu < \frac{1}{2}$, and $\bar{s} = \min(s_1, s_2)$, $\bar{t} = \min(t_1, t_2)$. The supremum is taken over the range $[4, \infty)$ for all variables.

One can manage to prove, from Eqs. (2.2)–(2.7) that, if v and ϱ_n^{el} both belong to the space defined by Eqs. (2.10)–(2.11), in particular if

$$\|v\| \leqq B\,, \tag{2.12}$$

and

$$\|\varrho_n^{\mathrm{el}}\| \leqq b\,, \tag{2.13}$$

then

$$\|\varrho_{n+1}^{\mathrm{el}}\| \leqq \Gamma(b+B)^2\,, \tag{2.14}$$

where Γ is a constant that depends only on the parameters μ and ε. If

$$B < (4\Gamma)^{-1}\,, \tag{2.15}$$

and

$$b_- \leqq b \leqq b_+\,, \tag{2.16}$$

where

$$b_\pm = \frac{1 - 2B\Gamma \pm [1 - 4B\Gamma]^{1/2}}{2\Gamma}\,, \tag{2.17}$$

you can easily check that the right-hand side of (2.14) is less than b. In other words, if $v(s,t)$ is such that

$$\|v\| < (4\Gamma)^{-1}\,, \tag{2.18}$$

and b satisfies (2.16), then we have proved, by induction, that every iterate, ϱ_n^{el}, lies in the ball (2.13).

To show that the iterates actually converge, consider

$$\varrho_{n+2}^{\text{el}} - \varrho_{n+1}^{\text{el}} = M[\varrho_{n+1}^{\text{el}}] - M[\varrho_n^{\text{el}}] . \tag{2.19}$$

One can show, by another detailed examination of Eqs (2.2)–(2.7), that

$$\|\varrho_{n+2}^{\text{el}} - \varrho_{n+1}^{\text{el}}\| \leqq 2\Gamma(b+B) \, \|\varrho_{n+1}^{\text{el}} - \varrho_n^{\text{el}}\| . \tag{2.20}$$

By iteration,

$$\|\varrho_{n+2}^{\text{el}} - \varrho_{n+1}^{\text{el}}\| \leqq \{2\Gamma(b+B)\}^{n+1} \, \|\varrho_1^{\text{el}}\| , \tag{2.21}$$

since $\varrho_0^{\text{el}} \equiv 0$. Hence, for any $m > n$

$$\begin{aligned} \|\varrho_m^{\text{el}} - \varrho_n^{\text{el}}\| &\leqq \sum_{r=n}^{m-1} \|\varrho_{r+1}^{\text{el}} - \varrho_r^{\text{el}}\| \\ &\leqq \frac{x^n - x^m}{1-x} \, \|\varrho_1^{\text{el}}\| , \end{aligned} \tag{2.22}$$

where $x = 2\Gamma(b+B)$. If $x < 1$, the left-hand side here tends to zero as $m, n \to \infty$, which shows immediately that $\{\varrho_n^{\text{el}}\}$ satisfies Cauchy's criterion of convergence. The space defined by Eqs. (2.10) and (2.11) is complete, so there is a limit,

$$\varrho_m^{\text{el}}(s,t) \xrightarrow[m \to \infty]{} \varrho^{\text{el}}(s,t) . \tag{2.23}$$

The final restriction on b is

$$b_- \leqq b < \frac{1}{2\Gamma} - B , \tag{2.24}$$

which is a sufficient condition for the limit (2.23) to exist. Moreover, one can prove rigorously that this limit function really does satisfy the Eqs. (2.2)–(2.7), and that it is the only solution in the ball (2.13). It is not difficult to see also that if one changes the "generating function", $v(s,t)$, in Eq. (2.7), then one necessarily changes the solution. I refer you to the published papers [6] for a proof of these statements, and for a more sophisticated statement of the Contraction Mapping Principle, which is one of the names of the Theorem I have just used.

The double spectral function, $\varrho(s,t)$, is such that a Mandelstam representation with no subtractions exists, and that crossing symmetry and elastic unitarity are satisfied. We have now to show that the inelastic inequalities for $s > 16$ can be met. The imaginary part of the partial-wave projection of the amplitude has the Froissart-Gribov representation

$$\operatorname{Im} F_l(s) = \frac{4}{\pi(s-4)} \int_4^\infty dt \, \varrho(s,t) \, Q_l\left(1 + \frac{2t}{s-4}\right) . \tag{2.25}$$

The elastic part of this comes from the first term in Eq. (2.7) and the inelastic remainder may be written

$$\operatorname{Im} F_l(s) - \left(\frac{s-4}{s}\right)^{1/2} |F_l(s)|^2 \\ = \frac{4}{\pi(s-4)} \int_4^\infty dt [\varrho^{el}(t,s) + v(s,t) + v(t,s)]\, Q_l\left(1 + \frac{2t}{s-4}\right). \tag{2.26}$$

It will be shown that one can arrange

$$\varrho^{el}(t,s) + v(s,t) + v(t,s) \geqq 0 \tag{2.27}$$

everywhere. Then, since the Legendre function in Eq. (2.26) is positive, it follows that the integral is non-negative, and this means that the inelastic unitarity constraint is satisfied.

For $t > 16$, one can easily manage (2.27) by requiring $v(s,t) + v(t,s)$ to be positive. This is because the limit of the iteration, (2.23), necessarily satisfies

$$\|\varrho^{el}\| \leqq b_- < 4\Gamma B^2 . \tag{2.28}$$

So by choosing a value of B that is small enough, one can be sure that the positive contribution from $v(s,t) + v(t,s)$ is greater than $|\varrho^{el}(t,s)|$.

It remains to prove that $\varrho^{el}(t,s)$ is positive for $t \leqq 16$. One first requires that the contribution of $v(,s\, t) + v(t,s)$ to the integral (2.5) gives a positive result for $4 \leqq s \leqq 16$ and all t. Then it is proved, by induction, that if

$$\varrho_n^{el}(s,t) \geqq 0 , \tag{2.29}$$

for $4 \leqq s \leqq 16$ and all t, then necessarily

$$\varrho_{n+1}^{el}(s,t) \geqq 0 , \tag{2.30}$$

for the same range of s. Then it follows that the same condition holds for the limit function, $\varrho^{el}(s,t)$; and this condition is just what we wanted, namely that $\varrho^{el}(t,s)$ be positive for $t \leqq 16$.

The induction proof follows from the observations that the kernel K in (2.2) is non-negative, and that $d^*(s,t_1)\, d(s,t_2)$ may be replaced by

$$\operatorname{Re} d(s,t_1) \operatorname{Re} d(s,t_2) + \varrho(s,t_1)\, \varrho(s,t_2) . \tag{2.31}$$

It may be shown that, with the given restrictions on v, the quantity (2.31) may be constrained to be positive for $s \leqq 16$, so that therefore the integral in (2.2) will also be positive for $s \leqq 16$. Since $\varrho_0^{el}(s,t) \equiv 0$ manifestly satisfies (2.29), the induction is complete.

3. Subtractions and CDD Poles

I would like now to sketch the latest developments in the construction of crossing-symmetric, unitary functions [6]. I will indicate how CDD poles and one subtraction may be introduced. It is easy to add as many CDD poles as you like, but there is an essential difficulty in having more then one subtraction. This facet of the problem has only been partially resolved.

One has to make a subtraction in Eq. (2.5), in which the subtraction function can be written in terms of the t-channel S-wave absorptive part, say $\operatorname{Im} F_0(t)$. If this latter function were known, the iteration for the double-spectral function would be much the same as before. However, it is not known, of course, but must itself be determined. This is done by writing N/D equations for the S-wave. However, these are not old-fashioned N/D equations in which the left-hand cut discontinuity is given as a divergent series of crossed-channel partial waves. Instead, the discontinuity is given in terms of the crossed-channel absorptive part, which is known explicitly in terms of the double-spectral function, and the S-wave absorptive part itself. Hence, in the end, one has a double, interlocking iteration for the double-spectral function and for the S-wave. One can find conditions for the convergence of the iteration, much as in the no-subtraction case.

In the N/D equations, one has the freedom to introduce explicit poles into the dispersion relation for the D-function. These are the CDD poles. One can show that the double iteration still converges, if the CDD pole residues are small enough. If one puts the CDD poles on the cut, one can also show that if the residue has a certain well-defined sign, which is determined by the inelastic part of the double-spectral function, then the D-function has no zeros on the physical sheet, so there are no ghosts. For the wrong sign of the residue, there is a ghost, i. e., a zero of D (a pole of the S-wave amplitude), near the CDD pole position. As one changes the CDD pole residue's sign from the wrong to the right one, this ghost slips from the physical to the unphysical sheet, and becomes a bona fide resonance pole. This has been proved rigorously, for small CDD pole residues [6].

I turn lastly to the question of having more than one subtraction. If one does not worry about the inelastic constraints above the four-pion threshold, but only about the elastic unitarity and crossing symmetry, then one can handle any number of subtractions. However, if one insists on the inelastic constraints, as indeed one should, the best that has been done so far is to allow inelastic generating functions that may have more than one subtraction, but which are such that their t-discontinuities only need one subtraction in the s-channel elastic region.

This is not as good as one would like, but it seems hard to go further. Nevertheless, one can obtain total cross-sections that have an asymptotic behaviour

$$\sigma(s) \underset{s\to\infty}{\approx} (\log s)^{-2-\varepsilon}, \tag{3.1}$$

which is not too far from being constant. As Wanders has told you, and I shall prove in a moment, analyticity, crossing and positivity imply that necessarily

$$\sigma(s) \lesssim (\log s)^2 , \tag{3.2}$$

which is the Froissart bound [1]. You see that in the construction program, we are presently $4+\varepsilon$ logarithms away from the maximal allowed asymptotics.

4. Froissart Bounds

I turn now to a consideration of the high-energy bounds that any amplitude which satisfies axiomatic field-theory must observe. We may call these inequalities "Froissart bounds", although I am going to report some modern developments that go considerably beyond the original result of *Froissart* [1].

The first result I am going to prove, following a method of *Martin* [8], is that

$$\overline{\lim_{s\to\infty}} \{(\log s)^{-2} \sigma_{\text{tot}}(s)\} \leqq \pi . \tag{4.1}$$

This is usually written, a little loosely, as

$$\sigma_{\text{tot}}(s) \underset{s\to\infty}{\lesssim} \pi \left[\log \frac{s}{s_0}\right]^2 \tag{4.2}$$

which is to be understood asymptotically, and where an unknown scale, s_0, has been thrown in, somewhat gratuitously.

The absorptive part of the amplitude is

$$[F(s,t)]_s = \sum_{l=0}^{\infty} (2l+1) \operatorname{Im} f_l(s) P_l\left(1+\frac{2t}{s-4}\right). \tag{4.3}$$

The total cross-section, $\sigma_{\text{tot}}(s)$, will therefore be, by the optical theorem,

$$\sigma_{\text{tot}}(s) = \frac{16\pi}{\varrho s} \sum_{l=0}^{\infty} (2l+1) \operatorname{Im} f_l(s), \tag{4.4}$$

where $\varrho(s) = [(s-4)/s]^{1/2}$.

Now let us use the fact that the singularity in the t-plane nearest to $t=0$ is at $t=4$, so that the series (4.3) certainly converges for $t=t_0<4$.

We will obtain a lower bound on $[F(s, t_0)]_s$. This is done by choosing those values of $\operatorname{Im} f_l(s)$ that minimize the sum (4.3) (for $t = t_0$), for clearly $[F(s, t_0)]_s$ will have to be not smaller than this minimum. The minimization must be done subject to two constraints. The first constraint is that the $\operatorname{Im} f_l(s)$ add up, in sum (4.4), to $\sigma_{\text{tot}}(s)$; and the second constraint is that each $\operatorname{Im} f_l(s)$ should lie in the interval

$$0 \leqq \operatorname{Im} f_l(s) \leqq 1/\varrho \,. \tag{4.5}$$

This is of course required by unitarity.

The solution of the conditional minimization, as I will prove in a moment, is as follows. Define the integer, L, by

$$L = \operatorname{int}\left[\left(\frac{\varrho s \sigma_{\text{tot}}(s)}{16\pi}\right)^{1/2}\right] - 1 \tag{4.6}$$

where int [ ] means the integral part of the number within the square brackets. The solution of the problem is to choose $\operatorname{Im} f_l(s) = 1/\varrho$ for $l = 0, 1, \ldots, L$, and $\operatorname{Im} f_l(s) = 0$ for $l \geqq L+2$, and finally to choose $\operatorname{Im} f_{L+1}(s)$ so that the sum (4.4) comes out right. This will give

$$\sigma_{\text{tot}}(s) = \frac{16\pi}{\varrho s} \sum_{l=0}^{L} (2l+1) + \frac{16\pi}{\varrho s} (2L+3) \operatorname{Im} f_{L+1}^{\min}(s) \,, \tag{4.7}$$

so that

$$\operatorname{Im} f_{L+1}^{\min}(s) = \frac{1}{2L+3} \left\{ \frac{\varrho s}{16\pi} \sigma_{\text{tot}}(s) - (L+1)^2 \right\}. \tag{4.8}$$

This is the solution which minimizes series (4.3), i. e. the minimum permissible $[F(s, t_0)]_s$ is

$$\begin{aligned} [F(s, t_0)]_s^{\min} = {} & \frac{1}{\varrho} \sum_{l=0}^{L} (2l+1) P_l\left(1 + \frac{2t_0}{s-4}\right) \\ & + (2L+3) \operatorname{Im} f_{L+1}(s) P_{L+1}\left(1 + \frac{2t_0}{s-4}\right). \end{aligned} \tag{4.9}$$

This conditional minimum, you see, amounts to filling up the $L+2$ lowest waves with the maximum allowed $\operatorname{Im} f_l(s)$, namely $1/\varrho$, except the last one, which is just right to ensure that the series (4.4) comes to $\sigma_{\text{tot}}(s)$. All the other $\operatorname{Im} f_l(s)$ are set to zero. The reason that this minimizes series (4.3) is that, for $s > 4$ and $t > 0$, the Legendre function is a strictly increasing function of l (see Appendix). Therefore the way to minimize series (4.3) is precisely to use up as much of the low l values as is consistent with the constraint (4.5). Let us prove this more rigorously. Take any sequence $\operatorname{Im} f_l(s)$ which gives a sum (4.4) equal to $\sigma_{\text{tot}}(s)$, and which does not violate the unitarity constraint (4.5). Define $[F(s,t)]_s$ by Eq. (4.3). Then we want to show in particular that

$$[F(s,t_0)]_s \geqq [F(s,t_0)]_s^{\min}. \tag{4.10}$$

Now

$$\begin{aligned}[F(s,t_0)]_s-[F(s,t_0)]_s^{\min} = &-\sum_{l=0}^{L}(2l+1)\left[\frac{1}{\varrho}-\operatorname{Im}f_l(s)\right]P_l\left(1+\frac{2t_0}{s-4}\right)\\ &-(2L+3)\left[\operatorname{Im}f_{L+1}^{\min}(s)-\operatorname{Im}f_{L+1}(s)\right]P_{L+1}\left(1+\frac{2t_0}{s-4}\right)\\ &+\sum_{l=L+2}^{\infty}(2l+1)\operatorname{Im}f_l(s)P_l\left(1+\frac{2t_0}{s-4}\right).\end{aligned} \tag{4.11}$$

We can obtain a lower bound on the last term here, the series from $l=L+2$ to $l=\infty$, by using the fact that both the general $\operatorname{Im}f_l(s)$, and the "minimal" set add up to the same $\sigma_{\text{tot}}(s)$ in Eq. (4.4). Hence we must have

$$\begin{aligned}\sum_{l=0}^{L}(2l+1)\left[\frac{1}{\varrho}-\operatorname{Im}f_l(s)\right]+(2L+3)\left[\operatorname{Im}f_{L+1}^{\min}(s)-\operatorname{Im}f_{L+1}(s)\right]&\\ -\sum_{l=L+2}^{\infty}(2l+1)\operatorname{Im}f_l(s)=0.&\end{aligned} \tag{4.12}$$

From this result, and the fact that $P_l(z)$ is a monotonic increasing function of l for $z>1$, we find

$$\begin{aligned}&\sum_{l=L+2}^{\infty}(2l+1)\operatorname{Im}f_l(s)P_l\left(1+\frac{2t_0}{s-4}\right)\\ &\quad\geqq P_{L+1}\left(1+\frac{2t_0}{s-4}\right)\sum_{l=L+2}^{\infty}(2l+1)\operatorname{Im}f_l(s)\\ &\quad=\sum_{l=0}^{L}(2l+1)\left[\frac{1}{\varrho}-\operatorname{Im}f_l(s)\right]P_{L+1}\left(1+\frac{2t_0}{s-4}\right)\\ &\qquad+(2L+3)\left[\operatorname{Im}f_{L+1}^{\min}(s)-\operatorname{Im}f_{L+1}(s)\right]P_{L+1}\left(1+\frac{2t_0}{s-4}\right).\end{aligned} \tag{4.13}$$

On substituting this lower bound into (4.11), we have

$$\begin{aligned}&[F(s,2)]_s-[F(s,2)]_s^{\min}\\ &\geqq\sum_{l=0}^{L}(2l+1)\left[\frac{1}{\varrho}-\operatorname{Im}f_l(s)\right]\left[P_{L+1}\left(1+\frac{2t_0}{s-4}\right)-P_l\left(1+\frac{2t_0}{s-4}\right)\right]\\ &\geqq 0,\end{aligned} \tag{4.14}$$

since $\mathrm{Im} f_l(s)$ satisfies (4.5) and $l < L+1$ in the sum. This concludes the proof that $[F(s, t_0)]_s^{\min}$ is indeed the minimum possible value of $[F(s, t_0)]_s$.

We need now a lower bound on $P_l(z)$, which is good when $z-1 \to 0_+$, since this corresponds to the large s-limit in Eq. (4.9). Such a formula can be obtained from the Laplace representation:

$$P_l(z) = \frac{1}{\pi} \int_0^\pi \mathrm{d}\theta [z + (z^2-1)^{1/2} \cos\theta]^l . \tag{4.15}$$

One certainly has, for $z \geqq 0$, and for any $\theta_0 \geqq 0$,

$$\begin{aligned} P_l(z) &\geqq \frac{1}{\pi} \int_0^{\theta_0} \mathrm{d}\theta [z + (z^2-1)^{1/2} \cos\theta]^l \\ &\geqq \frac{\theta_0}{\pi} \{1 + [2(z-1)]^{1/2} \cos\theta_0\}^l . \end{aligned} \tag{4.16}$$

Hence, from Eq. (4.9),

$$\begin{aligned} [F(s, t_0)]_s \geqq [F(s, t_0)]_s^{\min} &\geqq \frac{1}{\varrho} \sum_{l=0}^{L} (2l+1) P_l\left(1 + \frac{2t_0}{s-4}\right) \\ &> \frac{2\theta_0}{\varrho\pi} \sum_{l=0}^{L} l y^l \\ &= \frac{2\theta_0}{\varrho\pi} \left[\frac{L y^L}{y-1} + \frac{1-y^L}{(y-1)^2}\right] \\ &> \frac{2\theta_0}{\varrho\pi} \left[\frac{L}{y-1} - \frac{1}{(y-1)^2}\right] y^L \end{aligned} \tag{4.17}$$

where

$$y = 1 + \cos\theta_0 \left[\frac{4t_0}{s-4}\right]^{1/2} . \tag{4.18}$$

Recall now the definition of L [Eq. (4.6)]. For large s, one has

$$L \sim s^{1/2} \left\{\frac{\sigma_{\mathrm{tot}}(s)}{16\pi}\right\}^{1/2} . \tag{4.19}$$

So, from Eqs. (4.17) and (4.18),

$$\begin{aligned} [F(s, t_0)]_s \gtrsim \frac{2s\theta_0}{\pi} &\left\{\sec\theta_0 \left[\frac{\sigma_{\mathrm{tot}}(s)}{64\pi t_0}\right]^{1/2} - \frac{\sec^2\theta_0}{4t_0}\right\} \\ &\cdot \exp\left\{\cos\theta_0 \left[\frac{t_0 \sigma_{\mathrm{tot}}(s)}{4\pi}\right]^{1/2}\right\} . \end{aligned} \tag{4.20}$$

We have proved that $[F(s,t_0)]_s$ will be asympotically not less than the expression on the right-hand side of (4.20). The ingredients, you remember, were the assumptions that $[F(s,t)]_s$ is analytic in t up to $t=4$, and that the unitarity constraints (4.5) are not violated. We needed the analyticity, in order to be sure that the Legendre series (4.3) converges at $t=t_0$; and we needed the constraint (4.5) during our minimization procedure. We need a third ingredient in order to derive the Froissart bound, one that has not yet been used. This is the assumption that $[F(s,t_0)]_s$ is bounded asymptotically by a polynominal in s, say by s^N. This would, of course, be the case even if Regge trajectories rise indefinitely: we are not requiring a simultaneous and uniform polynomial bound for $F(s,t)$ is s for all t, but only for one t-value. The polynomial boundedness we need has in fact been proved in axiomatic field theory, as you have already learned from Wanders.

It is clear now that we must have

$$\sigma_{\text{tot}}(s) \lesssim C\left[\log\left(\frac{s}{s_0}\right)\right]^2, \tag{4.21}$$

where C and s_0 are constants, since any more divergent expression would give something worse than a power of s in Eq. (4.20), and so would violate the assumption of a polynomial bound. On inserting (4.21) into (4.20), one finds

$$[F(s,t_0)]_s \gtrsim \frac{s_0\theta_0}{4\cos\theta_0}\left(\frac{C}{\pi^3 t_0}\right)^{1/2}\left(\frac{s}{s_0}\right)^B \log\left(\frac{s}{s_0}\right), \tag{4.22}$$

where

$$B = 1 + \cos\theta_0\left(\frac{t_0 C}{4\pi}\right)^{1/2}. \tag{4.23}$$

To finish the demonstration, we need to evaluate C. To do this we have to examine the number of subtractions that a fixed-t dispersion relation for $F(s,t)$ needs for $0<t<4$. This can be done by repeating the above proof, but now taking $[F(s,t)]_s, 0<t<t_0$, to be specified, and getting, as before, a lower bound on $[F(s,t_0)]_s$. The minimum is again obtained by filling up the first $L'+1$ lowest waves, say, with the unitarity limit, in such a way that the sum for $[F(s,t)]_s, 0<t<t_0$, comes out right. An extension of the technique gives a bound also on $|F(s,t)|$, and not merely on $[F(s,t)]_s$. I do not have time to go into details, but one finds that if

$$|F(s,t_0)| \lesssim C_0 s^N, \tag{4.24}$$

then

$$|F(s,t)| \lesssim C_1 s^{1+(N-1)(t/t_0)^{1/2}+\varepsilon}, \tag{4.25}$$

where C_0 and C_1 are constants, and where $\varepsilon > 0$ can be as small as one likes.

From Eq. (4.25), one sees that there will be a finite domain of t, say $0 \leqq t \leqq t_1$, where $F(s,t)$ will satisfy a dispersion relation in s with only two subtractions. I will now show, following *Jin* and *Martin* [9], that in fact this twice subtracted dispersion relation continues to be valid for $t \leqq t_0$, where t_0 can be arbitrarily close to 4. For simplicity, I will pretend that there is only a right-hand cut. The proof with two cuts is very similar, but a little longer, and I refer you to the original paper if you want to see the demonstration in its full splendour. For $t \leqq t_1$, we need two subtractions:

$$F(s,t) = \alpha(t) + s\beta(t) + \frac{s^2}{\pi} \int_4^\infty \frac{\mathrm{d}s'}{s'^2(s'-s)} [F(s',t)]_s, \tag{4.26}$$

whereas for $t \leqq t_0 < 4$ we know that we certainly need no more than $N+1$ subtractions:

$$F(s,t) = \sum_{n=0}^{N} C_n(t)\, s^n + \frac{s^{N+1}}{\pi} \int \frac{\mathrm{d}s'}{s'^{N+1}(s'-s)} [F(s',t)]_s. \tag{4.27}$$

For $t \leqq t_1$, both (4.26) and (4.27) hold, so if $N \geqq 2$, we have

$$C_N(t) = \frac{1}{\pi} \int_4^\infty \frac{\mathrm{d}s'}{s'^{N+1}} [F(s',t)]_s. \tag{4.28}$$

Now $[F(s',t)]_s$ has a Taylor expansion

$$[F(s',t)]_s = \sum_{n=0}^{\infty} \frac{t^n}{n!} \left[\left(\frac{\partial}{\partial t} \right)^n [F(s',t)]_s \right]_{t=0} \tag{4.29}$$

which is certainly convergent for $t \leqq t_0 < 4$, uniformly with respect to s'. Moreover, according to Eq. (4.25), the convergence of the infinite integral in (4.28) will be uniform with respect to t, if $t \leqq t_1$. Hence we can insert the series (4.29) into (4.28) and interchange the order of summation and integration:

$$C_N(t) = \frac{1}{\pi} \sum_{n=0}^{\infty} \frac{t^n}{n!} \int_4^\infty \frac{\mathrm{d}s'}{s'^{N+1}} \left[\left(\frac{\partial}{\partial t} \right)^n [F(s',t)]_s \right]_{t=0}. \tag{4.30}$$

Now we know from unitarity that

$$\left[\left(\frac{\partial}{\partial t} \right)^n [F(s',t)]_s \right]_{t=0} \geqq 0, \tag{4.31}$$

so that the power series (4.30) has positive coefficients, and hence it must converge for all $|t| < 4$, since the nearest possible singularity of $C_N(t)$ is at $t = 4$.

Hence, *now for any* $t \leqq t_0 < 4$, we can reverse the summation and integration orders in (4.30) to recover (4.28), so we have proved that the integral (4.28) continues to converge for $t \leqq t_0 < 4$, and this means that one subtraction in (4.27) can be removed for these values of t. The whole process can be repeated until there are only two subtractions left, for then (4.27) just matches (4.26).

Therefore we can see from Eq. (4.22) that

$$B < 2\,, \tag{4.32}$$

since otherwise $F(s, t_0)$ would need more than two subtractions. Hence

$$C < \frac{4\pi}{t_0 \cos^2\theta_0} \tag{4.33}$$

Now we can take t_0 as close as we like to 4, and we can take θ_0 as small as we like (but not quite zero, for then the right-hand side of Eq. (4.20) would vanish, and the argument would be spoiled). In other words, the smallest value we may take for the coefficient C in Eq. (4.21) is

$$C = \pi\,. \tag{4.34}$$

If one were to put any larger value than this into Eq. (4.23), one could find a value of $t_0 < 4$, and a value of $\theta_0 > 0$, for which the right-hand side of (4.23) would be greater than 2, and this is not allowed. This concludes the proof of the Froissart bound, Eq. (4.2).

5. Yndurain's Bound

Next, I will describe a very recent extension of the Froissart bound by *Yndurain* [2], in a generalized form given by Common. I will use, however, a simpler estimate for the Legendre function than either of these authors, and I will introduce another modification.

One obtains a bound for an averaged cross-section, defined by

$$\bar{\sigma}_{\rm tot}^{(m)}(s) = \frac{m+1}{(s-4)^{m+1}} \int_4^s \mathrm{d}s'(s'-4)^m\, \sigma_{\rm tot}(s)\,, \tag{5.1}$$

rather than for $\sigma_{\rm tot}$ itself. For a sensible function, $\sigma_{\rm tot}(s)$, that does not oscillate infinitely, $\bar{\sigma}_{\rm tot}(s) \sim \sigma_{\rm tot}^{(m)}(s)$ as $m \to \infty$, but in fact the bounds we will obtain on $\bar{\sigma}_{\rm tot}(s)$ do not require $\sigma_{\rm tot}$ to be non-oscillatory.

The big advantage of the Yndurain bound is that there are no unknown constants. The unknown scale, s_0, in Eq. (4.21), will be replaced by unity, the square of the pion mass, and the inequality will be a rigorous one, valid for all energies. This is an important advance, because the original Froissart bound is only asymptotic, and one does not know, in Eq. (4.21), when one is supposed to have reached the asymptotic region.

From the definition (5.1) and the optical theorem (4.4), one has

$$\bar{\sigma}_{\text{tot}}^{(m)}(s) = 16\pi \frac{m+1}{(s-4)^{m+1}} \int_4^s \mathrm{d}s'(s'-4)^{m-1/2} s'^{-1/2} \sum_{l=0}^{\infty} (2l+1) \operatorname{Im} f_l(s') . \quad (5.2)$$

Now break up the sum into two pieces, A and B, respectively comprising the terms of the infinite sum up to $l = L$, and those for $l > L$, where L is defined by

$$L + 1 = N(s-4)^{1/2} \log s . \quad (5.3)$$

The constant N will be fixed later. The term A is majorized by replacing $\operatorname{Im} f_l(s')$ by its unitarity limit, namely $1/\varrho(s')$, thus giving

$$\begin{aligned} A &\leqq 16\pi \frac{m+1}{(s-4)^{m+1}} \sum_{l=0}^{L} (2l+1) \int_4^s \mathrm{d}s'(s'-4)^{m-1} \\ &= 16\pi \frac{m+1}{m} \frac{(L+1)^2}{s-4} \\ &= 16\pi N^2 \frac{m+1}{m} (\log s)^2 . \end{aligned} \quad (5.4)$$

Consider next the term B. The problem is to obtain an upper bound for

$$\int_4^s \mathrm{d}s'(s'-4)^{m-1/2} s'^{-1/2} \sum_{l=L+1}^{\infty} (2l+1) \operatorname{Im} f_l(s') . \quad (5.5)$$

The D-wave amplitude is given by

$$f_2(t) = \frac{4}{\pi} (t-4)^{-1} \int_4^{\infty} \mathrm{d}s [F(s,t)]_s \, Q_2 \left(1 + \frac{2s}{t-4}\right) . \quad (5.6)$$

The point of using the D-wave, rather than the S-wave, is that we know that this integral converges, for $t < 4$ whereas the corresponding integral for the S-wave could diverge. As $t \to 4_-$, on finds that

$$f_2(t) \sim \alpha_2 \left(\frac{t-4}{4}\right)^2 \quad (5.7)$$

where α_2 is the D-wave scattering length, given by

$$\alpha_2 = \frac{16}{15\pi} \int_4^{\infty} \mathrm{d}s' s'^{-3} [F(s',4)]_s . \quad (5.8)$$

The idea is to obtain a bound on the series (5.5) in terms of α_2. Since $[F(s,4)]_s$ is positive for $0<t<4$, one certainly has

$$\alpha_2 > \frac{16}{15\pi}\int_4^s ds' s'^{-3} \sum_{l=0}^{\infty} (2l+1)\,\mathrm{Im} f_l(s)\, P_l\left(1+\frac{8}{s'-4}\right) \tag{5.9}$$

for any s. Let us further minimize this expression by throwing away the terms from $l=0$ to L in this series, and then replace P_l by P_{L+1}. The result is

$$\alpha_2 > \frac{16}{15\pi}\int_4^s ds' s'^{-3} P_{L+1}\left(1+\frac{8}{s'-4}\right) \sum_{l=L+1}^{\infty} (2l+1)\,\mathrm{Im} f_l(s'). \tag{5.10}$$

With an eye on the term (5.5) that we are trying to bound, let us extract a factor

$$(s'-4)^{-m+1/2} s'^{-5/2} P_{L+1}\left(1+\frac{8}{s'-4}\right) \tag{5.11}$$

from the integral (5.10), by replacing this expression by its value at $s'=s$, its smallest value. This gives

$$\begin{aligned}\alpha_2 > \frac{16}{15\pi}(s-4)^{-m+1/2} s^{-5/2} P_{L+1}\left(1+\frac{8}{s-4}\right)\int_4^s ds'(s'-4)^{m-1/2}\,. \\ \cdot s'^{-1/2} \sum_{l=L+1}^{\infty} (2l+1)\,\mathrm{Im} f_l(s')\,.\end{aligned} \tag{5.12}$$

We have succeeded in constructing the expression (5.5). Hence we have the following bound for B:

$$B < \frac{15\pi^3}{\theta_0}\alpha_2(m+1)(s-4)^{-3/2} s^{5/2}\left[1+\frac{4\cos\theta_0}{(s-4)^{1/2}}\right]^{-N(s-4)^{1/2}\log s} \tag{5.13}$$

where the inequality 4.16 for the Legendre function has been used, and where the definition (5.3) of $L+1$ has been inserted. The last factor in (5.13) has the form

$$\left[1+\frac{a}{h}\right]^{-bn} \tag{5.14}$$

where $a=4\cos\theta_0$, $b=N\log s$, $n=(s-4)^{1/2}$. As $n\to\infty$ i.e. as $s\to\infty$, we know that this tends to e^{-ab}, but since we want a rigorous inequality for finite s, and not just an asymptotic one, we will use the inequality

$$\left[1+\frac{a}{n}\right]^{-bn} < \exp\left\{-ab+\frac{a^2 b}{2n}\right\} \tag{5.15}$$

which may be proved to hold for all $b>0$ and $0<a<n$. Hence, for $s>20$ one has

$$B<\frac{15\pi^3}{\theta_0}\alpha_2(m+1)\left(\frac{s}{s-4}\right)^{3/2}s^{1-4N\cos\theta_0}\exp\left\{\frac{8N\cos^2\theta_0\log s}{(s-4)^{1/2}}\right\}. \quad (5.16)$$

Now let us choose $N=1/4$ and θ_0 such that

$$\cos\theta_0=1-(\log s)^{-2\eta} \quad (5.17)$$

where η will be specified in a moment. One can prove that, for any real θ_0,

$$\frac{1}{\theta_0}<[2(1-\cos\theta_0)]^{-1/2}, \quad (5.18)$$

so one has, finally, combining (5.4) and (5.17),

$$\begin{aligned}\bar{\sigma}_{\text{tot}}^{(m)}(s)<\pi\frac{m+1}{m}(\log s)^2+\frac{15\pi^3}{2^{1/2}}\alpha_2(m+1)\left(\frac{s}{s-4}\right)^{3/2}(\log s)^{\eta}\,\cdot \\ \cdot\exp\left\{(\log s)^{1-2\eta}+\frac{2\log s}{(s-4)^{1/2}}\right\}.\end{aligned} \quad (5.19)$$

This bound holds rigorously for any $s>20$. If one takes $\frac{1}{2}\leqq\eta<2$, then the first term dominates the second as $s\to\infty$, for fixed m, and one has asymptotically,

$$\bar{\sigma}_{\text{tot}}^{(m)}(s)\lesssim\pi\frac{m+1}{m}(\log s)^2. \quad (5.20)$$

The sharpest asymptotic result is obtained by taking, say, $m=\log s$ and $\frac{1}{2}\leqq\eta<1$. Then the first term in (5.19) still dominates asymptotically, and one has

$$\bar{\sigma}_{\text{tot}}^{(\log s)}(s)\lesssim\pi(\log s)^2, \quad (5.21)$$

which corresponds precisely to the earlier result (4.2).

Common [2] has developed a rather more complicated minimization method, that I do not have time to describe, which gives numerically somewhat smaller bounds than the above. However, unfortunately the numerical bounds turn out to be very big. With a reasonable value of 16×10^{-5} for the *D*-wave scattering length (in units of the pion mass), *Common* found bounds for the total cross-section, at 5–10 GeV/*c* pion momentum, of two or three barns, which is really awfully big! However, the latest news is that a similar treatment of pion-nucleon scattering gives much more reasonable results. There are no published results yet.

6. Pomeranchuk Theorem

I want to conclude by mentioning some theoretical aspects of the Pomeranchuk theorem [3, 10]. Let $F_\pm(s,t)$ be the elastic scattering amplitudes for, say, $\pi^\pm$ scattering from some target particle, which might be another pion, or a nucleon, for example.

The forward amplitudes are analytic in s, with cuts $(M+1)^2 < s < \infty$ and $-\infty < s \leqq (M-1)^2$, and they satisfy the crossing condition

$$F_\pm(s+i\varepsilon, 0) = F_\mp(2M^2+2-s-i\varepsilon, 0), \tag{6.1}$$

where M is the mass of the target particle in units of the pion mass. The Pomeranchuk theorem states that the difference between the total cross-sections for $\pi^\pm$ scattering on the target should vanish as $s\to\infty$:

$$\sigma_+(s) - \sigma_-(s) \xrightarrow[s\to\infty]{} 0. \tag{6.2}$$

A sufficient set of conditions for this to hold is as follows:

1. $F_\pm(s,0)$ is polynomially bounded as $|s|\to\infty$.

2. $$\frac{F_+(s+i\varepsilon,0) - F_-(s+i\varepsilon,0)}{s\log s} \xrightarrow[s\to\infty]{} 0. \tag{6.3}$$

3. A limit $$\sigma_+(s) - \sigma_-(s) \xrightarrow[s\to\infty]{} C \tag{6.4}$$

exists. If these conditions hold, then necessarily $C=0$.

I will first give a proof, before discussing the credibility of the conditions. Define the symmetrical variable

$$\omega = \left[\frac{s-M^2-1}{2M}\right]^2, \tag{6.5}$$

and the new function

$$\zeta(\omega) = 16\pi[s-(M+1)^2]^{-1/2}[s-(M-1)^2]^{-1/2}[F_+(s,0)-F_-(s,0)]. \tag{6.6}$$

Then $\zeta(\omega)$ only has one cut $1\leqq\omega<\infty$, on which its imaginary part is

$$\operatorname{Im}\zeta(\omega) = \sigma_+(\omega) - \sigma_-(\omega). \tag{6.7}$$

In view of the conditions (1)–(3), we may write a once-subtracted dispersion relation for ζ:

$$\zeta(\omega) = \zeta(0) + \frac{\omega}{\pi}\int_1^\infty \frac{d\omega'}{\omega'(\omega'-\omega)} \operatorname{Im}\zeta(\omega'). \tag{6.8}$$

According to assumption (1) and (2), the analytic function $\zeta(\omega)/\log\omega$ is polynomially bounded in the ω-plane, and tends to zero as $\omega\to+\infty$

just above or just below the cut. The Phragmén-Lindelöf Theorem then tells us that

$$\frac{\zeta(\omega)}{\log\omega}\to 0 \tag{6.9}$$

in any complex direction, in particular as $\omega\to-\infty$. We will suppose that C in Eq. (6.4) is non-zero, and obtain a contradiction with Eq. (6.9).

There certainly exists a number, Λ, so large that

$$\operatorname{Im}\zeta(\omega)\geqq\frac{C}{2} \tag{6.10}$$

for all $\omega\geqq\Lambda$, where it has been assumed that $C>0$ for definiteness. Then, for $\omega<0$,

$$\left|\frac{\omega}{\pi}\int_{\Lambda}^{\infty}\frac{d\omega'}{\omega'(\omega'-\omega)}\operatorname{Im}\zeta(\omega')\right|\geqq\frac{C}{2\pi}\log\frac{|\omega|+\Lambda}{\Lambda}. \tag{6.11}$$

Since the subtraction constant and the piece of the integral in Eq. (6.8) from $\omega=1$ to $\omega=\Lambda$ are asymptotically bounded as $\omega\to-\infty$, if follows that

$$\zeta(\omega)\underset{\omega\to-\infty}{\gtrsim}\frac{C}{2\pi}\log\frac{|\omega|}{\Lambda}. \tag{6.12}$$

But this contradicts Eq. (6.9), and so we must have $C=0$, which is the Pomeranchuk theorem.

It should be clear to you that, although we would hate to have to drop condition (1), we might very well consider relaxing (2) and (3), now that we have some experimental reason for trying to break the Pomeranchuk theorem. What I will do is to relax condition (3) a little. Suppose that, instead of Eq. (6.4), one has, more generally,

$$|\sigma_+(s)-\sigma_-(s)|\geqq C(\log s)^a, \tag{6.13}$$

for all $s\geq\Lambda$, where $a\geq 0$. One can repeat the argument and find

$$\zeta(\omega)\underset{\omega\to-\infty}{\sim}\frac{C}{\pi(m+1)}\left[\log\frac{|\omega|}{\Lambda}\right]^{a+1}. \tag{6.14}$$

So one can force the conclusion $C=0$ by excluding (6.14) as a possibility. This can be done by requiring that

$$\frac{\operatorname{Re}F_+(s,0)-\operatorname{Re}F_-(s,0)}{s[\log s]^{a+1}}\xrightarrow[s\to\infty]{}0. \tag{6.15}$$

Now you see that this is guaranteed by the Froissart bound, which both F_+ and F_- must satisfy, if $a>1$, so one may as well take $a=1$ in (6.13) and (6.15). A more illuminating condition, which can be derived immediately from (6.13) and (6.15), is that

$$\frac{\operatorname{Re} F_+(s,0) - \operatorname{Re} F_-(s,0)}{\operatorname{Im} F_+(s,0) - \operatorname{Im} F_-(s,0)} \cdot \frac{1}{\log s} \xrightarrow[s\to\infty]{} 0 \tag{6.16}$$

To break the Pomeranchuk theorem most easily, we would have to violate (6.16), so that the forward amplitude $F_+ - F_-$ would have to become asymptotically real. Then one could avoid the conclusion $\sigma_+ - \sigma_- \to 0$.

Appendix

An amusing proof that for $z > 1$ and l a positive integer,

$$P_{l+1}(z) > P_l(z): \tag{A1}$$

Consider the recurrence relation

$$(l+1)\,P_{l+1}(z) - (2l+1)\,zP_l(z) + lP_{l-1}(z) = 0\,, \tag{A2}$$

which can be rewritten as

$$(l+1)\,[P_{l+1}(z) - zP_l(z)] = l[zP_l(z) - P_{l-1}(z)]\,. \tag{A3}$$

Hence

$$\begin{aligned}(l+1)\,[P_{l+1}(z) - P_l(z)] &> (l+1)\,[P_{l+1}(z) - zP_l(z)]\\ &= l[zP_l(z) - P_{l-1}(z)]\\ &> zl[P_l(z) - P_{l-1}(z)]\\ &\cdots\cdots\cdots\cdots\\ &> z^l[P_1(z) - P_0(z)]\\ &= z^l(z-1)\\ &> 0\,.\end{aligned} \tag{A4}$$

The theorem also holds for any real l (not necessarily integral), but the general proof is less enjoyable than this one.

References

1. *Froissart, M.*: Phys. Rev. **123**, 1053 (1961).
2. *Yndurain, F. J.*: Phys. Letters **31** B, 368 (1970).
   Common, A. K.: CERN preprint TH. 1145 (1970).
3. *Pomeranchuk, I.*: JETP **34**, 725 (1958).
4. *Allaby, J. V., et al.*: Phys. Letters **30** B, 500 (1969).
5. *Martin, A.*: Scattering Theory: Unitarity, Analyticity and Crossing. Berlin-Heidelberg-New York: Springer 1969.

6. *Atkinson, D.*: Nucl. Phys. B **7**, 375 (1968); B **8**, 377 (1968); B **13**, 415 (1969). CERN preprint TH. 1168 (1970).
Atkinson, D., Warnock, R. L.: Phys. Rev. **188**, 2098 (1969).
Kupsch, J.: Nucl. Phys. B **11**, 573 (1969); B **12**, 155 (1969); Nuovo Cimento **66** A, 202 (1970) CERN preprint TH. 1151 (1970).
7. *Mandelstam, S.*: Phys. Rev. **112**, 1344 (1958), **115**, 1741, 1752 (1959).
8. *Martin, A.*: Phys. Rev. **129**, 1432 (1963).
9. *Jin, Y. S., Martin, A.*: Phys. Rev. **135**, B 1375 (1964).
Martin, A.: CERN preprint TH 1075 (1969).
10. *Eden, R. J.*: UCR preprint 34 P 107–105 (1969).

Dr. *D. Atkinson*
Physikalisches Institut der Universität
D-5300 Bonn

Analyticity, Unitarity and Crossing-Symmetry Constraints for Pion-pion Partial Wave Amplitudes

G. Wanders

Contents

1. General Properties of the Pion-pion Scattering Amplitudes Resulting from Axiomatic Field Theory

The purpose of these notes is to show that, apart the usual dispersion relations, there are practical consequences which can be extracted from the "axiomatic" properties of scattering amplitudes. The relevance of the consequences I shall present here is due to the fact that they can be used either as tests of existing low-energy models or as basic conditions in the construction of new models.

In this Section I shall present a short review of some general properties of scattering amplitudes resulting from axiomatic field theory. I shall confine myself to the pion-pion case and to those properties which will be used in the following Sections.

By axiomatic field theory we mean field theory in the LSZ formalism. Using the asymptotic condition, a scattering amplitude $M(p_1, p_2, p_3, p_4)$ ($p_i =$ energy – momentum of the particles, $p_i^2 = m_i^2$, $p_{i0} \geqq 0$) is expressed as the Fourier transform of the vacuum expectation value of retarded or advanced products of field operators. This expression allows an off-shell continuation ($p_i^2 \neq m_i^2$) of $M(p_1, p_2, p_3, p_4)$ and the locality of the field operators implies analyticity properties of the scattering amplitude [1]. Furthermore, I shall assume that the retarded or advanced products are tempered distributions; this implies that $M(p_1, p_2, p_3, p_4)$ is polynomially bounded as the p_i's tend to infinity.

If one uses the method developed by *Bros*, *Epstein* and *Glaser*, one first determines a domain of holomorphy Δ of $M(p_1, p_2, p_3, p_4)$ as a function of the vectors p_i. Then one determines the (non empty!) intersection D of Δ with the complex mass shell ($p_i^2 = m_i^2$). D can be described in terms of the usual invariants s, t, and u ($s=(p_1+p_2)^2$, $t=(p_1-p_3)^2$, $u=(p_1-p_4)^2$; $s+t+u=4$ in units where the pion mass $m_\pi=1$). Let $T^I(s,t,u)$ be the isospin I s-channel pion-pion amplitude. The relevant properties of D are contained in the following statements:

a) Every physical point of the s-channel ($s \geqq 4$, $t \leqq 0$, $u \leqq 0$) is on the boundary of D. $T^I(s,t,u)$ is the boundary value of an analytic function $F^I(s,t,u)$, holomorphic in D:

$$T^I(s,t,u) = \lim F^I(s+i\varepsilon, t, u-i\varepsilon)\,.$$

b) D ensures crossing symmetry. There is a path in D connecting any physical point of the s-channel to any physical point of the t- or u-channel. One has the crossing relation:

$$\lim_{\varepsilon\to 0+} F^I(s, t+i\varepsilon, u-i\varepsilon) = \sum_{I'} C_{II'} T^{(I')}(s,t,u) \quad \text{if} \quad t \geqq 4, s \leqq 0, u \leqq 0\,.$$

$T^{(I)}(s,t,u) = T^I(t,s,u) = t$-channel, isospin I amplitude.

$$C_{II'} = \tfrac{1}{6}\begin{pmatrix} 2 & 6 & 10 \\ 2 & 3 & -5 \\ 2 & -3 & 1 \end{pmatrix}.$$

c) $F^I(s,t,u)$ verifies fixed-t dispersion relations for $-t_0 \leqq t \leqq 0$. This means:

$$\{s,t \,|\, t=t_1, -t_0 \leqq t_1 \leqq 0, s \neq 4+\lambda, s \neq -t_1-\lambda, \lambda \geqq 0\} \subset D\,.$$

d) for $s \geqq 4$, $F^I(s,t,u)$ is holomorphic in the Lehmann ellipse $E(s)$; i.e.:

$$\{s,t \,|\, s=s_1, s_1 \geqq 4, t \in E(s_1)\} \subset D\,.$$

$E(s)$ = ellipse of the t-plane, foci at $t=0$ and $t=-(s-4)$, semi major axis $a(s) = \dfrac{1}{2}(s-4)\left[1+\dfrac{4^4}{s(s-4)}\right]^{\frac{1}{2}}$.

e) $F^I(s,t,u)$ is holomorphic in s and t in a neighbourhood of every point of the domain described in c):

$$\{s,t \,|\, |s-s_1| < \varrho(s_1,t_1), |t-t_1| < \varrho(s_1,t_1), -t_0 \leqq t_1 \leqq 0,$$
$$s_1 \neq 4+\lambda, s_1 \neq -t_1-\lambda, \lambda \geqq 0\} \subset D\,.$$

The properties listed until now are consequences of locality and of the mass spectrum of the pion-pion system. As it was shown by *Martin*, the holomorphy of $F^I(s, t, u)$, as a function of the two complex variables s and t, in an arbitrarily small neighbourhood of the points $(s_1 t_1)$ leads to a finite (and large) domain of holomorphy in s and t. This result is obtained from the positivity properties implied by unitarity.

These positivity properties are readily seen if we write the (s-channel) partial-wave expansion of $T^I(s, t, u)$:

$$F^I(s, t, u) = \sum_{l=0}^{\infty} (2l+1)\, F_l^I(s)\, P_l\left(1 + \frac{2t}{s-4}\right).$$

As a consequence of unitarity:

$$A^I(s) = \operatorname{Im} T_l^I(s + i\varepsilon) = \sqrt{\frac{s-4}{s}}\, |T_l^I(s)|^2$$

for $s \geqq 4$, we have: $A_l^I(s) > 0$. Furthermore:

$$\frac{\partial^n}{\partial z^n} P_l(z) > 0 \quad \text{for} \quad z \geqq 1\,.$$

Therefore:

$$\frac{\partial^n}{\partial t^n} A^I(s, t) \geqq 0 \qquad A^I(s, t) = \operatorname{Im} F^I(s + i\varepsilon, t, u - i\varepsilon)$$

for $s > 4$ and those positive values of t for which the partial-wave expansion converges (i.e. at least for t in the Lehmann ellipse $E(s)$).

Starting from these positivity properties and the holomorphy of $F^I(s, t, u)$ in a neighbourhood of $(s = s_0, t = 0)$, $0 \leqq s_0 \leqq 4$, *Martin* succeeded in proving the following result:

$F^I(s, t, u)$ is a holomorphic function of both variables s and t in the domain:

$$\bar{D} = D_{st} \cup D_{tu} \cup D_{us}$$

$$D_{st} = \{s, t \mid |t| < 4,\ s \neq 4 + \lambda,\ s \neq -t - \lambda,\ \lambda \geqq 0\}\,.$$

$\bar{D}$ is not a natural domain of holomorphy. Its holomorphy envelope is smaller than the domain corresponding to the Mandelstam representation.

Now I give a list of the particular implications of the existence of $\bar{D}$ which I shall use later.

1. Existence of fixed-t dispersion relations for $|t| < 4$.

2. Convergence of the partial wave expansion in an ellipse $E_1(s)$ with foci at $t = 0$ and $t = -(s-4)$ and semi-major axis $a(s) = \frac{1}{2}(s+4)$. $(E(s) \subset E_1(s))$.

3. As a consequence of (2), the positivity properties extend to the interval $0 \leqq t < 4$:

$$\frac{\partial^n A^I(s, t)}{\partial t^n} \geqq 0 \quad \text{for} \quad s \geqq 4,\ 0 \leqq t < 4,\ n = 0, 1, 2, \ldots.$$

4. (2) and unitarity $[A_l^I(s) \leqq [s/(s-4)]^{\frac{1}{2}}]$ imply the Froissart bound:

$$|T(s,0)| \lesssim \text{Const.}\, s \left(\log \frac{s}{s_0}\right)^2 \qquad s \geqq 4$$

and a bound valid for $0 \leqq t < 4$:

$$|T(s,t)| \lesssim \text{Const.}\, s^{(1+(t/4)^{1/2}+\varepsilon)}, \varepsilon > 0, s \geqq 4 .$$

5. From (4) and (5) we deduce that the number of subtractions needed in fixed-t dispersion relations is at most two if $0 \leqq t < 4$.

Notice that this number of subtractions has no connection with the number of subtractions in a possible Mandelstam representation.

If we write down fixed-t dispersion relations for the amplitudes $T^{(I)}(s,t,u)$ having a definite isospin in the t-channel, the effective number of subtractions is reduced. This is due to the symmetry of these amplitudes under $s \leftrightarrow u$ exchange. For $I=0$ and 2 we have only one subtraction:

$$\begin{aligned} T^{(I)}(s,t,u) &= T^{(I)}(s_0,t,u_0) + \frac{1}{\pi}\int_4^\infty \mathrm{d}x\, A^{(I)}(x,t) \\ &\times \left[\frac{1}{x-s} - \frac{1}{x-s_0} + \frac{1}{x-u} - \frac{1}{x-u_0}\right] \quad 0 \leqq t < 4,\ I=0,2 . \end{aligned} \tag{1.1}$$

6. Pomeranchuk theorem. If:

$$\lim_{s\to\infty} \frac{1}{s\log s}\left[T_{AB\to AB}(s,t) - T_{A\bar{B}\to A\bar{B}}(s,t)\right] = 0 .$$

($T_{AB\to AB}(s,t)$ is the absorptive part for the process $AB \to AB$, and $A_{AB\to AB}$ its absorptive part) and if the quantity

$$\frac{1}{s}\left[A_{AB\to AB}(s,t) - A_{A\bar{B}\to A\bar{B}}(s,t)\right]$$

has a limit as $s \to \infty$, this limit vanishes.

Proofs of the results (4–6) are given in Atkinson's lecture notes.

For pion-pion scattering, $A = B = \pi^+$ is the only choice leading to a result which is not already implied by isospin variance. As:

$$T_{\pi^+\pi^+\to\pi^+\pi^+}(s,t) - T_{\pi^+\pi^-\to\pi^+\pi^-}(s,t) = A^{(1)}(s,t) ,$$

the Pomeranchuk theorem tells us that

$$\lim_{s\to\infty} \frac{1}{s} A^{(1)}(s,t) = 0 .$$

If we make the slightly stronger assumption

$$\int^\infty \mathrm{d}s \frac{1}{s^2} A^{(1)}(s,t) < \infty$$

we get an unsubtracted fixed-t dispersion relation for $T^{(1)}(s,t,u)$

$$T^{(1)}(s,t,u)=\frac{1}{\pi}\int_4^\infty \mathrm{d}x\, A^{(1)}(x,t)\left[\frac{1}{x-s}-\frac{1}{x-u}\right]. \tag{1.2}$$

7. The Froissart-Gribov representation for $T_l^I(s)$ is valid for $l\geqq 2$, $I=0,2,0\leqq s\leqq 4$.

This result is obtained if one inserts the expression of

$$T^I(s,\tfrac{1}{2}(4-s)(1-z),\tfrac{1}{2}(4-s)(1+z))$$

given by a once-subtracted, fixed-s dispersion relation, into

$$T_l^I(s)=2\int_{-1}^{+1}\mathrm{d}z\, T^I(s,\tfrac{1}{2}(4-s)(1-z),\tfrac{1}{2}(4-s)(1+z))\,P_l(z). \tag{1.3}$$

The Froissart-Gribov representation reads:

$$T_l^I(s)=\frac{2((-1)^l+(-1)^I)}{\pi(4-s)}\int_4^\infty \mathrm{d}t\, A^{(I)}(t,s)\,Q_l\left(\frac{2t}{4-s}-1\right),\quad l\geqq 2,\ 0<s<4. \tag{1.4}$$

If we assume the validity of the unsubtracted dispersion relation (1.2), we get a Froissart-Gribov representation for $T_1^1(s)$.

8. Analyticity properties of the partial wave amplitudes. $T_l^I(s)$ is regular at s if $\bar{D}$ or its holomorphy envelope contains a path joining the points $t=0$ and $t=-(s-4)$. It is readily seen that $\bar{D}$ itself contains such a path if s is in the circle $|s-4|<4$ provided with the cut $s=4+\lambda$, $0\leqq\lambda<4$.

In fact the holomorphy domain of $T_l^I(s)$ resulting from the envelope of $\bar{D}$ is much larger. It is of the form:

$$\{s\,|\,s\neq\lambda,\ 4\leqq\lambda<\infty,\ -\infty<\lambda\leqq 0\}\cap\mathscr{D}$$

$\mathscr{D}$ is a bounded domain, containing the interval $(-28,+120)$ of the real axis.

2. Crossing-Symmetry Constraints on the Partial Waves

This Section and the three following ones are devoted to the properties of the partial wave amplitudes on the unphysical interval $0\leqq s\leqq 4$. The search of such properties is justified by the hope that they may lead to limitations on the low energy behaviour of the phase shifts.

In this Section I discuss the constraints resulting from crossing symmetry alone. This type of constraints has been discovered by *Balachandran* and *Nuyts* [2]. I follow the simplified derivation due to *Roskies* [3] and *Basdevant*, *Cohen-Tannoudji*, and *Morel* [4].

Let $Q(s, t, u)$ be a polynomial in s, t and u, symmetric (or antisymmetric) under the exchange $s \leftrightarrow u$. We have, trivially:

$$\int_{\Delta} \mathrm{d}s\,\mathrm{d}u\, T^{(I)}(s,t,u)\, Q(s,t,u) = 0 \tag{2.1}$$

where Δ is the triangle $\{s, t, u \mid s \geqq 0,\ t \geqq 0,\ u \geqq 0\}$ and where we choose $Q(s, t, u)$ antisymmetric if $I = 0.2$ and symmetric if $I = 1$. If we expand $Q(s, t, u)$ in the Legendre polynomials $P_l(z_s)$ $\left(z_s = 1 + \frac{2t}{s-4}\right)$:

$$Q(s,t,u) = \sum_{l=0}^{N} q_l(s)\, P_l(z_s)$$

(2.1) becomes

$$\begin{aligned}\sum_{l=0}^{N} \int_0^4 \mathrm{d}s(4-s)\, q_l(s) \int_{-1}^{+1} \mathrm{d}z_s\, T^{(I)}(s,t,u)\, P_l(z_s)\\ = \sum_{l=0}^{N} \sum_{I'} C_{II'} \int_0^4 \mathrm{d}s(4-s)\, q_l(s)\, T_l^{I'}(s) = 0\,.\end{aligned} \tag{2.2}$$

Remembering that the $q_l(s)$ are polynomials in s, we see that (2.2) is a linear relation between moments of a finite number of partial waves over the unphysical interval $[0, 4]$.

If we require (2.2) for a set of symmetric (I odd) or antisymmetric (I even) polynomials which is complete over the triangle Δ we get an infinite set of necessary and sufficient conditions the partial waves have to satisfy in order to ensure crossing symmetry. A nice feature of these conditions is that the number of conditions involving only the elements of a finite set $\{T_l^I(s) \mid I = 0, 1, 2;\ l \leqq L\}$ of partial waves is finite. Thus, there are two (and only two) conditions involving S waves only, and three (and only three) conditions involving both S and P waves only:

$$\begin{aligned}
&\int_0^4 \mathrm{d}s(4-s)(3s-4)(T_0^0(s) + 2T_0^2(s)) = 0\\
&\int_0^4 \mathrm{d}s(4-s)(2T_0^0(s) - 5T_0^2(s)) = 0\\
&\int_0^4 \mathrm{d}s(4-s)\, s(2T_0^0(s) - 5T_0^2(s)) = -3\int_0^4 \mathrm{d}s(4-s)^2\, T_1^1(s)\\
&\int_0^4 \mathrm{d}s(4-s)^2\, s(2T_0^0(s) - 5T_0^2(s)) = -3\int_0^4 \mathrm{d}s(4-s)^2\, s\, T_1^1(s)\\
&\int_0^4 \mathrm{d}s(4-s)^3\, s(2T_0^0(s) - 5T_0^2(s)) = -3\int_0^4 \mathrm{d}s(4-s)^2\, s(3s-4)\, T_1^1(s)
\end{aligned} \tag{2.3}$$

As it has been shown by *Balachandran* and *Nuyts*, it is convenient to expand $T_l^I(s)$ in Jacobi polynomials $P_n^{(2l+1,0)}(\frac{1}{2}s-1)$:

$$T_l^I(s)=\sum_{n=0}^{\infty} f_{l,n}^I P_n^{(2l+1,0)}(\tfrac{1}{2}s-1)\,. \tag{2.4}$$

This series converges for $0<s<4$. The crossing conditions relate the coefficients $f_{l,n}^I$ with constant $(l+n)$:

$$f_{l,n}^I=\sum_{l'=0}^{l+n}\sum_{I'} B_{l,l'}^{n,I,I'} f_{l',l+n-l'}^{I'} \tag{2.5}$$

3. Local Conditions on the Partial Waves Resulting from Crossing Symmetry and the Positivity of the Absorptive Parts

The investigations of the constraints imposed by positivity and crossing symmetry on the S and P waves in the interval $[0,4]$ were initiated by *Martin* [5] in 1967. The methods he developed with his coworkers ([6–11]) lead to three types of inequalities:

– inequalities giving the sign of the derivatives of the $\pi^0\pi^0$ S wave at given points of $[0,4]$,

– inequalities comparing the values of the S and P waves (or their first order derivative) at two given points of $[0,4]$,

– inequalities relating the value of a linear combination of S and P waves at a given point of $[0,4]$ to an integral over a linear combination of S and P waves extending to the interval $[0,4]$.

We call these conditions "local" because they involve the values of the partial waves at given points. As we shall see in the next Section, there are also "non local" conditions, which involve only integrals over partial waves.

Two general remarks before I describe the local properties a little more in detail.

1. If $\{T^I(s,t,u)\,|\,I=0,1,2\}$ is a crossing symmetric set of functions satisfying the positivity conditions, the set $\{c\,T^I(s,t,u)\}$ satisfies the same requirements, if c is a positive constant. Furthermore, the set of linear functions

$$t^0=2a(s-\tfrac{4}{3})+5b\,,$$
$$t^1=a(t-u)\,,$$
$$t^2=-a(s-\tfrac{4}{3})+2b$$

is crossing symmetric and verifies twice subtracted dispersion relations (with vanishing absorptive parts). Therefore if $\{T^I(s,t,u)\}$ is consistent with crossing symmetry and positivity, the same is true for $\{c\,T^I(s,t,u) + t^I(s,t,u)\}$. This means that our conditions do not distinguish a set of partial waves

$$T_0^0(s),\quad T_1^1(s),\quad T_0^2(s)$$

from:

$$c\,T_0^0(s)+2a(s-\tfrac{4}{3})+5b,\qquad c\,T_1^1(s)+\tfrac{1}{3}a(s-4),\qquad c\,T_0^2(s)-a(s-\tfrac{4}{3})+2b$$

with $c>0$.

2. As the higher partial waves ($l\geqq 2$) are given by the Froissart-Gribov representation (1.4), it is possible to get quite rapidly some restrictions on these partial waves. For instance $Q_l(z)>0$ for $z\geqq 1$ implies

$$T_l^{(I)}(s)>0\,,\qquad I=0,2\,,$$

$$T_l^{(1)}(s)<0\,,$$

$$\text{if}\quad l\geqq 2\quad\text{and}\quad 0\leqq s\leqq 4\,.$$

We have no Froissart-Gribov representation for the S-waves and the Froissart-Gribov representation for $T_1^1(s)$ gives no connection with positivity because it contains the absorptive part $A^{(1)}(t,s)$ whose sign is not definite. Therefore the S- and P-waves require more refined techniques than the higher waves.

It is hard to give an exhaustive discussion of the methods leading to the local properties à la Martin. There are many tricks and, in general, one is unable to predict if the trick one is using will provide us with a new result or with a result already implied by the conditions obtained from other tricks.

I shall restrict myself to a description of the technique used in one of the last papers on the subject, due to *Auberson, Brander, Mahoux* and *Martin* [9]. We consider an amplitude $T(s,t,u)$ which is a combination of T^I's with no symmetry properties under $s\leftrightarrow u$ exchange. This amplitude satisfies a twice subtracted fixed-t dispersion relation. Subtracting at the point $s=u=\frac{1}{2}(4-t)$, this dispersion relation can be written:

$$T(s,t,u)=\alpha(t)+\beta(t)\,z_t+\frac{1}{\pi}\int\limits_{x_0(t)}^{\infty}\mathrm{d}x\left\{A_s(s(x),t)\left(\frac{1}{x+z_t}-\frac{1}{x}+\frac{z_t}{x^2}\right)\right.$$
$$\left.+A_u(s(x),t)\left(\frac{1}{x-z_t}-\frac{1}{x}-\frac{z_t}{x^2}\right)\right\}\tag{3.1}$$

where $z_t=1+\dfrac{2s}{t-4}$, $\quad s(x)=\frac{1}{2}(4-t)(1+x)$, $\quad x_0(t)=\dfrac{4+t}{4-t}$.

A_s and A_u are the s- and u-channel absorptive parts. Using the Darboux-Christoffel formula:

$$\frac{1}{x-z_t}=\sum_{l=0}^{L-1}(2l+1)\,Q_l(x)\,P_l(z_t)+L\,\frac{Q_{L-1}(X)\,P_L(z_t)-Q_L(x)\,P_{L-1}(z_t)}{x-z_t} \tag{3.2}$$

for $L=2$ we get:

$$\begin{aligned}T(s,t,u)=\alpha(t)+\beta(t)\,z_t+\frac{1}{\pi}\int\limits_{x_0(t)}^{\infty}\mathrm{d}x\Bigg\{&A_s(s(x),t)\\ \times&\left[\left(Q_0(x)-\frac{1}{x}\right)-\left(Q_1(x)-\frac{1}{x^2}\right)z_t\right]+A_u(s(x),t)\\ \times&\left[\left(Q_0(x)-\frac{1}{x}\right)+\left(Q_1(x)-\frac{1}{x^2}\right)z_t\right]\Bigg\}+R_t(s,t,u)\,.\end{aligned} \tag{3.3}$$

The integral written explicitely and the rest $R_t(s,t,u)$ converge separately. Furthermore, as the S- and P-wave projections of $R_t(s,t,u)$ vanish, $(T(s,t,u)-R_t(s,t,u))$ coincides with the sum of the t-channel S- and P-wave parts of $T(s,t,u)$

$$T(s,t,u)=T_0(t)+3\,T_1(t)\,z_t+R_t(s,t,u),\quad (s,t,u)\in\Delta \tag{3.4}$$

where:

$$\begin{aligned}R_t(s,t,u)=\frac{1}{\pi}\int\limits_{x_0(t)}^{\infty}\mathrm{d}x\Bigg\{&A_s(s(x),t)\,\frac{Q_1(x)\,P_2(-z_t)-Q_2(x)\,P_1(-z_t)}{x-z_t}\\ &+A_u(s(x),t)\,[z_t\to -z_t]\Bigg\}\,.\end{aligned} \tag{3.5}$$

Similarly, we get from the fixed-s dispersion relation:

$$T(s,t,u)=T_0(s)+3\,T_1(s)\,z_s+R_s(s,t,u),\quad (s,t,u)\in\Delta\,. \tag{3.6}$$

A_s and A_u are linear combinations of the positive A^I's and $Q_l(x)>0$ for $x\geqq 1$. For suitable linear combinations the integrand of (3.5) is positive for $x_0<x<\infty$, at some points (s_0,t_0,u_0) of the triangle Δ. Therefore $R_t(s_0,t_0,u_0)>0$ and

$$T(s_0,t_0,u_0)>T_0(t_0)+3\,T_1(t_0)\left(1-\frac{2s_0}{4-t_0}\right). \tag{3.7}$$

Now it may happen that the integrand of $R_s(s,t,u)$ is negative at (s_0,t_0,u_0);

$$T(s_0,t_0,u_0)<T_0(s_0)+3\,T_1(s_0)\left(1-\frac{2t_0}{4-s_0}\right). \tag{3.8}$$

Eliminating $T(s_0, t_0, u_0)$ from (3.7) and (3.8) we get an inequality relating the values of S- and P-wave amplitudes at the points s_0 and t_0 of the interval $[0, 4]$. Examples of such inequalities are:

$$\begin{aligned} &T_0^{00}(0) > T_0^{00}(3.155) \\ &T_0^{00}(0.2134) > T_0^{00}(2.9863) \end{aligned} \tag{3.9}$$

$$\begin{aligned} &1.844\, T_1^1(0.2937) + 3.765\, T_1^1(2.4226) \\ &\quad < T_0^0(0.2937) - T_0^0(2.4226) - T_0^{00}(0.2937) + T_0^{00}(2.4226) \\ &0.6146\, T_1^1(0.2937) + 2.510\, T_1^1(2.4226) \\ &\quad > T_0^{00}(2.4226) - T_0^0(0.2937) + \tfrac{2}{3}\, T_0^{00}(0.2937) \end{aligned} \tag{3.10}$$

$T_l^{00}(s) = \frac{1}{3}\, T_l^0(s) + \frac{2}{3}\, T_l^2(s)$ are the $\pi^0 - \pi^0$ amplitudes.

If we choose $L = 3$ in the Darboux-Christoffel formula (3.2) we get, for example:

$$\begin{aligned} 1.494\, T_2^{00}(0.537) - 1.623\, T_2^{00}(2.363) &< T_0^{00}(0.537) - T_0^{00}(2.363) \\ &< 1.510\, T_2^{00}(0.537) - 1.622\, T_2^{00}(2.363). \end{aligned} \tag{3.11}$$

The meaning of these inequalities is that $T_0^{00}(0.537)$ is equal to $T_0^{00}(2.363)$, up to D-wave corrections.

Other techniques lead to inequalities involving derivatives of S- and P-wave amplitudes. For instance, the set of inequalities:

$$\begin{aligned} &\frac{d}{ds}\, T_0^{00}(s) < 0 \quad \text{for} \quad 0 \leqq s \leqq 1.127\,, \\ &\frac{d}{ds}\, T_0^{00}(s) > 0 \quad \text{for} \quad 1.7 \leqq s \leqq 4\,, \\ &\frac{d^2}{ds^2}\, T_0^{00}(s) > 0 \quad \text{for} \quad 0 \leqq s \leqq 1.7 \end{aligned} \tag{3.12}$$

shows that $T_0^{00}(s)$ has a unique minimum in the interval $[0, 4]$ located between $s = 1.127$ and $s = 1.7$. If we combine (3.12) with inequalities like

$$\begin{aligned} &T_0^{00}(3.155) < T_0^{00}(0) < T_0^{00}(4), \\ &T_0^{00}(2.9863) < T_0^{00}(0.2134) < T_0^{00}(3.205) \end{aligned} \tag{3.13}$$

we get a fairly good idea of the shape of the $\pi^0\pi^0$ S-wave in the interval $[0, 4]$.

4. Constraints on Moments of the Partial Waves Resulting from Positivity

The constraints I discuss in this Section are of another type than the local conditions described in the preceding section. They are inequalities relating moments of the partial wave amplitudes over the interval [0, 4]. I give an outline of the method used in [12].

Define the following moments of the partial waves:

$$\mu_{l,n}^{I} = \frac{1}{4^{l+n+2}} \int_{0}^{4} \mathrm{d}s (4-s)^{l+1} s^{n} T_{l}^{I}(s) . \tag{4.1}$$

If $l \geqq 2$, $T_l^I(s)$ can be replaced by its Froissart-Gribov representation (1.4) and $\mu_{l,n}^I$ becomes a linear combination of quantities like

$$d_{l,n}^{I} = \frac{1}{4^{l+n+2}} \int_{4}^{\infty} \mathrm{d}x \int_{0}^{4} \mathrm{d}s (4-s)^{l+1} s^{n} Q_{l}\left(\frac{2x}{4-s}-1\right) A^{I}(x,s) . \tag{4.2}$$

Now, if we replace $A^I(x,s)$ by its partial wave expansion, we get:

$$d_{l,n}^{I} = \sum_{k=0}^{\infty} \int_{4}^{\infty} \mathrm{d}x \, R_{n}^{l}(k,x) \, A_{k}^{I}(x), \qquad l \geqq 2 \tag{4.3}$$

with:

$$R_{n}^{l}(k,x) = \frac{(2k+1)}{4^{l+n+2}} \int_{0}^{4} \mathrm{d}s (4-s)^{l+1} s^{n} Q_{l}\left(\frac{2x}{4-s}-1\right) P_{k}\left(1+\frac{2s}{x-4}\right). \tag{4.4}$$

We ask for the properties of the $d_{l,n}^I$'s resulting from the fact that the absorptive parts $A_l^I(x)$ of the partial waves are positive. This problem is a generalized moment problem. For practical purposes it is useful to know the solution of the following reduced problem. Given a finite set $\{d_{l,n}^I \,|\, l+n \leqq \sigma, l \geqq 2\}$, we ask whether it is possible to construct a sequence of positive $A_l^I(x)$'s reproducing this set via (4.3). Evidently this is not possible for an arbitrary set, and we are looking for the necessary and sufficient conditions the set $\{d_{l,n}^I \,|\, l+n \leqq \sigma, l \geqq 2\}$ has to satisfy in order to be given by positive parts $A_l^I(x)$.

In order to get a compact statement of the conditions we are looking for, we have to develop our notation. Let $\boldsymbol{d}^I$ be the vector of $\mathbb{R}^{\frac{1}{2}\sigma(\sigma-1)}$ whose components are the elements of the set $\{d_{l,n}^I \,|\, l+n \leqq \sigma, l \geqq 2\}$. Furthermore let $\boldsymbol{b} \in \mathbb{R}^{\frac{1}{2}\sigma(\sigma-1)}$ ($\boldsymbol{b} = \{b_{l,n} \,|\, l+n \leqq \sigma, l \geqq 2\}$) be such that:

$$R_{\boldsymbol{b}}(k,x) \equiv \sum_{\substack{l,n \\ l+n \leqq \sigma \\ l \geqq 2}} b_{l,n} R_{n}^{l}(k,x) \geqq 0 \tag{4.5}$$

for $x \geqq 4$ and

$$k = 0, 2, 4, \ldots \quad \text{if} \quad I = 0 \text{ or } 2\,,$$

$$k = 1, 2, 3, \ldots \quad \text{if} \quad I = 1\,.$$

The vectors $\boldsymbol{b}$ satisfying (4.5) belong to a convex cone B^I_σ. It is immediately seen that a necessary condition for $\boldsymbol{d}^I$ is

$$(\boldsymbol{d}^I, \boldsymbol{b}) \equiv \sum_{\substack{l,n \\ l+n \leqq \sigma \\ l \geqq 2}} d^I_{l,n} b_{l,n} = \sum_{k=0}^{\infty} \int_4^{\infty} \mathrm{d}x\, R_{\boldsymbol{b}}(k, x)\, A^I_k(x) \geqq 0\,. \tag{4.6}$$

$\forall\, \boldsymbol{b} \in B^I_\sigma$. It has been shown [13] that (4.6) provides also the sufficient conditions for $\boldsymbol{d}^I$, ensuring the existence of positive $A^I_l(x)$ producing this vector through (4.3). (4.6) defines a convex cone D^I_σ, the dual cone of B^I_σ.

Therefore the necessary and sufficient condition we are looking for is

$$\boldsymbol{d}^I \in D^I_\sigma\,. \tag{4.7}$$

Now we can convert this condition into a condition for the "physical" moments $\mu^I_{l,n}$ or, equivalently, for the $f^I_{l,n}$'s introduced in Section 2.

Notice that, because of Bose statistics, $f^I_{l,n}$, as well as $\mu^I_{l,n}$, is defined only for $I = 0{,}2$ if l is even and $I = 1$ if l is odd. However, the expression of $f^I_{l,n}$ or $\mu^I_{l,n}$ in terms of the $d^I_{l,n}$'s contains $d^I_{l,n}$'s with $I = 0, 1, 2$ for every l. As a consequence the number $\frac{3}{2}\sigma(\sigma-1)$ of elements of the set $\{d^I_{l,n} | I = 0, 1, 2;\ l+n \leqq \sigma, l \geqq 2\}$ is larger than the number $\nu(\sigma)$ of elements of the "physical" set $\{f^I_{l,n} | l+n \leqq \sigma, l \geqq 2; I = 0{,}2 \text{ for } l \text{ even}, I = 1 \text{ for } l \text{ odd}\}$. This number is given by:

$$\nu(\sigma) = \begin{cases} \frac{1}{2}\sigma(\frac{3}{2}\sigma - 1) & \text{if } \sigma \text{ is even,} \\ \frac{1}{2}\sigma(\frac{3}{2}\sigma - 1) - \frac{1}{4} & \text{if } \sigma \text{ is odd.} \end{cases}$$

Let $\boldsymbol{f}$ be the vector of $\mathbb{R}^{\nu(\sigma)}$ whose components are the elements of the set $\{f^I_{l,n} | l+n \leqq \sigma,\ l \geqq 2;\ I = 0,\ 2 \text{ for } l \text{ even}, I = 1 \text{ for } l \text{ odd}\}$. The linear relations between the components of $\boldsymbol{f}$ and the components of the vectors $\boldsymbol{d}^I (I = 0, 1, 2)$ define a projection of $\mathbb{R}^{\frac{3}{2}\sigma(\sigma-1)}$ on $\mathbb{R}^{\nu(\sigma)}$. The condition (4.7) implies that $\boldsymbol{f}$ has to belong to the projection of $D_\sigma = D^0_\sigma \times D^1_\sigma \times D^2_\sigma$. This projection is a convex cone F_σ:

$$\boldsymbol{f} \in F_\sigma = \text{projection of } D_\sigma \text{ onto } \mathbb{R}^{\nu(\sigma)}\,. \tag{4.8}$$

The crossing constraints of Section 2 impose linear relations between the components of $\boldsymbol{f}$ and the S- and P-wave coefficients $f^I_{0,n}$ $(I=0,2)$ and $f^1_{1,n}(n \leqq \sigma)$. These coefficients are the components of a new vector $\boldsymbol{g}$ belonging to a space of S and P-wave coefficients. The restriction of $\boldsymbol{f}$ to the cone F_σ and the crossing relations imply that $\boldsymbol{g}$ has to be in a cone K_σ:

$$\boldsymbol{g} \in K_\sigma. \tag{4.9}$$

It is obvious from our derivation that (4.9) is a necessary condition for $\boldsymbol{g}$. As we shall see, (4.9) is not sufficient (in spite of the fact that the condition (4.7) for $\boldsymbol{d}^I$ is necessary and sufficient).

The outlined program has been carried through in detail in the simplest non trivial case $\sigma=3$ [12]. In this case, the points of K_3 are determined in terms of S waves only. (This is no longer true if $\sigma>3$.) Another simplifying feature of this case is that the cones B^I_3, D^I_3, F_3 and K_3 are bounded by hyperplanes. The restrictions we get are expressed by a finite number of linear inequalities. We have:

$$\begin{aligned}
&D^I_3: \begin{cases} d^I_{3,0} \geqq 0, \\ d^I_{2,0} - d^I_{2,1} \geqq \frac{5}{3} d^I_{3,0}, & I=0,2, \\ -d^I_{2,0} - d^I_{2,1} \geqq -d^I_{3,0}, \end{cases} \\
&D^1_3: \begin{cases} d^1_{3,0} > 0, \\ d^1_{2,0} - d^1_{2,1} \geqq \frac{7}{3} d^1_{3,0}, \\ -d^1_{2,0} + 7 d^1_{2,1} \geqq \frac{1}{3} d^1_{3,0}. \end{cases}
\end{aligned} \tag{4.10}$$

The cone K_3 is defined by 7 inequalities of the form:

$$\alpha_p f^0_{0,2} + \beta_p f^0_{0,3} + \gamma_p f^2_{0,2} + \delta_p f^2_{0,3} \geqq 0, \quad p=0,1,\dots 7. \tag{4.11}$$

The coefficients $\alpha_p, \dots, \delta_p$ are given in the following table:

p	α_p	β_p	γ_p	δ_p
1	12	5	24	10
2	9	−40	18	−80
3	162	−370	999	1510
4	36	71	72	−110
5	−6	1	33	5
6	−2	5	11	−10
7	−18	−32	99	−160

The known results dealing with the case $\sigma = 4$ concern only the $\pi^0 - \pi^0$ amplitudes [13, 14]. The boundary of the cone F_4^{00} is not formed by hyperplanes and has not been determined exactly. The following inequalities define a cone $\bar{F}_4^{00}$ which contains F_4^{00} [13]:

$$\begin{vmatrix} \mu_{2,0}^{00} & \mu_{2,1}^{00} \\ \mu_{2,1}^{00} & \mu_{2,2}^{00} \end{vmatrix} \geqq 0,$$

$$\begin{aligned} -4\,\mu_{2,1}^{00} + \ 25\,\mu_{2,2}^{00} &\geqq 0, \\ -5\,\mu_{2,0}^{00} + 128\,\mu_{2,1}^{00} - 300\,\mu_{2,0}^{00} &\geqq 0, \\ \mu_{2,0}^{00} - \ 18\,\mu_{2,1}^{00} + \ 45\,\mu_{2,2}^{00} &\geqq 0, \\ \mu_{2,0}^{00} - \ 16\,\mu_{2,1}^{00} + \ 36\,\mu_{2,2}^{00} &\geqq 0. \end{aligned} \tag{4.12}$$

The crossing relations define a section $\bar{K}_4^{00}$ of $\bar{F}_4^{00}$. The projection of $\bar{K}_4^{00}$ onto the space of the S-wave coefficients $f_{0,n}^{00}$ with $n \leqq 3$ is given by:

$$\begin{aligned} f_{0,2}^{00} &\geqq 0, \\ 9\,f_{0,2}^{00} - 40\,f_{0,3}^{00} &\geqq 0, \\ 3.36\,f_{0,2}^{00} + \ 5\,f_{0,3}^{00} &\geqq 0. \end{aligned} \tag{4.13}$$

We notice that the last inequality (4.13) is stronger than the first inequality (4.11) $p = 1$. This proves the fact mentioned before that the conditions obtained in the case $\sigma = 3$ are not the sufficient conditions the set $\{f_{l'n}^I | I = 0, 1, 2;\ l + n \leqq 3,\ l \geqq 2\}$ has to satisfy.

Conclusion: we have a method allowing the construction of an infinite set of constraints resulting from the positivity of the absorptive parts $A_l^I(s)$. This set is complete: it provides the necessary and sufficient conditions the $f_{l,n}^I$'s have to satisfy in order to secure this positivity of the absorptive parts. However we are not able to extract from this set of conditions the necessary and sufficient conditions a finite set of $f_{l,n}^I$'s has to satisfy. In other words the reduced problem for the "physical" quantities $f_{l,n}^I$ remains unsolved.

Finally I mention that *Balachandran* and *Blackmon* [15] and *Pennington* [16] have an elegant technique to exploit directly the positivity of some combinations of higher partial waves in the interval [0, 4]. However, whereas this posivity results from the positivity of the absorptive parts, it is not equivalent to it. Therefore the method I have indicated should give stronger results.

5. Threshold Behaviour at $s=0$

As it is well known, the left hand cut discontinuity of a partial wave amplitude $T_l^I(s)$ is determined by the crossed channel partial waves at physical energies if $-s_0<s<0$. If the Mandelstam representation is valid, $s_0=32$. In particular the threshold behaviour of $\operatorname{Im} T_l^I(s)$ at $s=0$ is rigorously related to the behaviour of the crossed channel partial waves at $s=4$. These relations are useful in our context and I shall discuss them briefly in this short Section [23].

If $s\leqq 0$ and sufficiently small:

$$T_l^I(s)=\frac{2}{4-s}\int_{\frac{1}{2}(4-s)}^{(4-s)} \mathrm{d}t\, T^I(s,t,u)\, P_l\left(1-\frac{2t}{4-s}\right) \tag{5.1}$$

and

$$\operatorname{Im} T_l^I(s+i\varepsilon)=\frac{2}{4-s}\int_4^{4-s} \mathrm{d}t\, \operatorname{Im} T^I(s+i\varepsilon, t-i\varepsilon, u)\, P_l\left(1-\frac{2t}{4-s}\right)$$

because $\operatorname{Im} T^I(s+i\varepsilon, t-i\varepsilon, u)=0$ for $\frac{1}{2}(4-s)<t<4$.

The relation can be written

$$\operatorname{Im} T_l^I(s+i\varepsilon)=-\frac{2}{4-s}\sum_{I'} C_{II'}\int_4^{4-s} \mathrm{d}t\, A^{I'}(t,s)\, P_l\left(1-\frac{2t}{4-s}\right) \tag{5.2}$$

In this expression, $A^{I'}(t,s)$ can be replaced by its partial wave expansion:

$$A^{I'}(t,s)=\sum_{l=0}^{\infty}(2l+1)\, A_l^{I'}(t)\, P_l\left(1+\frac{2s}{t-4}\right). \tag{5.3}$$

If we assume a normal threshold behaviour at the elastic threshold, $A_l^{I'}(t)$ is given, for t sufficiently small, by the expansion

$$A_l^{I'}(t)=(t-4)^{2l+\frac{1}{2}}\left(\alpha_l^{I'}+\beta_l^{I'}(t-4)+\ldots\right). \tag{5.4}$$

After successive insertion of (5.4) into (5.3) and of (5.3) into (5.2), we get an expansion of $\operatorname{Im} T_l^I(s+i\varepsilon)$ in powers of $\sqrt{s}$. This shows that $T_l^I(s)$ has a square-root type branch point at $s=0$. There is no $s^{\frac{1}{2}}$-term in the expansion of $\operatorname{Im} T_l^I(s+i\varepsilon)$ and we have, restricting ourself to S- and P-waves [29]:

$$\begin{aligned}\operatorname{Im} T_0^I(s+i\varepsilon)=\sum_{I'=0,2} C_{II'}\Bigg[&-\frac{1}{6}(a_{I'})^2(-s)^{\frac{3}{2}}\\ &+\left(\frac{13}{240}(a_{I'})^2-\frac{1}{10}\left[\frac{\mathrm{d}}{\mathrm{d}s}|T_0^{I'}(s)|^2\right)_{s=0}\right)(-s)^{\frac{5}{2}}\Bigg]+O((-s)^{\frac{7}{2}}),\end{aligned} \tag{5.5}$$

$$\begin{aligned}\operatorname{Im} T_1^1(s+i\varepsilon) = \sum_{I'=0,2} C_{1I'} \Bigg[& -\frac{1}{6}(a_{I'})^2(-s)^{\frac{3}{2}} \\ & + \left(\frac{7}{80}(a_{I'})^2 - \frac{1}{10}\left(\frac{d}{ds}|T_0^{I'}(s)|^2\right)_{s=0}\right)(-s)^{\frac{5}{2}}\Bigg] + O((-s)^{\frac{7}{2}}) .\end{aligned} \tag{5.6}$$

a_I is the isospin I scattering length $(I=0, 2)$: $a_I = T_0^I(4)$. We see that up to $(-s)^{\frac{7}{2}}$-terms the threshold behaviour at $s=0$ of the S- and P-wave left hand cut discontinuities is determined by the behaviour of the S-waves at the elastic threshold.

6. Constructing Low-energy Models of Pion-pion Scattering

As it was stated at the beginning of these lectures, the rigorous properties of the partial waves in the interval [0, 4] are likely to impose restrictions on the possible behaviour of the low-energy phase shifts. Therefore, they should be taken into account in the construction of low-energy models.

A first way of applying our conditions is to test if models for S and P waves obtained from other sources are consistent with positivity and crossing symmetry. Examples of such models are the partial waves obtained by *Lovelace* [17] through unitarization of the Veneziano amplitudes. The unitarization procedure destroys crossing symmetry. The sum rules (2.3) allow an evaluation of the extent to which crossing symmetry is violated. This has been done [18] and the violation is claimed to be significant.

A second application of the rigorous constraints is to use them in order to improve existing models. For instance they can be used to determine a unitarization procedure of the Veneziano amplitudes which minimizes the violation of crossing symmetry and positivity [18, 19].

The current algebra models constitute another type of models which can be improved with the help of the rigorous constraints. The original Ansatz of *Weinberg* [20] corresponds to a set of amplitudes, linear in s, t and u, like those encountered on page 28. The current algebra conditions determine the constants a and b. One of the improved current algebra models which have been proposed is due to *Iliopoulos* [21]. He constructs a third order expansion of $T^I(s, t, u)$ in powers of the momenta $k_s = \frac{1}{2}\sqrt{4-s}$, $k_t = \cdots$, $k_u = \cdots$. Imposing the current algebra conditions and elastic unitarity (up to third order) at threshold, he gets four solutions. The local conditions discussed in Section 3 [9] as well as the conditions on moments of Section 4 [12] reject three of these solutions. The remaining solution produces the scattering lengths predicted by *Weinberg*. Another unitarization procedure using crossing and positivity constraints is applied to the Weinberg amplitudes in [22].

A third possibility of applying the rigorous conditions is to use them as basic conditions determining the parameters of a model. Such a model can be parametrized in such a way that it exhibits exact elastic unitarity. It should be emphasized immediately that our constraints do not define a unique model, in spite of the fact that, because we are imposing unitarity, the conditions on the parameters are necessarily non linear. As we may learn from Atkinson's lecture notes, a unitary amplitude satisfying the Mandelstam representation is far from being uniquely determined; its double spectral function contains an arbitrary function. Such an amplitude satisfies all our constraints. Therefore we need some kind of extra input.

This input can be in the type of parametrization choosen. This has been experienced in the case of neutral theory models submitted to the local conditions [23, 24].

The S-wave can be written

$$T_0(s) = \frac{1}{K(s) - i\varrho(s)}, \qquad \varrho(s) = \sqrt{\frac{s-4}{s}}, \tag{6.1}$$

where $K(s)$ is regular at $s=4$. A specific model corresponds to a specific parametrization. In all the cases which have been studied it appears that the number of practically independent parameters is drastically reduced by the constraints. One finds always pairs of inequalities which define rather small windows for the values of some parameters.

At the time the first charged theory models were constructed [25, 26], only the first positivity constraints for the $\pi^0 - \pi^0$ S-wave were known. Therefore, it was not possible to construct simultaneously a model for the $I=0$ and the $I=2$ S-wave. An $I=2$ S-wave satisfying elastic unitarity was choosen as input in accordance with our rough experimental knowledge (negative, slowly decreasing phase shift, $\delta_0^2(m_\varrho^2) \sim -20^0$).

An $I=0$ S-wave fulfilling elastic unitarity was constructed from this input through the positivity constraints on the $(\frac{1}{3} T_0^0(s) + \frac{2}{3} T_0^2(s))$ combination and the threshold conditions at $s=0$. The results of *Auberson, Piguet*, and *Wanders* [25] and of *Bonnier* [26] coincide practically at low energies (below the mass of the ϱ-meson), in spite of the use of the different parametrizations. The $I=0$ S-wave phase shift is rapidly rising and reaches values near 90^0 in the ϱ-region without going through this value. A different $I=2$ input has been proposed by *Castoldi* [27]; this does not alter substantially the $I=0$ S-wave. Similar phase shifts have been obtained by applying the Padé method to the σ-model [28].

As we have seen in the preceding Sections, at present we know conditions which involve other S-wave combinations than the $\pi^0-\pi^0$ combination; we have also constraints containing the $I=1$ P-wave. This allows the construction of S-wave and P-wave models with the information of the existence of the ϱ-meson as input. A first investigation in this direction has been performed by *Bonnier* and *Gauron* [29]. They display three solutions; all of them exhibit an $I=0$ S-wave resonance slightly below the ϱ-meson mass. One of their solutions is compatible with the current algebra scattering lengths but produces an $I=2$ S-wave phase shift which contradicts the experimental results. Models are presently being constructed along similar lines by *Le Guillou*, *Morel*, and *Navelet* [30] and *Piguet* [31].

References

1. Proofs and references concerning the results presented in Section 1 can be found in: *Martin, A.*: Scattering Theory: Unitarity, Analyticity and Crossing. Lecture Notes in Physics, Vol. 3. Berlin-Heidelberg-New York: Springer 1969.
2. *Balachandran, A. P., Nuyts, J.*: Phys. Rev. **172**, 1821 (1968).
3. *Roskies, R.*: Nuovo Cimento **65** A, 467 (1970).
4. *Basdevant, J. L., Cohen-Tannoudji, G., Morel, A*: Nuovo Cimento **64** A, 585 (1969).
5. *Martin, A.*: Nuovo Cimento **47** A, 265 (1967).
6. *Common, A. K.*: Nuovo Cimento **53** A, 946 (1968).
7. *Martin, A.*: Nuovo Cimento **58** A, 303 (1968).
8. — Nuovo Cimento **63** A, 167 (1969).
9. *Auberson, G., Brander, O., Mahoux, G., Martin, A.*: Nuovo Cimento **65** A, 743 (1970).
10. *Cheung, F. F. K., Chen Cheung, F. S.*:Nuovo Cimento **65** A, 347 (1970).
11. *Auberson, G.*: Saclay preprint (1970).
12. *Piguet, O., Wanders, G.*: Phys. Letters **30** B, 418 (1969).
13. — unpublished.
14. *Roskies, R.*: Yale University preprint (1969).
15. *Balachandran, A. P., Blackmon, M. L.*: Phys. Letters **31** B, 655 (1970).
16. *Pennington, M. R.*: University of London, Westfield College preprint 1970.
17. *Lovelace, C.*: Proceedings of the Argonne Conference on π and K Interactions (Argonne, Ill. USA 1969) p. 562.
18. *Lipinski, H. M.*: Wisconsin University preprint COO 264 (1969).
19. *Baier, R., Kühnelt, H., Widder, F.*: Lettere al Nuovo Cimento **3**, 594 (1970).
20. *Weinberg, S.*: Phys. Rev. Letters **17**, 616 (1966).
21. *Iliopoulos, J.*: Nuovo Cimento **53** A, 552 (1968).
22. *Baier, R., Kühnelt, H., Widder, F.*: preprint (1970). To appear in Acta Phys. Aust.
23. *Wanders, G., Piguet, O.*: Nuovo Cimento **56** A, 417 (1968).
24. *Bonnier, B.*: Paris preprint IPNO/TH 161 (1969).
25. *Auberson, G., Piguet, O., Wanders, G.*: Phys. Letters **28** B, 41 (1968).
26. *Bonnier, B.*: Nucl. Phys. B **10**, 467 (1969).

27. *Castoldi, P.*: Nucl. Phys. B **12**, 567 (1969).
28. *Basdevant, J. L., Lee, B. W.*: Saclay preprint (1970).
29. *Bonnier, B., Gauron, P.*: Paris preprint IPNO/TH 176 (1970).
30. *Le Guillou, J. C., Morel, A., Navelet, H.*: private communication.
31. *Piguet, O.*: in preparation.

Dr. G. *Wanders*
Institut de Physique Théorique,
Université de Lausanne
CH-1005 Lausanne

New Methods in the Analysis of π—N Scattering*

J. HAMILTON

Contents

I. Scope of the Lectures

These lectures deal with recent developments in the analysis of $\pi N\to\pi N$ scattering and other experimental data in order to obtain information on the structure of the πN interaction and to study related topics such as the existence of the σ-meson. The emphasis is on the new methods which are now available, and in particular on the modern methods of using analytic continuation. Predictions of πN scattering will not be discussed here.

* I am indebted to the Theoretical Studies Division of CERN for facilities and hospitality while these notes were prepared.

Briefly the new methods are:

a) Determining the πN partial wave amplitudes (p. w. a.) $f_{l\pm}(s)$ in new unphysical regions in terms of experimental data. These are on the real axis near $s=0$, and on the cut $-(M^2-\mu^2)\leqq s<0$.

b) Determination of the p. w. a. discontinuity across the t-channel cut, i. e. across the circle $|s|=M^2-\mu^2$, from experimental data on $\pi N\to\pi N$ by analytic continuation methods. In a similar fashion information about the cut $-\infty<s<(-M^2+\mu^2)$ in the p. w. a. can be obtained.

c) The relation of *backward* $\pi N\to\pi N$ to $\pi\pi\to N\bar{N}$ using correct analytic continuation methods. A further new method here makes it possible to extract the helicity amplitudes $f^J_\pm(t)$ for $\pi\pi\to N\bar{N}$ by paying due attention to the convergence of the Legendre series.

d) From the application of these techniques we obtain consistant descriptions of the $T=0\,J=0$ (σ-meson), and $T=1\,J=1$ (ϱ-meson), $\pi\pi\to N\bar{N}$ amplitudes $f^J_\pm(t)$. The results of the σ-case in particular will be discussed and compared with other methods.

II. Partial Wave Amplitudes in New Regions

II.1 Introduction

We use the s-plane where

$$s=W^2=[(M^2+q^2)^{\frac{1}{2}}+(\mu^2+q^2)^{\frac{1}{2}}]^2$$

M,μ being the nucleon and pion masses and q the momentum in the πN c. m. s. Partial wave amplitudes have the form

$$f_{l\pm}(s+i\varepsilon)=\frac{e^{i\alpha}\cdot\sin\alpha}{q}\qquad(\alpha\text{ real})$$

on the elastic portion of the physical cut.

The cut structure of $f_{l\pm}(s)$ is shown in Fig. 1 where the causes of the cuts are also indicated. For further explanation of how the cuts arise and of the various exchange process see, for example, Ref. [1]. Here we shall not worry about the dynamical aspects for the moment — we shall instead use any relevant physical data to determinate $f_{l\pm}(s)$ on as much as possible of the unphysical real axis cuts. Notice that we want $f_{l\pm}(s)$ itself, not merely $\operatorname{Im} f_{l\pm}(s)$.

This problem was solved long ago [2] for the "crossed physical cut" $0<s<(M-\mu)^2$. We can usefully illustrate the method by using Fig. 2.

The p. w. a. are defined by

$$f_{l\pm}(s)=\tfrac{1}{2}\int_{-1}^{+1}dx_s\{f_1(s,x_s)\,P_l(x_s)+f_2(s,x_s)\cdot P_{l\pm1}(x_s)\}\tag{1}$$

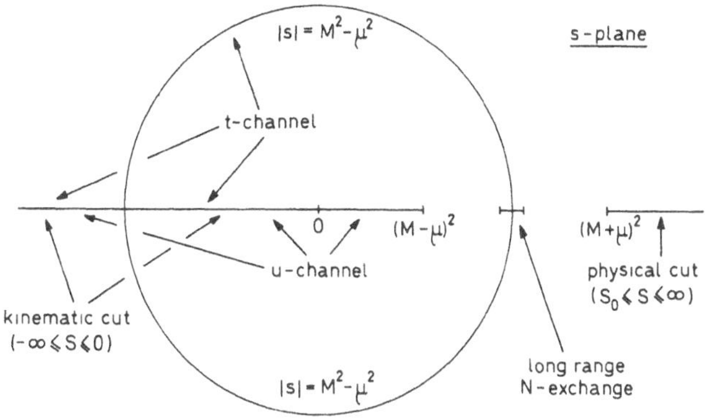

Fig. 1. The cuts of the πN partial wave amplitude $f_{l\pm}(s)$ and a description of the exchange processes etc. which cause them

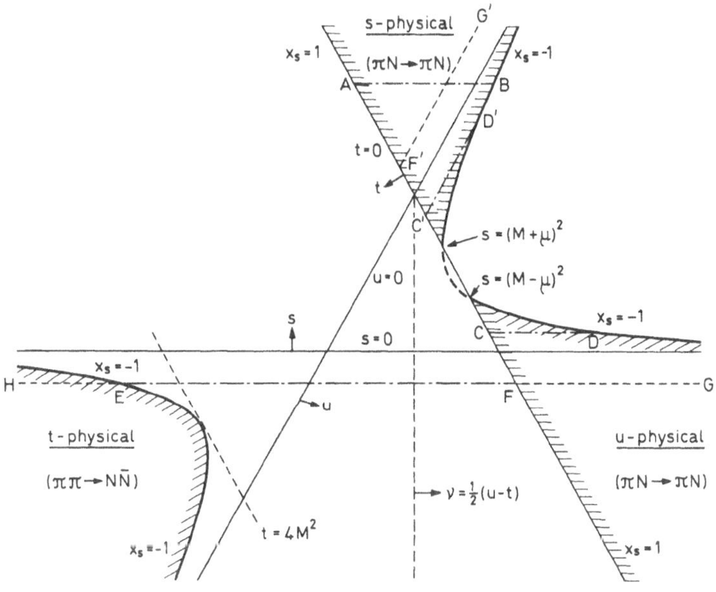

Fig 2. The boundaries of the three physical regions

when x_s is the cosine of the $\pi N \to \pi N$ scattering angle and the πN helicity amplitudes are given by*

$$f_{1,2}(s, x_s) = \frac{(W \pm M)^2 - \mu^2}{16\pi s} \{\pm A(s, t) + (W \mp M) B(s, t)\} . \qquad (2)$$

Here $A(s, t)$, $B(s, t)$ are the invariant amplitudes.

* For subscript 1 (2) use the top (bottom) signs.

Using the well known formulae

$$s+t+u=2M^2+2\mu^2$$
$$t=-2q^2(1-x_s) \tag{2a}$$
$$4q^2s=s^2-2s(M^2+\mu^2)+(M^2-\mu^2)^2$$

we find that the curve $x_s=-1$ which gives one edge of the s-physical region can be written

$$s\cdot u=(M^2-\mu^2)^2\,. \tag{3}$$

The other boundary of the s-physical region is $x_s=+1$ (i. e. $t=0$). These curves are shown in Fig. 2.

Because Eq. (3) and the relation $t=0$ are unaltered under the exchange of s and u, the curve (3) and $x_s=+1$ also give the boundaries of the u-physical region. In the channel $\pi\pi\to N\bar{N}$, the nucleon and pion momenta in the c. m. s. are given by

$$p_t^2=t/4-M^2, \qquad q_t^2=t/4-\mu^2\,.$$

Also

$$-s=p_t^2+q_t^2-2p_tq_t\cdot\cos\vartheta_t \tag{2b}$$

where $\cos\vartheta_t$ gives the scattering angle. Simple algebra shows that the boundary of the physical region for $\pi\pi\to N\bar{N}$, i. e. $\cos^2\vartheta_t=1$ is again given by Eq. (3).

Thus we have the particularly simple result that the line $x_s=+1$ and the hyperbola $x_s=-1$ give the boundaries of all three physical regions [3].

For physical s the integration in Eq. (1) is equivalent to integration along the line AB in Fig. 2. For $0<s<(M-\mu)^2$ it is the equivalent to integration along the line CD in Fig. 2. By crossing symmetry ($s\leftrightarrow u$, $t\leftrightarrow t$) the values of the invariant amplitudes $A(s,t)$, $B(s,t)$ at any point of CD are simply related to their values at the corresponding point on the line C′D′. Thus by using Eq. (1), (2) we can easily find $f_{l\pm}(s\pm i\varepsilon)$ for $0<s<(M-\mu)^2$ in terms of physical $\pi N\to\pi N$ scattering data.

II.2 Near $s=0$

From Eq. (2) it follows that as $s\to 0$

$$f_{1,2}(s,x)\simeq\pm\frac{M^2-\mu^2}{16\pi s}\{A(s,t)-MB(s,t)\}$$

and this may give rise to some singular behaviour in $f_{l\pm}(s)$ at $s=0$ (Eq. 1). Moreover we see from Fig. 2 that as s passes through zero we have to give up the prescription of integrating along the line CD; also $D \to \infty$ as $s \to +0$ so amplitudes at infinite energy are involved as $s \to +0$. Clearly we expect $f_{l\pm}(s)$ to be singular [2] at $s=0$, and in the early work various models [4, 5] were used to estimate the behaviour near $s=0$. The nature of the singularity has now been given by *Jakob* and *Steiner* [6] and by *J. Lyng Petersen* [7].

As $s \to +0$ the line C′D′ (Fig. 2) is for almost the whole of its length close to the curve $x = -1$. Therefore $f_{l\pm}(s)$ near $s=0$ is determined [4] by high energy backward $\pi N \to \pi N$ scattering. This backward scattering is dominated by Δ-exchange (Fig. 3), and fitting the appropriate Regge formulae [8] one finds that near $s=0$ the S-wave has the singularity

$$\begin{aligned} f_0(s) &\simeq C \cdot s^{-\alpha_\Delta(0) - \frac{1}{2}} \\ &\simeq C \cdot s^{-0.69} \end{aligned} \tag{4}$$

where C is a complex constant.
The numerical value of C is determined by the Regge parameters [9].

In the accurate work to be described in § III the Δ and N trajectories were included in determining $f_0(s)$ in the neighbourhood of $s=0$.

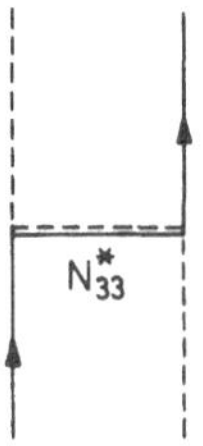

Fig. 3

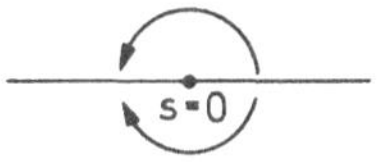

Fig. 4

Fig. 3. An example of the Δ-exchange which is important in near backward $\pi N \to \pi N$ at high energies

Fig. 4. Analytic continuation of Eq. (4) around the singular point $s=0$

It is clear that Eq. (4) (or the more accurate expression) can be continued to give $f_0(s)$ for small negative values of s. The continuation is made separately in $\operatorname{Im} s > 0$ and in $\operatorname{Im} s < 0$ (Fig. 4). Using the Δ trajectory approximation (Eq. 4) it is easy to see that

$$\operatorname{Re} f_0(s + i\varepsilon) = \operatorname{Re} f_0(-s + i\varepsilon)$$

$$\operatorname{Im} f_0(s + i\varepsilon) = -\operatorname{Im} f_0(-s + i\varepsilon)$$

for s small and positive [6]. This analytic continuation agrees with the more general result [7] in the next section.

II.3 On $-(M^2-\mu^2)<s<0$

Again returning to Fig. 2 we see that to evaluate $f_{l\pm}(s)$ on $-\infty<s<0$ we must in Eq. (1) integrate along the line EF. We cannot use a Legendre expansion in any channel to find $A(s,t)$, $B(s,t)$ on EF; such an expansion only converges for $\operatorname{Im}A$ and $\operatorname{Im}B$ when $s\gtrsim -3\mu^2$ and because of the kinematic factor W in Eq. (2) we also require $\operatorname{Re}A$ and $\operatorname{Re}B$ even to get $\operatorname{Im}f_{l\pm}(s)$ when $s<0$.

This difficulty was avoided by *Lyng Petersen* [7] by replacing the integration from E to F by an integration from F to G, around the point at infinity, and finally from H to E. This integration around the point at infinity is in the spirit of the similar procedure in deriving finite energy sum rules (FESR) [10].

In order to be able to replace the integration over x_s from -1 to $+1$ by a closed contour in this way, *without further complications*, it is in the first place necessary that for $\operatorname{Im}s=\varepsilon$ (ε being small and positive) the t-channel and the u-channel cuts are both above or both below $-1\leqq x_s\leqq +1$. By Eqs. (2a)

$$x_s=1+t/2q^2$$

$$4\frac{dq^2}{ds}=(1-R^2/s^2)\qquad (R=M^2-\mu^2).$$

Since $q^2<0$ on $-\infty<s<0$, the t-channel cut runs from $-\infty$ up to*

$$x_\mu=1+4\mu^2/q^2<+1.$$

For $-R<s<0$, $\operatorname{Im}(q^2)<0$ for $\operatorname{Im}s>0$, so $\operatorname{Im}(1/q^2)>0$, and this cut lies in $\operatorname{Im}x_s>0$.
By Eqs. (2a)

$$u=R^2/s-2q^2(1+x_s) \tag{5}$$

so

$$x_s=-1+(\tfrac{1}{2}q^2)\cdot(R^2/s-u)\,. \tag{5a}$$

The u channel singularities are the nucleon pole at $u=M^2$ and the cut $(M+\mu)^2\leqq u\leqq\infty$. The pole gives $x_M>-1$, and the cut runs to $x_s\to+\infty$ (Fig. 5). Also $\operatorname{Re}x_s=+1$ for

$$u=2\cdot(M^2+\mu^2)-s>2(M^2+\mu^2) \tag{5b}$$

By using Eq. (5a) we can show that if $\operatorname{Im}s>0$, then $\operatorname{Im}x_s>0$ for

$$(M+\mu)^2\leqq u<2R^2\cdot\frac{M^2+\mu^2-s}{R^2-s^2} \tag{5c}$$

---

* Remembering that the t-channel cut is $4\mu^2\leqq t<\infty$.

Moreover the upper value of u in Eq. (5c) exceeds the value in Eq. (5b). Therefore both the t and the u cuts lie above $-1 \leqq x_s \leqq +1$ if $\operatorname{Im} s > 0$ and $-R < s < 0$ (Fig. 5).

Finally we notice that for $-\infty < s < -R$, the t-channel cut lies on the opposite side of $-1 \leqq x_s \leqq +1$, and the simple method to be described below will not be valid.

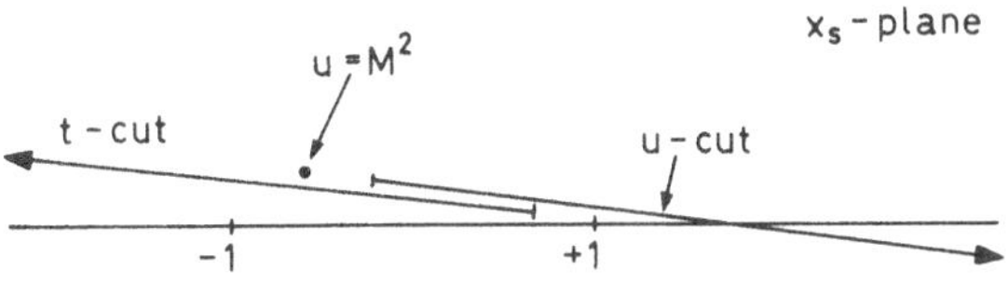

Fig. 5. Location of the t and u-channel singularities in the x_s plane for $\operatorname{Im} s > 0$ and $-R < s < 0$.

The Integration

It is convenient to replace x_s by

$$v = \tfrac{1}{2}(u - t) \tag{6}$$

By Eq. (5)

$$v = -R^2/2s + 2q^2 x_s . \tag{6a}$$

On $-R < s < 0$, v increases as x_s increases. The points E and F are

$$v_1 = R^2/s - (M^2 + \mu^2) + \tfrac{1}{2} \cdot s < 0$$
$$v_2 = M^2 + \mu^2 - \tfrac{1}{2} s$$

respectively. The integration in Eq. (1) can be written

$$f_{l\pm}(s) = \int_{v_1}^{v_2} \mathrm{d}v\, f(v, s) \tag{7}$$

where $f(v, s)$ depends on the invariant amplitudes A, B and various kinematic factors. Of course s is fixed on $-R < s < 0$ (and $\operatorname{Im} s > 0$).

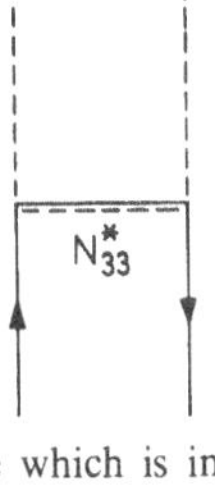

Fig. 6. An example of the Δ-exchange which is important in near forward $\pi\pi \to N\bar{N}$ at high energies.

If we were to continue the integration along the line FG (Fig. 2) that would involve the amplitudes along the line F′ G′. At the upper end of such a path we are again close to backward scattering in $\pi N \to \pi N$. Thus Δ exchange is again dominant (Fig. 3) and it is easy to see that for ν large, $f(\nu, s)$ behaves like

$$\nu^{(\alpha_\Delta(s) - \frac{1}{2} + l)} \quad \text{for} \quad f_{l-}(s)$$

$$\nu^{(\alpha_\Delta(s) + \frac{1}{2} + l)} \quad \text{for} \quad f_{l+}(s).$$

We now assume, in the usual way, that $f(\nu, s)$ behaves in this fashion for $|\nu|$ large. Of course this implies that Δ-exchange is dominant in near forward $\pi\pi \to N\bar{N}$ at high energies, ie. near H in Fig. 2.

For the S-wave $f(\nu, s)$ behaves as $\nu^{(\alpha_\Delta(s)+\frac{1}{2})}$ for large $|\nu|$. Since $\alpha_\Delta(0) \approx -0.19$, the exponent is positive for most of the range (for $-40\mu^2 \lesssim s < 0$, in fact). *Lyng Petersen* then writes Eq. (7) as

$$f_0(s) = \int_{+\nu_1}^{(-\nu_1)} d\nu f(\nu, s) - \int_{+\nu_2}^{(-\nu_1)} d\nu f(\nu, s). \tag{8}$$

For small $|s|$, $(-\nu_1)$ is very large and the first integral in Eq. (8) can be evaluated as in the FESR method, ie. it is evaluated by integration along a large semi-circle of radius $|\nu_1|$ in the ν-plane*. When s comes near to $-40\mu^2$, $(-\nu_1)$ is barely in the range where the Regge-method is applicable and we would only get a rough result (see however below).

The second integral in Eq. (8) (from ν_2 to $-\nu_1$) is in the physical $\pi N \to \pi N$ range. It consists of a portion of the line F′G′ staring at F′. It presents no problem in principle, and with the results of modern phase shift analyses and the Regge analysis it can be evaluated with fair or good accuracy.

For $s \gtrsim -R$ a different technique can be used [7]. Here $-1 < \alpha_\Delta(s) + \frac{1}{2} < 0$, and if $f^R(\nu, s)$ denotes the Δ and N trajectories, leading contribution to $f(\nu, s)$ we then expect (on reasonable assumptions about the Δ daughter trajectory) that $(f(\nu, s) - f^R(\nu, s))$ is super convergent, ie. its integral over the semi-circle of infinite radius will vanish. Thus we can write

$$\begin{aligned} f_0(s) = &- \int_{-\infty}^{\nu_1} d\nu (f(\nu, s) - f^R(\nu, s)) - \int_{\nu_2}^{\infty} d\nu (f(\nu, s) - f^R(\nu, s)) \\ &+ \int_{\nu_1}^{\nu_2} d\nu f^R(\nu, s). \end{aligned} \tag{9}$$

(What has been done here is to integrate $(f(\nu, s) - f^R(\nu, s))$ around a closed contour comprising the real ν axis and the semi-circle $|\nu| \to \infty$).

* See Ref. [7] for more details.

Eq. (9) is in principle expressed in terms of quantities derived from physical scattering processes. However in practice there is some difficulty about giving an accurate evaluation of the right-hand side of Eq. (9). The first integral requires knowledge of the $\pi\pi \rightarrow N\bar{N}$ amplitude at energies $t^{\frac{1}{2}}$ not much above the threshold $2M$ whenever $s \simeq -R$. No doubt the Regge expressions are not accurate down to such low values of t, and we have little direct information on the amplitudes for $\pi\pi \rightarrow N\bar{N}$.

Finally we should remark that the value of $f_0(s)$ for small negative s as determined by FESR method is just the analytic continuation of the value obtained for small positive s in § II, as it should be.

Summary

By the new methods discussed in § II it is possible in principle to determine the πN p. w. a. $f_{l\pm}(s)$ near $s=0$ and on $-(M^2-\mu^2)<s<0$. Near $s=0$ the p. w. a. are determined by high energy backward $\pi N \rightarrow \pi N$ or forward $\pi\pi \rightarrow N\bar{N}$. On $-(M^2-\mu^2)<s<0$ a FESR type method is used, and we no longer have the simple interpretation in terms of various exchange forces. However given sufficient experimental data we can find $f_{l\pm}(s)$.

These new methods have been used by *Henry Nielsen*, *J. Lyng Petersen* and *E. Pietarinen* [11] to find $f_0(s)$ on $-(M^2-\mu^2)<s\lesssim 8\mu^2$. As was seen above the evaluation is poor for $s \simeq -(M^2-\mu^2)$ on account of poor experimental data.

Because of lack of space new work by *Baacke* and *Steiner* [12] on determining πN p. w. a. in certain regions of real positive s from fixed t dispersion relations cannot be discussed.

III. Discrepancy Method and Analytic Continuation

III.1 Introduction

Some years ago the discrepancy method was invented [4] to extract information on the dynamics of the πN interaction, including the t-channel effects, from the experimental data on πN p. w. a. In the final form of that early work there was clear evidence [1] for a strong interaction in the $T=0$ $J=0$ $\pi\pi \rightarrow N\bar{N}$ amplitude as well as the ϱ-exchange term (in the $T=1$ $J=1$ $\pi\pi \rightarrow N\bar{N}$ amplitudes). The analysis was carried out for the S-wave and the P-wave πN amplitudes [13] and they gave self consistent results.

The interpretation of the discrepancy function in the early work was to some extent intuitive, and it seemed desirable to find the solution given by analytic continuation techniques, while at the same time making use of the wealth of data now available on πN phase shifts and Regge parameters etc. Here we report on this programme as applied to the πN s-wave amplitudes $f_0^{(\pm)}(s)$ by the authors of Ref. [11] (to be referred to as NPP).

III.2 The Discrepancy Function

The charge notation for any $\pi N \to \pi N$ amplitude T is as follows:

$$T^{(\pm)} = \tfrac{1}{2}(T_- \pm T_+)$$

where $T_\pm$ are the amplitudes for $\pi^\pm p \to \pi^\pm p$. On going to the t-channel the $(+), (-)$ amplitudes are related respectively to isospin $T=0$, $T=1$ $\pi\pi \to N\bar{N}$ scattering.

The authors NPP [11] use the discrepancy functions $\Delta^{(\pm)}(s)$ defined by

$$\begin{aligned} \Delta(s) = f(s) - \frac{1}{\pi}\Bigg\{ \int_{(M+\mu)^2}^{s_p} ds' \frac{\operatorname{Im} f(s')}{s'-s} + \int_{R^2/s_p}^{(M-\mu)^2} ds' \frac{\operatorname{Im} f(s')}{s'-s} \\ + \int_{-R}^{R^2/s_p} ds' \frac{\operatorname{Im} f(s')}{s'-s} + \int_{s_p}^{\infty} ds' \frac{\operatorname{Im} f(s')}{s'-s} \Bigg\}. \end{aligned} \tag{10}$$

Here $R = M^2 - \mu^2$. For simplicity we write $f(s) \equiv f_0^{(\pm)}(s)$, the two s-wave amplitudes. The value s_p is the highest value for which the πN phase shifts are available.

The term $\operatorname{Re} f(s)$ is obtained from the data on $(M+\mu)^2 \leqq s \leqq s_p$, and by the crossing method of § II.1 above on $R^2/s_p < s \leqq (M-\mu)^2$. Similarly the first two integrals on the right of Eq. (10) can be evaluated accurately. The phase shift analyses go up to about 2 GeV (lab. pion energy) so $s_p \approx 250\,\mu^2$ (μ = pion mass) and $R^2/s_p \simeq 8\,\mu^2$.

The values of the $\operatorname{Im} f(s')$ required in the third integral are obtained by the methods of § II.2 and § II.3 above. These will be approximate values; however, the inclusion of this integral is important. Since the $f(s')$ values join smoothly at R^2/s_p there is no logarithmic singularity in $\Delta(s)$ at that point. Moreover including the Δ and N-trajectories removes the divergence which would otherwise exist in $\Delta(s)$ at $s=0$.

The fourth integral in Eq. (10) is evaluated by assuming a somewhat arbitrary high energy behaviour for $\operatorname{Im} f(s')$, going as $(s')^{-\frac{1}{2}}$ for large s' and joining smoothly onto the phase shift value at s_p. This prevents any logarithmic singularity in $\Delta(s)$ at s_p. In Eq. (10) there is no long range Born term because these terms are very small [1].

With these input data $\Delta^{(\pm)}(s)$ are evaluated by NPP [11] on $(M+\mu)^2 \leqq s \lesssim 180\,\mu^2$ and $8\,\mu^2 \lesssim s \leqq (M-\mu)^2$. (Figs. 7a, 7b). One could in principle go further towards negative s or to higher positive s, but it is feared that the accuracy of the FESR method and of the conjectured high energy form of $\operatorname{Im} f(s')$ would not justify that.

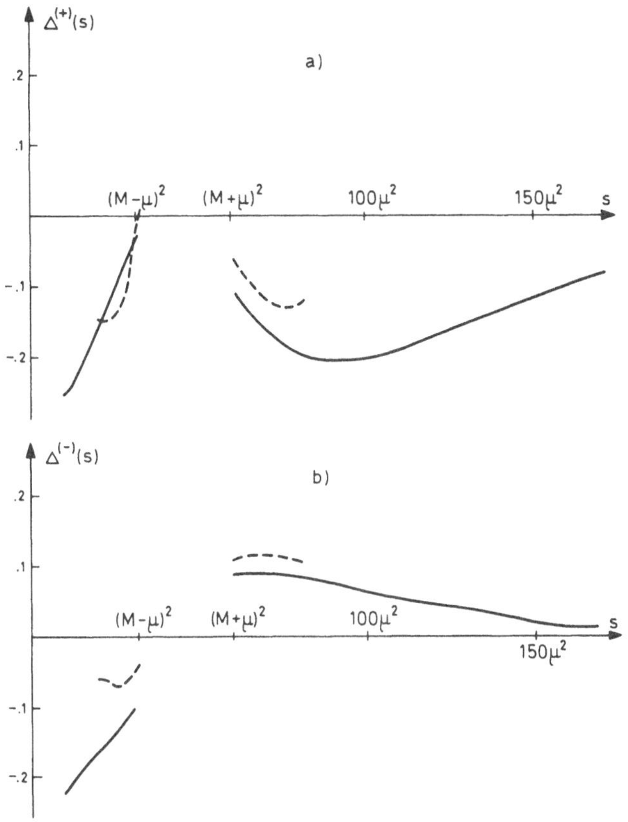

Fig. 7a and b. The values of the S-wave πN discrepancies $\Delta^{(\pm)}(s)$ due to NPP (Ref. [11]) are the solid lines. The broken lines show the old values of HMOV (Ref. [13]).

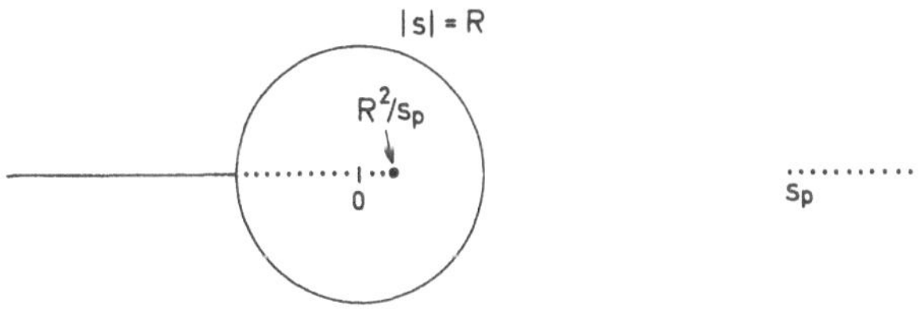

Fig. 8. The singularity structure of the discrepancy functions $\Delta^{(\pm)}(s)$ defined by Eq. (10). The discontinuity across the dotted lines is small.

The dispersion relation for $f(s)$ and Eq. (10) show that $\Delta^{(\pm)}(s)$ has no cut along $8\mu^2 < s < s_p$. However, the FESR and Regge approximations used for $-R < s < 8\mu^2$ and the high energy extrapolation ($s > s_p$) are certainly not exact, so the method of NPP gives $\Delta^{(\pm)}(s)$ having small discontinuities across $-R < s < 8\mu^2$, $s_p < s < \infty$. The singularity structure of $\Delta^{(\pm)}(s)$ is shown in Fig. 8.

In Figs. 7a, 7b the old values of $\Delta^{(\pm)}(s)$ due to HMOV (Ref. [13]) are shown for comparison. The difference between these and the NNP values is almost entirely due to the better treatment of the far off singularities in the work of NPP.

III.3 Analytic Continuation: The Convergence Problem

We would like to continue the analytic functions $\Delta^{(\pm)}(s)$ into the complex plane from the values on segments of the real axis given in Figs. 7a, b. If this could be done, we could for example explicitly demonstrate the discontinuity in $\Delta^{(\pm)}(s)$ across the t-channel cut $|s| = R$. NPP [11] have shown that this can indeed be done. We must first examine *the convergence problem and the error problem* in analytic continuation. For a thorough discussion the works of *Cutkosky* [14] and *Cuilli* [15] and their collaborators should be consulted.

We shall be brief at the risk of oversimplifying the matter. Let $z = x + iy$, x and y being real, and suppose $f(z)$ is an analytic function of z which is regular in some domain containing the line $-1 \leqq x \leqq 1$ and $f(z)$ is real on that line. Suppose further that we are given the numerical values $f(x)$ of the function on $-1 \leqq x \leqq +1$ and we want to continue it into the complex plane.

We cannot use the Taylor series about $z = 0$ because that requires the value of all derivatives $f^{(p)}(x)$ at $x = 0$ $(p = 1, 2 \ldots)$. A practical method is to approximate $f(x)$ by some sequence of polynominals $h_N(x)$ $(N = 0, 1, 2, \ldots)$.

If we require that $h_N(x)$ are chosen so that

$$\int_{-1}^{+1} \mathrm{d}x (f(x) - h_N(x))^2$$

be minimum, then $h_N(x)$ is of the form

$$a_0 + a_1 \cdot P_1(x) + a_2 \cdot P_2(x) + \cdots + a_N \cdot P_N(x)$$

where a_i are real coefficients and $P_i(x)$ are Legendre polynomials.* If we require that

$$\int_{-1}^{+1} \mathrm{d}x |f(x) - h_N(x)|$$

be minimum, then $h_N(x)$ is a sum of Tschebycheff polynomials. In any case, the larger N, the better is the fit of $h_N(x)$ to $f(x)$, on $-1 \leqq x \leqq 1$.

* We must use orthogonal polynominals to express $h_N(x)$ in order that the coefficients $a_i (i \leqq N)$ do not depend on N.

Having found the sequence of polynomials $h_N(x)$ we can replace x by z and use the sequence of polynomials $h_N(z)$ to approximate the continued function $f(z)$. It is well known that in both cases above, the sequence $h_N(z)$ converges to $f(z)$ as $N \to \infty$ provided z lies *inside* an ellipse Γ. The ellipse is the largest ellipse having -1 and $+1$ as foci which does not contain a singular point of $f(z)$ (Fig. 9). Thus Γ touches at least one singularity of $f(z)$.

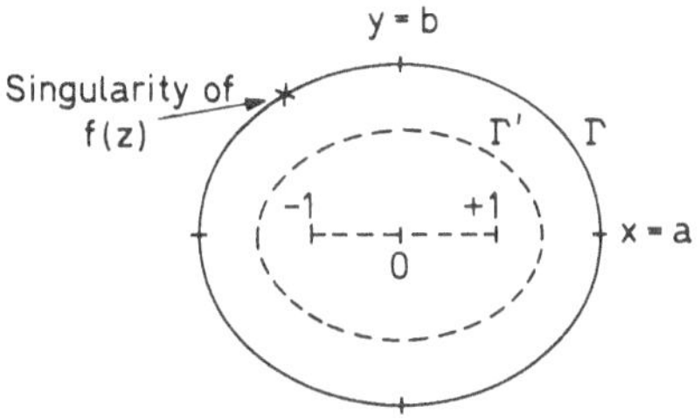

Fig. 9. The ellipse Γ and a confocal ellipse Γ'.

We now examine the rate of convergence. Suppose that $f(z)$ has a finite number of singular points on Γ and that at all points z, z_0 on Γ a Hölder condition.

$$|f^{(p)}(z) - f^{(p)}(z_0)| \leqq \text{const} \cdot |z - z_0|^\alpha$$

is obeyed. Here $f^{(p)}(z)$ is the pth derivative.

For example if there is a single pole on Γ then $p + \alpha = -1$, if there is a logarithmic singularity $p + \alpha = 0$, while if there is no more than a square root branch point then $p + \alpha = \frac{1}{2}$.

Then on Γ the convergence (or lack of it) is given by

$$|f(z) - h_N(z)| \leqq \frac{A}{N^{p+\alpha}} \quad \text{as} \quad N \to \infty \tag{11}$$

where A is a constant. Clearly if there is no more than a square root branch point there is a convergence on Γ.

Let a, b be the semi-major and minor axes of the ellipse Γ and with $R = a + b$. Let ϱ be the analogue of R for any confocal ellipse Γ' inside Γ (Fig. 9). At any point z on Γ' the convergence is given by

$$|f(z) - h_N(z)| \leqq \frac{A}{N^{p+\alpha}} \left(\frac{\varrho}{R}\right)^N.$$

It is worth noticing that the conformal transformation

$$z \to C \quad \text{where} \quad z = \tfrac{1}{2}(C + C^{-1})$$

transforms the line $-1 \leqq x \leqq 1$ (both sides of it) into the circle $|C|=1$, and transforms the ellipse Γ into the circle $|C|=R$. In this way the problems can be related to properties of Fourier series, or Laurent expansions [15].

The Conformal Transformations

We shall apply these ideas to the problem of extrapolating the discrepancy function from the line $R^2/s_p < s < (M-\mu)^2$ to the circle $|s|=R$. First we shall ignore the small discontinuity in $\Delta(s)$ across $-R < s < R^2/s_p$. We know the numerical value of the function on the segment AB (cf. Fig. 10). By the Legendre polynominal method we can then find $\Delta(s)$ at any point inside the ellipse, having A and B as foci, which touches $|s|=R$. Clearly this ellipse will touch the circle at $s=R$, and it follows that this methods *does not allow us to extrapolate to any points near the circle except for $s \simeq R$.*

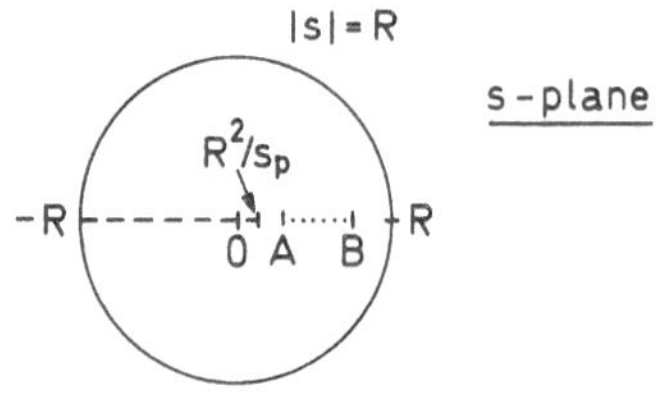

Fig. 10. Extrapolation inside $|s|=R$.

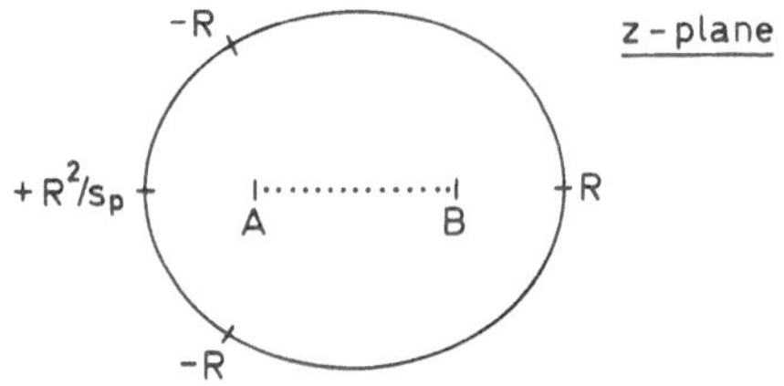

Fig. 11. The interior region in Fig. 10 has been transformed into the interior of an ellipse in the z-plane, and $s=\mathrm{A}$, $s=\mathrm{B}$ are transformed to the foci of the ellipse. Other transformed points (eg. $s=R$) are marked on the figure.

If we do not ignore the small discontinuity in $\Delta(s)$ across $-R < s < R^2/s_p$, then we can choose the point A to be somewhat more positive than R^2/s_p and again B is $(M-\mu)^2$. Clearly this does not improve the situation.

The way out of this difficulty is to make a conformal transformation $z=z(s)$ to a new variable z. This transformation is chosen so that the boundary ($|s|=R$ plus $-R < s < R^2/s_p$) of the domain of regularity in

Fig. 10 becomes an ellipse in the z plane, *and* so that the points A, B in the s-plane become the foci of the ellipse (Fig. 11). For the details of how this is done see NPP [11]. Notice that the points $s=R$, $s=R^2/s_p$ are transformed into the ends of the major axis of the ellipse and points $R^2/s_p < s < R$ lie on the major axis. The two sides of the line $-R < s < R^2/s_p$ are transformed into different arcs of the ellipse.

Thus one proceeds as follows [11]. The point $s=\mathrm{A}$ is chosen to be a few units μ^2 to the right of $R^2/s_p \simeq 8\mu^2$; the results do not depend noticably on the size of this gap. Knowing the values of $\Delta^{(\pm)}(s)$ on AB we have the values of $\Delta^{(\pm)}(z)$ on the line joining the foci of the ellipse (Fig. 11). Then $\Delta^{(\pm)}(z)$ are fitted by Legendre series

$$\sum_{n=0}^{N} b_n \cdot P_n(z)$$

(having arranged that the foci are $z=\pm 1$), and this series then gives $\Delta^{(\pm)}(z)$ inside the ellipse. On $|s|=R$ and $-R \leqq s \leqq R^2/s_p$, $\Delta^{(\pm)}(s)$ has square root branch points so $p+\alpha=\frac{1}{2}$ and the series will converge almost everywhere on the ellipse.

III.4 Constraints and the Discontinuity Across the Circle

In this section we mention certain technical features which add to the strength of the method. The singularity structure of $\Delta^{(\pm)}(s)$ was shown in Fig. 8. Let $\Delta(s)$ now denote either $\Delta^{(+)}(s)$ or $\Delta^{(-)}(s)$. Consider the functions

$$\delta_{\pm}(s) = \Delta(s) \pm \Delta(R^2/s) \tag{12}$$

for $|s| \leqq R$. The functions $\delta_{\pm}(s)$ have for $|s| \leqq R$ the singularity structure shown in Fig. 10, and by Eq. 10 we know the values of $\delta_{\pm}(s)$ on the segment AB.

Since $\Delta(s)$ and $f(s)$ obey the reality condition ($\Delta^*(s) = \Delta(s^*)$, $f^*(s) = f(s^*)$) it follows from Eq. (12) that on $|s|=R$

$$\begin{aligned} \operatorname{Re}\delta_{-}(s) &= \operatorname{Re}\Delta f(s) \\ \operatorname{Im}\delta_{+}(s) &= \operatorname{Im}\Delta f(s) \end{aligned} \tag{12a}$$

where

$$\Delta f(s) = f(s(1-\varepsilon)) - f(s(1+\varepsilon)), \quad \varepsilon > 0, \quad |s| = R$$

is the discontinuity of the p. w. a. across the circle $|s|=R$. Therefore by applying the method of the preceeding section to the functions $\delta_{\pm}(s)$ we directly find the discontinuity $\Delta f(s)$ on $|s|=R$.

In analytic continuation it is desirable to employ any simple constraints which may be available, since this will improve the accuracy of the result, particulary near the position where the constraint is imposed.

Because the helicity amplitudes $f_{\pm}^{J}(t)$ for $\pi\pi \to N\bar{N}$ have the property

$$\operatorname{Im} f_{\pm}^{J}(t) \propto (t-4\mu^2)^{(\frac{1}{2}+J)} \quad \text{for} \quad t \gtrsim 4\mu^2$$

it follows that the circle discontinuity obeys (cf. Eq. (15) below)

$$\Delta f(s) \propto (s-R)^3 \quad \text{near} \quad s=R.$$

This gives

$$\frac{d}{ds}\delta_+(s=R)=0, \qquad \delta_-(s=R)=0,$$

$$R\cdot\frac{d^2}{ds^2}\delta_-(s=R) = -\frac{d}{ds}\delta_-(s=R).$$

These constraints are applied be NPP [11] in the analytic continuation of $\delta_{\pm}(s)$ by using a device which we shall not describe here.

III.5 The Error Problem

In discussing practical methods for analytic continuation in § III.3 we assumed that the function $f(x)$ was known on $-1 \leqq x \leqq 1$. In practice, however, experimental measurements will give the values of $f(x)$ at a number of points on $-1 \leqq x \leqq 1$, but these values will contain some errors. This introduces a somewhat severe *stability problem* in practical analytic continuation [14, 15].

A simple example for a real variable is shown in Fig. 12. The measured points seem to be scattered about a straight line $f_1(x)$ and we think of $f_1(x)$ as a sensible extrapolation function to give an estimate of $f(x)$ for real x outside $-1 \leqq x \leqq 1$. However we could fit the data points exactly by using a polynominal of high order $f_2(x)$. Using $f_2(x)$ as an extrapolation function will give a very different estimate of $f(x)$ at points outside $-1 \leqq x \leqq 1$.

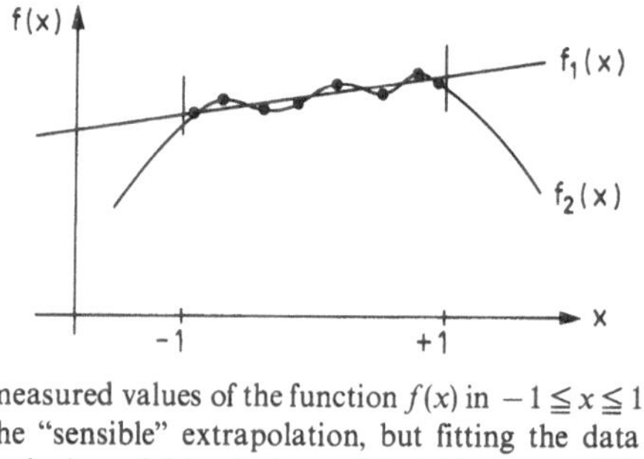

Fig. 12. The dots are measured values of the function $f(x)$ in $-1 \leqq x \leqq 1$. The extrapolation function $f_1(x)$ gives the "sensible" extrapolation, but fitting the data with a sufficiently high order polynominal gives $f_2(x)$ which would yield a very different extrapolation.

Before leaving Fig. 12 it should be noted that $f_1(z)$ and $f_2(z)$ would also give very different values if they were used for the analytic continuation of $f(x)$. The oscillations in $f_2(x)$ on $-1 \leqq x \leqq 1$ would, in most cases, be considerably amplified as z moves into the complex plane above or below $-1 \leqq x \leqq 1$.

Cuilli [15] has shown how to deal with the stability problem in analytic continuation in a simple way. Suppose that the data points measured on $-1 \leqq x \leqq 1$ lie within a band of width ε around the true function $f(x)$. One fits these (finite number of) data points with polynominals

$$h_N(x) = a_0 + a_1 \cdot P_1(x) + \cdots + a_N \cdot P_N(x) \qquad (N = 0, 1, 2, \ldots)$$

determining the coefficients a_i by least mean squares. It can be shown [15] that on the ellipse Γ (Fig. 9) we have

$$|f(z) - h_N(z)| < \frac{A}{N^{p+\alpha}} + B \cdot \varepsilon \cdot R^N , \tag{13}$$

and on the ellipse Γ'

$$|f(z) - h_N(z)| < \left(\frac{A}{N^{p+\alpha}} + B \cdot \varepsilon \cdot R^N\right)\left(\frac{\varrho}{R}\right)^N \tag{13a}$$

where B is $O(1)$ and positive and independent of N. Eq. (13a) replaces Eq. (11).

Now $R = a + b > 1$, so the right-hand side of Eq. (13) becomes large as N becomes large, in contrast with the behaviour in Eq. (11) (for $p + \alpha > 0$). There will be an optimum value of N which gives the least value for the right-hand side of Eq. (13). This optimum N gives the best continuation function $h_N(z)$ for continuing $f(z)$ to Γ.

This can be carried out in a simple way. In the case of no errors (i. e. $\varepsilon = 0$) the coefficient a_n in the Legendre series $h_N(z)$ will behave *for large n* like

$$|a_n| \lesssim \frac{A}{R^n} \cdot \frac{1}{n^{p+\alpha+1}} \tag{14}$$

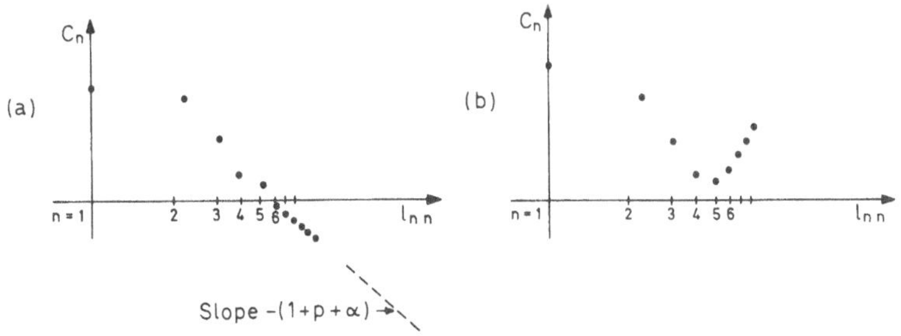

Fig. 13. a) Plot of $C_n = \ln(|a_n| R^n)$ versus $\ln n$ in the ideal case $\varepsilon = 0$. b) Plot of C_n in an actual case showing the divergence for larger n caused by the errors in the data

where A is the same constant which appears in Eqs. (11) and (13). Thus for $\varepsilon=0$ the plot of

$$C_n = \ln(|a_n| \cdot R^n)$$

would look like Fig. 13a. Because $\varepsilon>0$, the actual plot of will look like Fig. 13b. Terminating the series at the value of n giving the least C_n will give the best $h_N(z)$.

The Resolving Power

Suppose that in a practical case we have terminated the Legendre series at N (i.e. $h_N(z)$). This will put a limit on the grain of any structure of $f(z)$ which can be reproduced in the analytic continuation. In § III.3 the conformal transformation $z \to C$ was used to relate the family of confocal ellipses to a family of concentric circles. The data is now given on $|C|=1$ and Γ becomes $|C|=R$. The series are now essentially Fourier series. The *resolving power* is defined as

$$\delta_N = \pi/N$$

Structure features in $f(C)$ on $|C|=R$ which extends over angular intervals much less than δ_N will not be reproduced in the analytic continuation.

However, the *average* of $f(C)$ over an interval

$$\Delta(\arg C) \simeq m \cdot \delta_N \qquad (m>1)$$

is expected [11] to be reproduced with an error which is smaller than the maximum local error by a factor $(m\pi)^{-1}$.

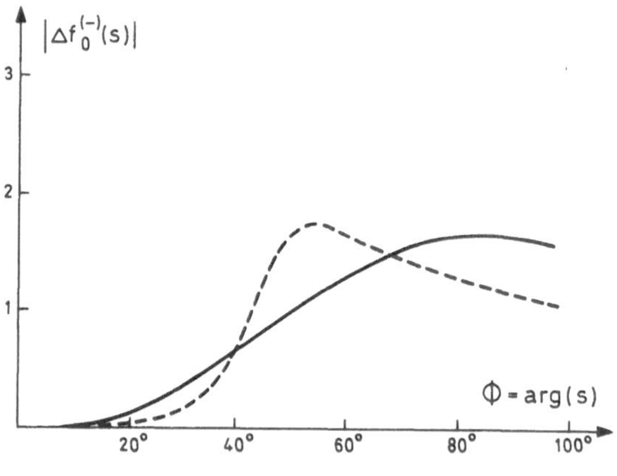

Fig. 14. The solid curve is the magnitude of the discontinuity in $f_0^{(-)}(s)$ across $|s|=R$ obtained by analytic continuation of $\Delta^{(-)}(s)$. The broken curve is the value deduced from the results of *Höhler* et al. (Ref. [16])

In the calculations of NPP [11] the cutoff value of N is 3, 4 or 5. Thus the accuracy of the presently available phase shift data does not permit the reproduction of narrow features on continuing the S-wave discrepancies $\Delta^{(\pm)}(s)$ to the circle $|s| = R$. Thus in the case of $\Delta^{(-)}(s)$ the magnitude of the discontinuity $|\Delta f_0^{(-)}(s)|$ across $|s| = R$ agrees with that which is deduced from the results of *Höhler et al.* [16], but the shape is smoother in the continuation method (Fig. 14). (If we look at arg $(\Delta f_0^{(-)}(s))$ which does vary slowly in the model of *Höhler et al.* the accuracy of the analytic continuation results is impressive).

III.6 The Circle Discontinuty and $f_+^0(t)$

With the methods outlined in the preceeding sections $\Delta^{(+)}(s)$ is analytically continued to give the discontinuity $\Delta f_0^{(+)}(s)$ across the circle $|s| = R$.

The result [11] is shown in Fig. 15.

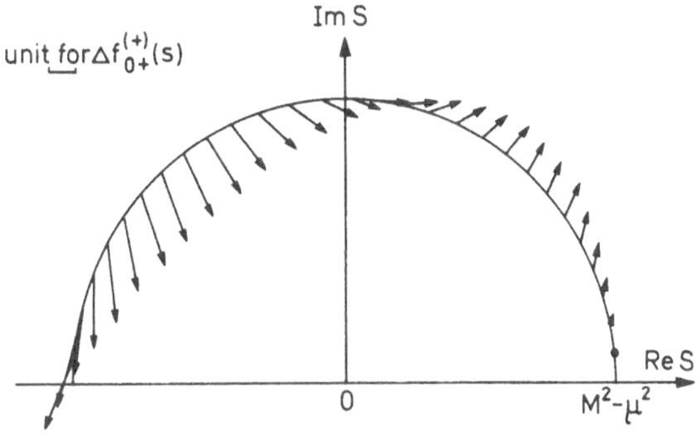

Fig. 15. The discontinuity $\Delta f_0^{(+)}(s)$ across $|s| = R$

If $f_\pm^J(t)$ are the helicity amplitudes for $\pi\pi \to N\bar{N}$, then in the notation of § II.1

$$A(s,t) = -\frac{8\pi}{p_t^2} \sum_{J=0}^{\infty} (J+\tfrac{1}{2})(p_t q_t)^J \left\{ P_J(\cos\vartheta_t) f_+^J(t) - \frac{M\cos\vartheta_t P_J'(\cos\vartheta_t)}{\sqrt{J(J+1)}} \cdot f_-^J(t) \right\} \tag{15a}$$

$$B(s,t) = 8\pi \sum_{J=1}^{\infty} \frac{(J+\tfrac{1}{2})}{\sqrt{J(J+1)}} (p_t q_t)^{J-1} P_J'(\cos\vartheta_t) \cdot f_-^J(t). \tag{15b}$$

For $A^{(\pm)}, B^{(\pm)}$ only $J = $ (even/odd) values occur in these Eqs.

For $0 \leqq |\arg s| \lesssim 66°$ the partial wave expansions in Eq. (15) are valid for calculating the discontinuities in $A(s,t), B(s,t)$ across $|s| = R$. For larger $|\arg s|$ the series do not converge, so we can only use Fig. 15 to obtain information on $\operatorname{Im} f^J_\pm(t)$ from the "front" of the circle.

The $\pi\pi \to \pi\pi$ $T=0, J=2$ amplitude is expected to be small for $4\mu^2 \leqq t \lesssim 50\mu^2$, so $\operatorname{Im} f^2_\pm(t)$ is small in this range. It follows from Eq. (1), (2) and (15) that

$$\arg(s^{\frac{1}{2}} \cdot \Delta f_0^{(+)}(s))$$

should be close to $\pi/2$ for $0 \leqq \arg s \lesssim 66°$. This is indeed well obeyed by the values [11] in Fig. 15, and we have here a useful test of the validity and the accuracy of the continuation method.

From $|\Delta f_0^{(+)}(s)|$ NPP [11] can directly determine $\operatorname{Im} f^0_+(t)$ in the range $4\mu^2 \leqq t \lesssim 50\mu^2$ (Fig. 16a). Since N was effectively 4 in the extrapolation, $\delta_N = \pi/4$. This means that around $t = 40\mu^2$ fine structure which is much narrower than $\Delta t \simeq 20\mu^2$ would not show up. For smaller t the resolving power is better.

Determination of $\operatorname{Re} f^0_+(t)$ *for* $t > 4\mu^2$

From fixed t dispersion relations one can calculate $\operatorname{Re} f^0_+(t)$ on $-26\mu^2 \lesssim t < 4\mu^2$ in terms of πN scattering data*. The partial wave expansion in the πN channel is sufficient to determine $\operatorname{Im} f^0_+(t)$ in the same range.

NPP [11] now use another discrepancy method to determine $\operatorname{Re} f^0_+(t)$ on $4\mu^2 \leqq t \lesssim 50\mu^2$. They write

$$f^0_+(t) = \frac{1}{\pi} \int_{-26\mu^2}^{0} \frac{\operatorname{Im} \tilde{f}^0_+(t')}{t'-t} dt' + \frac{1}{\pi} \int_{4\mu^2}^{50\mu^2} \frac{\operatorname{Im} f^0_+(t')}{t'-t} dt' + f^0_+(t)_{\text{BORN}} + \Delta^0_+(t) \tag{16}$$

where

$$f^0_+(t) = \tilde{f}^0_+(t) + f^0_+(t)_{\text{BORN}} \,.$$

Here $\Delta^0_+(t)$ is a real analytic function, regular in the plane cut along $-\infty < t \leqq -26\mu^2$, $50\mu^2 \leqq t < \infty$. $\Delta^0_+(t)$ represents the "far away" contributions to the dispersion relation Eq. (16). Notice that Eq. (16) can be used as it stands, or it can be used in subtracted form.

The method consists in evaluating $\Delta^0_+(t)$ on $-26\mu^2 \lesssim t < 4\mu^2$ and then continuing it analytically to $4\mu^2 \leqq t \lesssim 50\mu^2$. The analytic continuation is made with the help of a conformal transformation, and the method is similar to that described for $\Delta^{(\pm)}(s)$.

* A similar calculation has been carried out by *Höhler et al.* (Ref. [16]) for $f^1_\pm(t)$. The method was suggested by *Frazer* and *Fulco*, Ref. [2].

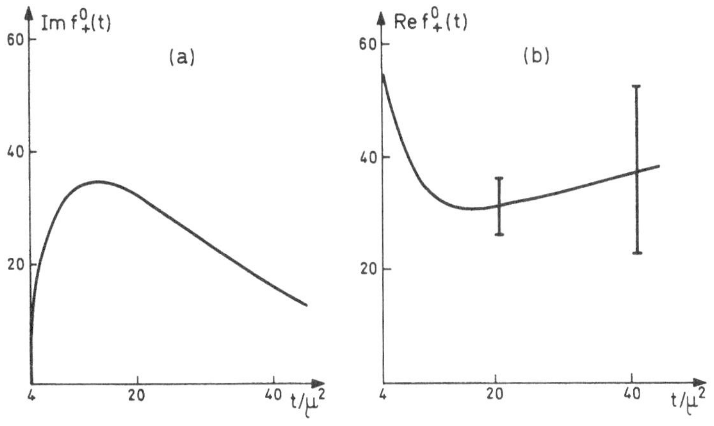

Fig. 16. $f_+^0(t)$ deduced from $\Delta_0^{(+)}(s)$

$\Delta_+^0(t)$ is evaluated on $-26\mu^2 \lesssim t < 4\mu^2$ by using the values of $f_+^0(t)$ on $-26\mu^2 \lesssim t < 4\mu^2$ calculated in the way we have indicated, and using the values of $\mathrm{Im} f_+^0(t)$ on $4\mu^2 \leqq t \lesssim 50\mu^2$ given in Fig. 16a. The errors in the analytic continuation over most of $4\mu^2 \leqq t \lesssim 50\mu^2$ will not be great since these points are inside the region of regularity. The errors are estimated using Eq. (13) and the values of $\mathrm{Re} f_+^0(t)$ with errors are shown in Fig. 16b (the errors shown do not include errors arising from the input values of $\mathrm{Im} f_+^0(t)$ on $4\mu^2 \leqq t \lesssim 50\mu^2$).

At present we merely comment that the values of $\mathrm{Im} f_+^0(t)$ in Fig. 16a do not differ greatly from the results of the earlier and much less sophisticated analysis by HMOV [13]. Also notice that $\mathrm{Re} f_+^0(t)$ does not go through zero, as it would if the $T=0\, J=0\, \pi\pi$ phase passed through $\pi/2$.

IV. Backward $\pi N \to \pi N$ and $f_+^0(t)$

IV.1 Introduction

The other standard method of determining $\pi\pi$ scattering from elastic πN scattering is by using backward πN scattering [17], this method has been applied by various authors*. It is useful to compare the results of this method with the partial wave method of § III, and NPP [11] have done so by introducing two improvements in the backward πN method: a) a good analytic continuation technique, b) proper treatment of a convergence problem which appears in extracting $f_\pm^0(t)$ (or $f_\pm^1(t)$) from the backward $\pi\pi \to N\bar{N}$ amplitude.

* See for example Ref. [18].

First we introduce some notation. From Eq. (2) we have

$$f_1 - f_2 = \frac{1}{4\pi W}\{E\cdot A(s,t) + M\cdot\omega B(s,t)\} \tag{17}$$

where $E=(M^2+q^2)^{\frac{1}{2}}$, $\omega=(\mu^2+q^2)^{\frac{1}{2}}$, q being the momentum in the πN c. m. s. Now

$$f_1(s, x_s=-1) - f_2(s, x_s=-1) \tag{17a}$$

is the backward πN amplitude, and its physical values (even though they are computed from phase shift analyses) should be known more accurately than the values of amplitudes which are only found by phase shift analysis.

Always keeping $x_s=-1$ the amplitude can be written as a function of $\nu \equiv q^2(s)$. A one-to-one relation between ν and s is obtained by introducing a two sheet Riemann surface for ν, the join being the line $-M^2 \leqq \nu \leqq -\mu^2$. We shall now stay on the top sheet where

$$s=[(M^2+\nu)^{\frac{1}{2}}+(\mu^2+\nu)^{\frac{1}{2}}]^2 \tag{18}$$

the roots being defined positive for $\nu>0$. By Eq. (2a)

$$t=-4\nu$$

Comparing Eqs. (2b) and (18) it is clear* that for $-\infty<\nu<-\mu^2$ we have $\cos\vartheta_t=-1$. Thus the amplitude in Eq. (17a) gives backward $\pi N \to \pi N$ for $0 \leqq \nu < \infty$, and is related to backward $\pi\pi \to N\bar{N}$ for $-\infty \leqq \nu \leqq -\mu^2$.

Another important property [17] is that for $x_s=-1$ the u-channel singularities of A and B do not appear on the top sheet. From Eq. (3) it follows that these singularities obey $0<s<M^2-\mu^2$ so that they lie on the second ν sheet.

We must be careful to avoid kinematic singularities in the relation to backward $\pi\pi \to N\bar{N}$. For this purpose we define

$$F^{(+)}(t) = \frac{1}{M}A^{(+)}(\nu, x_s=-1) + \frac{\omega}{E}\cdot B^{(+)}(\nu, x_s=-1) \tag{19}$$

By Eqs. (15)

$$F^{(+)}(t) = -\frac{8\pi}{Mp_t^2}\cdot \sum_{\substack{(J=0)\\(J\ \mathrm{even})}}^{\infty} (J+\tfrac{1}{2})(p_t q_t)^J \cdot f_+^J(t)\,. \tag{20}$$

The properties [19] of the helicity amplitudes $f_\pm^J(t)$ show that $F^{(+)}(t)$ is finite at $t=4\mu^2$ and $t=4M^2$ and that $F^{(+)}(t)$ has no kinematic singularities as a function of t.

* Remember that $\operatorname{Im} t$ and $\operatorname{Im} \nu$ have opposite signs.

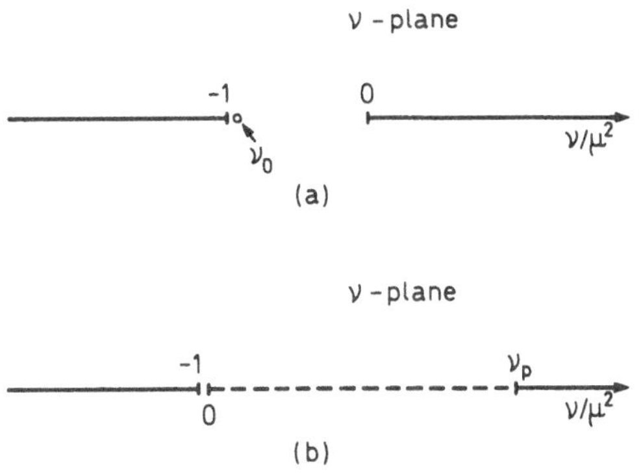

Fig. 17. a) shows the pole and cuts of $F^{(+)}(t)$ where $t = -4\nu$
b) shows the singularity structure of $\Delta_F(\nu)$ (Eq. (21)).

The function $F^{(+)}(t)$ has the hidden advantage that the D-wave $\pi\pi$ contribution to the absorptive part in the t channel is small compared with the D-wave contribution to other amplitudes which might have been chosen. The πN S-wave $f_0^{(+)}(s)$ has a similar advantage.

The singularity structure of $F^{(+)}(t)$ is shown in Fig. 17a. In addition to the two cuts there is a Born term pole $\xi/(\nu_0 - \nu)$ where

$$\nu_0 = -\mu^2 + \mu^4/4M^2$$

and $\xi = G_r^2 \cdot (\mu^4/4\mathrm{M}^2)$. This comes from the pole $(s - M^2)^{-1}$ in B.

IV.2 Analytic Continuation

Analytic continuation from the $\pi N \to \pi N$ cut $0 \leqq \nu \leqq \infty$ to the other cut will give $F^{(+)}(t)$ (Eq. (20)) for $4\mu^2 \leqq t < \infty$. NPP [11] carry this out as follows. The discrepancy function $\Delta_F(\nu)$ is defined by*

$$\Delta_F(\nu) = F^{(+)}(\nu) - \frac{\xi}{\nu_0 - \nu} - \frac{1}{\pi}\int_0^{\nu_p} d\nu' \frac{\operatorname{Im} F^{(+)}(\nu')}{\nu' - \nu} - \frac{1}{\pi}\int_{\nu_p}^{\infty} d\nu' \frac{\operatorname{Im} F_R^{(+)}(\nu')}{\nu' - \nu} \tag{21}$$

where ν_p is the highest value for which there are good phase shift data. $F_R^{(+)}(\nu)$ is the Regge approximation to the backward πN amplitude $F^{(+)}(\nu)$, as discussed in § I. The singularity of $\Delta_F(\nu)$ is shown in Fig. 17b. The cut $\nu_p < \nu < \infty$ is included to make it possible to allow for the approximate nature of $F_R^{(+)}(\nu)$ $(\nu > \nu_p)$.

* We have used $F^{(+)}(\nu)$ to mean $F^{(+)}(t = -4\nu)$.

By evaluating Eq. (21) the value of $\Delta_F(\nu)$ is found on a long segment $0 \leqq \nu \leqq \bar{\nu}$ where $\bar{\nu}$ is somewhat less than ν_p and $\nu_p \simeq 41\mu^2$. Now the conformal transformation and analytic continuation methods of § III are used to find the value of $F^{(+)}(\nu)$ (both $\operatorname{Re} F^{(+)}(\nu)$ and $\operatorname{Im} F^{(+)}(\nu)$) on $-\infty < \nu \leqq -\mu^2$.

IV.3 The Convergence Problem

Having obtained the value of $F^{(+)}(t)$ for $4\mu^2 \leqq t < \infty$ it is necessary to extract $f_+^0(t)$, using Eq. (20). In the range $4\mu^2 \leqq t \leqq 16\mu^2$ we have the unitarity relation

$$f_\pm^J(t) = |f_\pm^J(t)| \cdot e^{i\delta_J(t)} \tag{22}$$

where $\delta_J(t)$ is the $T=0$ or $T=1$ $\pi\pi$ phase for angular momentum J. In fact we shall assume that Eq. (22) is valid in the range $4\mu^2 \leqq t \lesssim 50\mu^2$; we do so because there is little experimental evidence for the process $2\pi \to 4\pi$ in this energy range.

There is reason to believe that the D-wave phase $\delta_2(t)$ is small [20] in $4\mu^2 \leqq t \lesssim 50\mu^2$ (note that the particle f^0 has $t \simeq 80\mu^2$) and the higher phase shifts will be very small. Thus, due to Eq. (22), $\operatorname{Im} f_\pm^J(t)$ will be small for $J \gtrsim 2$. However there is no reason to assume that the terms $\operatorname{Re} f_+^J(t)$ having $J \geqq 2$ in Eq. (20) can be ignored*, as various authors** have done.

Indeed NPP [11] show that the series in Eq. (20) converges poorly in much of the energy range. Consider the convergence of the series in Eqs. (15) in the $\cos\vartheta_t$ plane. The Born term singularity occurs at

$$\cos\vartheta_t = i\, b_N(t)$$

where

$$b_N(t) = \frac{\tfrac{1}{2}t - \mu^2}{2 \cdot (M^2 - t/4)^{\frac{1}{2}} \cdot (t/4 - \mu^2)^{\frac{1}{2}}}\,.$$

Thus the series for $B^{(+)}$ will only converge inside an ellipse having foci $\cos\vartheta_t = \pm 1$ and semi minor axis $b_N(t)$. The semi major axis is $a_N(t) = (1 + b_N^2(t))^{\frac{1}{2}}$.

The series in Eq. (20) for $F^{(+)}(t)$ will thus converge approximately like a geometric series $\sum_J R^{-J}$ where $R = a_N + b_N$. In the interval $4.2\mu^2 \leqq t < 25\mu^2$ we have $b_N(t) < 0.4$ and, in order to get 10% accuracy it is necessary to include terms up to $J = 6$ at least.

The singularities in $\cos\vartheta_t$ associated with the s and u channel thresholds give $b_{\mathrm{Th}} > 0.9$ in $4\mu^2 \leqq t \lesssim 50\mu^2$ and a rough estimate is that to get 10% accuracy it is sufficient to neglect terms $J > 2$ in Eq. (20).

* This remark is due to *G. C. Oades*.
** For example Ref. [18].

Since the Born term is the dominant cause of the poor convergence, NPP treat this term explicitly. Thus

$$F^{(+)}(t) = F^{(+)}(t)_{\mathrm{BORN}} + \tilde{F}^{(+)}(t) \tag{23}$$

where $F^{(+)}(t)_{\mathrm{BORN}}$ is the Born term contribution to $F^{(+)}(t)$. Using

$$B^{(+)}(s,t)_{\mathrm{BORN}} = -G^2 \cdot \left(\frac{1}{s - M^2} - \frac{1}{u - M^2} \right)$$

gives

$$F^{(+)}(t)_{\mathrm{BORN}} = -\frac{G^2}{M^4} \cdot \frac{t - 4\mu^2}{t - (4\mu^2 - \mu^4/M^2)} \,. \tag{23a}$$

In a similar way the helicity amplitudes are written

$$f_+^J(t) = f_+^J(t)_{\mathrm{BORN}} + \tilde{f}_+^J(t) \tag{23b}$$

where the Born terms $f_+^J(t)_{\mathrm{BORN}}$ are evaluated in the standard way using only $B^{(+)}(s,t)_{\mathrm{BORN}}$.

Replacing $f_+^J(t)$ in Eq. (20) by $f_+^J(t)_{\mathrm{BORN}}$ and $\tilde{f}_+^J(t)$ gives the series for $F^{(+)}(t)_{\mathrm{BORN}}$ and $\tilde{F}^{(+)}(t)$ respectively. It is easy to verify that the series for $F^{(+)}(t)_{\mathrm{BORN}}$ converges poorly in the range $4.2\mu^2 < t \lesssim 50\mu^2$.

The Born terms are of course real. The convergence of the imaginary part of Eq. (20) is expected to be very good.

IV.4 Extracting $f_+^0(t)$

For the imaginary part we just use

$$\mathrm{Im}\, f_+^0(t) = -\frac{M p_t^2}{4\pi} \,\mathrm{Im}\, F^{(+)}(t) \tag{24}$$

This should be accurate, and it gives the result in Fig. (18a). The values do not differ greatly from the results of the partial wave method of § III (Fig. 16a). If we make the same approximation (i. e. ignoring terms with $J \geqq 2$) in Eq. (20) to get $\mathrm{Re}\, f_+^0(t)$ the result is curve B in Fig. 18b. We have seen that this is bound to be a bad approximation. Its badness can be confirmed by using the values of $\mathrm{Im}\, f_+^0(t)$ given by Eq. (24) (Fig. 18a) in the method discussed in § III.6. This gives curve D in Fig. 18b for $\mathrm{Re}\, f_+^0(t)$.*

* The errors are explained in § III.6.

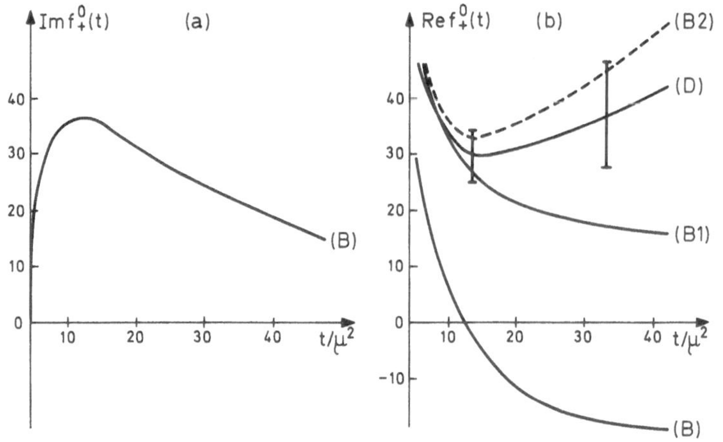

Fig. 18. a) $\operatorname{Im} f_+^0(t)$ from backward $\pi N \to \pi N$
b) $\operatorname{Re} f_+^0(t)$ from backward $\pi N \to \pi N$ (see the text).

NPP next use Eqs. (20), (23), (23 b) to get the relation

$$\operatorname{Re} f_+^0(t) = -\frac{M p_t^2}{4\pi} [\operatorname{Re} F^{(+)}(t) - F^{(+)}(t)_{\text{BORN}}] + f_+^0(t)_{\text{BORN}} \\ - \sum_{\substack{J=2 \\ (J\,\text{even})}}^{\infty} (2J+1)(p_t q_t)^J \cdot \operatorname{Re} \tilde{f}_+^J(t)\,. \tag{25}$$

This relation avoids the serious convergence trouble from the Born terms. It should be a reasonable first approximation to ignore the last term on the right. The result is curve B 1 in Fig. 18b. Clearly this is a considerable improvement.

Finally NPP approximate the next term $-5(p_t q_t)^2 \cdot \operatorname{Re} \tilde{f}_+^2$. This is done by neglecting the $J=2$ $\pi\pi$ phase and putting

$$\operatorname{Im} f_+^2(t) = \operatorname{Im} \tilde{f}_+^2(t) = 0\,, \qquad 4\mu^2 \leqq t \lesssim 50\mu^2\,.$$

Now using the continuation method of § III.6 they find $\operatorname{Re} \tilde{f}_+^2(t)$. The result for $\operatorname{Re} f_+^0(t)$ is shown by curve B 2 in Fig. 18b.

The general conclusion is that once the convergence problem has been understood and allowed for, there is reasonable agreement between the values of $\operatorname{Re} f_+^0(t)$ on $4\mu^2 \leqq t \lesssim 50\mu^2$ as deduced from analysing a) πN S-waves, b) πN backward scattering.

Some further discussion of the possible overall errors will be given in § V.

IV.5 The $T = 0$ $\pi\pi$ D-wave

Clearly with these new methods one can do more than finding $f_+^0(t)$. The limitation at present is rather in the accuracy of the input data, particularly for πN scattering below 250 MeV lab pion energy.

However NPP [11] have shown that the method used in § IV.1–3 gives reasonable values for the $\pi\pi$ phase δ_2^0. They use the backward amplitude

$$\begin{aligned} G^{(+)}(t) &\equiv \frac{1}{E\omega} B^{(+)}(\nu, x_s = -1) \\ &= 4\pi \sum_{\substack{J=2 \\ (J \text{ even})}}^{\infty} (J + \tfrac{1}{2}) \cdot (J(J+1))^{\frac{1}{2}} \cdot (p_t q_t)^{J-2} \cdot f_-^J(t) \,. \end{aligned} \tag{26}$$

This is convenient since it has no $J = 0$ term.

Analytic continuation gives* $G^{(+)}(t)$ for $t \geqq 4\mu^2$. It is then a good approximation to use

$$\operatorname{Im} f_-^2(t) = \frac{\sqrt{6}}{60\pi} \operatorname{Im} G^{(+)}(t)$$

to get $\operatorname{Im} f_-^2(t)$ on $4\mu^2 \leqq t \lesssim 50\mu^2$. For the real part they find

$$\operatorname{Re} G^{(+)}(t) \simeq G^{(+)}(t)_{\text{BORN}} \qquad (4\mu^2 \leqq t \lesssim 50\mu^2).$$

So they have $\operatorname{Re} f_-^2(t) \simeq f_-^2(t)_{\text{BORN}}$. Using Eq. (22) this gives the results in Fig. 19. The values of δ_2^0 shown are quite consistent with the existence of the f^0 resonance at $t \simeq 80\mu^2$.

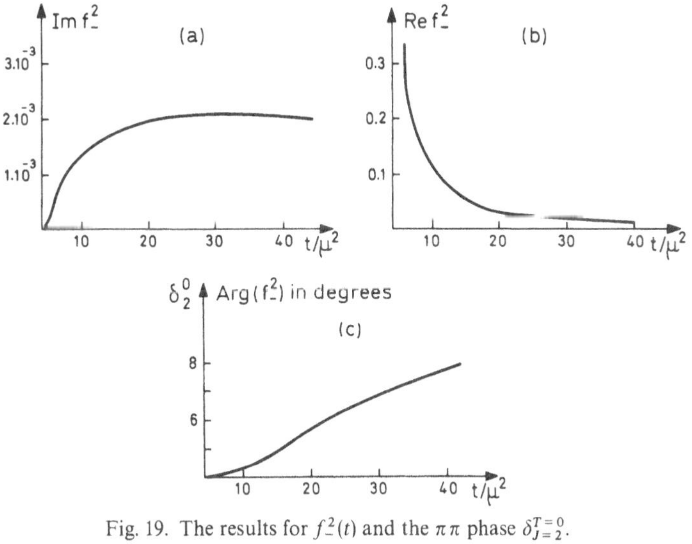

Fig. 19. The results for $f_-^2(t)$ and the $\pi\pi$ phase $\delta_{J=2}^{T=0}$.

---

* The constraint $\operatorname{Im} G^{(+)}(t) \propto (t - 4\mu^2)^{5/2}$ $(t \gtrsim 4\mu^2)$ has been used in the analytic continuation.

The analytic continuation method must give an accurate average value of $\operatorname{Im} f_-^2(t)$ in the range $4\mu^2 \leqq t \lesssim 50\mu^2$. Remembering that the constraint

$$\operatorname{Im} f_-^2(t) \propto (t-4\mu^2)^{5/2} \qquad (t \gtrsim 4\mu^2)$$

has been imposed, it is reasonable to believe that the D-wave phases which result are of the right order of magnitude.

These D-wave phases agree reasonably well with those of *Morgan* and *Shaw* [20a].

V. The σ-Meson

From Eq. (22)

$$\operatorname{Re} f_+^0(t) = \cos\delta_0^0(t) \cdot |f_+^0(t)|$$

and $\operatorname{Re} f_+^0(t)$ will pass through zero if $\delta_0^0(t)$ goes through $\pi/2$. In Fig. 20 the curve S shows the value of δ_0^0 given by the curves $\operatorname{Re} f_+^0$ and $\operatorname{Im} f_+^0$ in Fig. 16; this is the πN partial wave result. The curve F in Fig. 20 shows the result from backward $\pi N \to \pi N$ using curve D in Fig. 18b. In neither case does $\delta_0^0(t)$ reach near $\pi/2$.

NPP [11] consider possible errors, and in particular errors in $\operatorname{Im} f_+^0(t)$ on $4\mu^2 \leqq t \lesssim 50\mu^2$. The $(-)$ case can be compared with the results of *Höhler et al.* [16] so as to check on the error estimates. In order to see how near the analytic continuation can get to predicting a δ-resonance about $t_R \simeq 25\mu^2$, NPP [11] use model values of $\operatorname{Im} f_+^0(t)$ having their maximum at $t_R = 25\mu^2$. The area in the Cuilli integral

$$\int_{4\mu^2}^{50\mu^2} \operatorname{Im} f_+^0(t) \cdot \frac{d(\operatorname{Arg} C)}{dt}\, dt \tag{27}$$

(cf. § III.5) should be affected very little by errors. Two model values of $\operatorname{Im} f_+^0(t)$ having maxima at $t_R = 25\mu^2$ and having reasonably chosen widths are used. In (i) the area in Eq. (27) is unaltered, in (ii) the area is increased by 50%. This latter is regarded as unlikely to be possible. The corresponding values of δ_0^0 are the curves (1) and (2) in Fig. 20. Not even in curve (2) does δ_0^0 go through $\pi/2$.

The sort of result one would obtain if the situation is intermediate between curve F and curve (1) (or curve 2) is easily envisaged. This is the range in which the analytic continuation results almost certainly lie*.

Also shown in Fig. 20 are the regions where the experimental results from $\pi N \to \pi\pi N$ lie. The hatched regions are from $\pi^- p \to \pi^+ \pi^- n$ experiments [21] with the up/down ambiguity both above and below the resonance. The horizontal band is from a $\pi^- p \to \pi^0 \pi^0 n$ experiment [22].

* See Ref. [11] for further details.

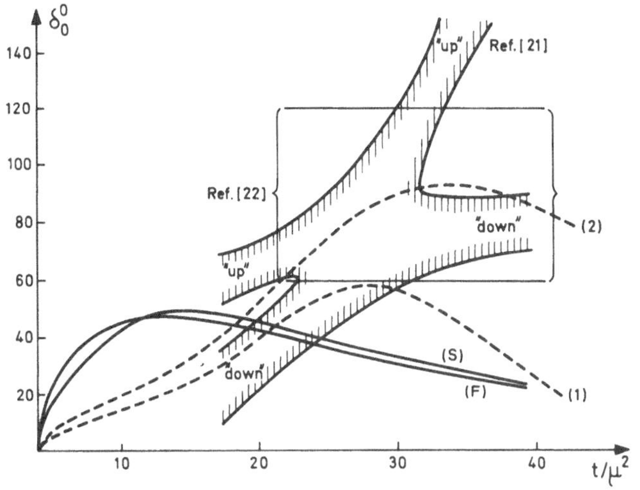

Fig. 20. Predictions and experiment for $\delta_0^0(t)$ (see text).

We can conclude:

(i) the analytic continuation results [11] are consistent with a broad σ-"resonance"

(ii) they do not allow δ_0^0 to increase above $\pi/2$ at energies above the resonance (at least not up to 900 MeV)

(iii) the analytic continuation results suggest that δ_0^0 may not even reach $\pi/2$. This type of behaviour would be consistent with both sets of experiments.

With better experimental results for low energy πN scattering the errors in the continuation work could be reduced considerably.

I am indebted to *H. Nielsen*, *J. Lyng Petersen* and *E. Pietarinen* for permission to quote freely from their work, and I am indebted to *S. Ciulli* for numerous discussions.

References

1. Review article "Pion-Nucleon Interactions" by *J. Hamilton* in High Energy Physics, Vol. I, (Ed. E. Burhop). New York: Academic Press 1967.
2. *Frazer, W. R., Fulco, J. R.*: Phys. Rev. **119**, 1402 (1960).
   Hamilton, J., Spearman, T. D.: Ann. Phys. (N. Y.) **12**, 172 (1961).
3. *Kibble, T. W. B.*: Phys. Rev. **117**, 1159 (1960).
4. *Hamilton, J., Spearman, T. D., Woolcock, W. S.*: Ann. Phys. (N. Y.) **17**, 1 (1962).
5. *Freedman, D. Z., Wang, J. M.*: Phys. Rev. Letters **17**, 569 (1966).
6. *Jakob, H. P., Steiner, F.*: Z. Physik **228**, 353 (1969).
7. *Lyng Petersen, J.*: Nucl. Phys. **1315**, 549 (1970).

8. *Singh, V.*: Phys. Rev. **129**, 1889 (1963).
9. *Barger, V., Cline, D.*: Phys. Rev. Letters **21**, 392 (1968).
10. See for example: *Dolen, R., Horn, D., Schmid, C.*, Phys. Rev. **166**, 1768 (1968).
11. *Henry Nielsen, Lyng Petersen, J., Pietarinen, E.*: Nordita preprint (april 1970) and Nucl. Phys. B **22**, 525 (1970).
12. *Baacke, J., Steiner, F.*: Fortschr. Physik **18**, 67 (1970).
13. *Hamilton, J., Menotti, P., Oades, G. C., Vick, L. L. J.*: Phys. Rev. **128**, 1881 (1962).
14. *Cutkosky, R., Deo, B. B.*: Phys. Rev. **174**, 1859 (1968), Phys. Rev. Letters **20**, 1272 (1968).
15. *Cuilli, S.*: Nuovo Cimento **61** A, 787 (1969) and **62** A, 301 (1969).
   Cuilli, S., Cutkosky, R.: Nordita Preprint (1969), to be published.
16. *Höhler, G., Strauß, R., Wunder, H.*: Karlsruhe Preprint (1968).
17. *Atkinson, D.*: Phys. Rev. **128**, 1908 (1962).
18. *Lovelace, C., Heinz, R. M., Donnachie, A.*: Phys. Letters **22**, 332 (1963).
19. *Frazer, W. R., Fulco, J. R.*: Phys. Rev. **117**, 1603 (1960).
20. *Oades, G. C.*: Phys. Rev. **132**, 1277, (1963).
20a. *Morgan, D., Shaw, R.*: Nucl. Phys. B **10**, 261 (1969).
21. *Scharenguivel, J. R., et al.*: Phys. Rev. **18** b, 1387 (1969).
22. *Deinet, W., et al.*: Phys. Letters **30** b, 359 (1969).

Prof. Dr. *J. Hamilton*
Nordita, Blegdamsvei 17
2100 Copenhagen/Danmark
and
CERN
CH-1211 Genf 23

Regge-Pole Phenomenology

EUAN J. SQUIRES

Contents

Abstract

After nine years, not without their difficulties, the Regge pole model, surrounded by her many friends and playing a central part in all theories exciting the most current interest, is happy and thriving.

I. Introduction

Regge theory was born in 1961 and grew rapidly during 1962. Towards the end of 1962 it was affected by a certain lack of shrinkage, a disease which quickly spread so that by 1964 it was seriously ill and apparently dying. The orbituary writers were eager. A conference in 1964 was summarised in 6 points, the fifth being "The Regge-pole model is dead"; a review of a book on the subject remarked "The years in which Regge poles were the 'cat's pyjamas' will be preserved for posterity. If Benjamin, Inc., were not so quick about publishing, rather less might be preserved." A count of papers in Phys. Rev. Letters and Phys. Letters on Regge-pole theory shows that there were 26 in 1962, 48 in 1963, but by 1964 there were only 7 stragglers left (all but one in the first half).

It is worth noting that at this stage there was no experimental data which could not be simply fitted by the model. Since then a great deal of new data has been available and there are several features which are not compatible with simple forms of the model. However, the dead theory came to life and in 1969 the above mentioned journals contained 121 papers in which Regge theory was an essential ingredient.

How do we explain this – apart from the (untenable?) assumption that elementary particle theorists are crazy? One possibility, that the theory has in some way retreated and its assumptions become less restrictive, whilst true in a few points, is in general clearly false. Indeed the range of applicability of the theory has been greatly extended, and assumptions of simplicity have been made which one would not have made in 1962. To demonstrate this we shall, in these lectures, state the 'assumptions' of the Regge model, as used today, in, according to my estimation, order of decreasing certainty, and then see how far these assumptions are justified-both on the basis of experimental evidence (to accord with my title) and our theoretical understanding (which I am supposed to know more about).

The amount of space devoted in these lectures to various topics is not a measure of their importance, since I shall only mention briefly those things that are well established, in order to concentrate on newer points and on those that I find most interesting. To keep within manageable limits I shall exclude photoproduction and discuss only hadron physics, the original domain of the model. For convenience I shall refer quite often to *Collins* and *Squires* [1] for further details and references (not because this reference contains the best treatment, but because I am easily able to find things in it).

II. The Assumptions of Regge Theory

Ass. 1. The Sommerfeld-Watson Transform

Before we can start we have to assume that there exists an analytic continuation of partial-wave amplitudes into the J-plane and that this continuation permits the Sommerfeld-Watson transformation to be made, with a contour which can be moved to the left of $\operatorname{Re} J = -1/2$.

There is no direct experimental verification of this of course, but without it the successes mentioned below are inexplicable. Much of what is required follows theoretically from the assumptions of S-matrix theory, e. g. Chapter I–VII of Ref. [1].

Ass. 2. Regge Trajectories

We need to assume that there exist Regge trajectories $\alpha(t)$ which can be continued smoothly from $t > 0$, where they show up as t-channel resonances (or bound states), to $t<0$ where they can be seen in s-channel high energy behaviour.

The classic case here is that of the ϱ in $\pi^- p \to \pi^0 n$. If one writes the differential cross-section for this process in the form

$$\frac{d\sigma}{dt} = f(t)\omega^{2\alpha(t)-2} \tag{1}$$

where ω is the π energy in the lab. system, then $\alpha(t)$ can be obtained from experiment for values of t from zero to about 1.5 $(\mathrm{GeV}/c)^2$; see Fig. 1 (taken from Ref. [2]). The points lie on a straight line which extrapolates almost exactly through the position of the ϱ-meson. Since this is the only known meson that can be exchanged in this process we have here a dramatic success of the theory, and one which has no alternative explanation.

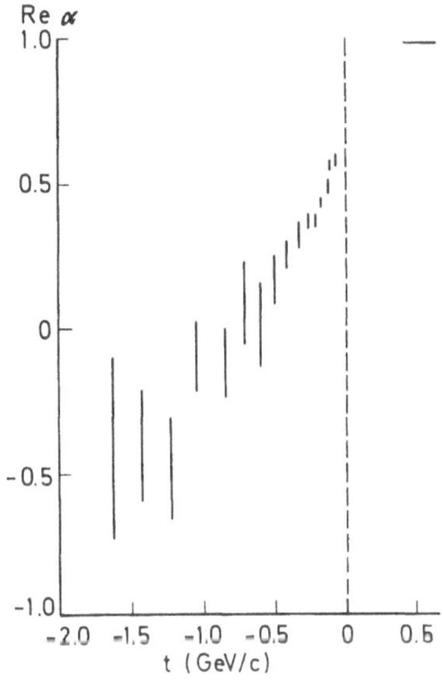

Fig. 1. ϱ trajectory from a fit to $\pi^- p \to \pi^0 n$

This result, and others like it*, is actually already rather nicer than one could have predicted theoretically, especially if one remembers that $\alpha(t)$ has a singularity at $t = 4m_\pi^2$ so that the extrapolation goes through a singular point (see chapter II of Ref. [1]). Indeed it was once claimed [3] that this singularity would completely destroy any connection between

* A particularly good example, of the $\varrho + A_2$ contribution to $K-N$ scattering, has recently been obtained by *C. Michael* and *C. Schmid* and is reported in the lectures at this school by *C. Schmid*.

the ϱ-meson and its trajectory in the $t<0$ region. Detailed calculations [4] in potential models and in dispersion models, however, do tend to confirm the smallness of the effect of the threshold singularity on the trajectory.

Ass. 3. No Elementary Particles

Since bound states clearly lie on Regge trajectories whilst CDD poles, in particular partial-waves, give rise to Kronecker delta singularities in $a(J,t)$, e. g. $\delta_{JJ_0}(t-m^2)^{-1}$, Regge theory offers a precise way of distinguishing between 'composite' and 'elementary' particles, and therefore of testing the idea of nuclear democracy that there are no elementary hadrons. The difference can readily be detected since a Kronecker delta singularity as above yields a high energy behaviour proportional to s^{J_0}, where J_0 is the spin of the elementary particle, whilst a Regge trajectory gives a behaviour $s^{\alpha(t)} \approx s^{J_0+\alpha'(t-m^2)}$. For no particles, except possibly the π, is the s^{J_0} behaviour seen and in general the behaviour has the Regge form with $\alpha' \sim 1(\mathrm{GeV})^{-2}$. The nucleon is a good example here (for no very obvious reason this is probably the most likely candidate for an 'elementary' particle); its contribution can be isolated in πN backward scattering for which one obtains [5] $\alpha_N(0) \simeq -0.34$. For the π the situation is not so clear since the difference between $\alpha(0)$ and $\alpha(m_\pi^2)$ is not likely to be appreciable. However the π is the only particle that can conceivably be 'elementary' in the above sense.

We recall (essentially because of the Froissart bound but the argument is rather involved, e. g. Chapter VIII of Ref. [1]) that there are strong theoretical reasons for believing that all particles of spin $\geqq 1$ are composite, so the experiments provide confirmation of our theoretical assumptions for $J_0 \geqq 1$ and of the nuclear-democracy idea for $J_0 < 1$.

It is worth noting here that the relationship of this result to the idea of CDD poles is not in fact clear. It is known that some particles, e. g. the nucleon (see § VII, 3 of Ref. [1]), are CDD poles in single channel calculations (even with correct inelasticity factors), and it may be that they remain CDD poles when all channels are included. If this is the case then presumably they require CDD poles in partial waves for all J values, i. e. are CDD trajectories. Evidence that the mesons may also be of the type is given by *Collins*, *Johnson*, and *Squires* [6].

Ass. 4. No other Important Singularities

Although it has long been known, on theoretical grounds, that there must be singularities in the J-plane other than Regge poles (in particular moving cuts and fixed poles, see Chapter 5 of Ref. [1]), these other singularities are not readily correlated with properties of particles so in order to have a useful theory it has been hoped that poles dominate. There is a considerable amount of evidence that this is the case. If one studies the energy dependence of total cross-sections for all $2 \to 2$ particle reactions one can fit them to the expression const. $s^{2\alpha-2}$ where the values of α lie in the following ranges:

(A) $\alpha \sim 1$, where there is zero quantum number exchange in the crossed channel.

(B) $\alpha \sim 1/2$, where only non-strange mesons, excluding those of case (A), can be exchanged.

(C) $\alpha \sim 0$, where only strange mesons or the π can be exchanged.

(D) $\alpha \sim -1/2$, where only baryons can be exchanged.

(E) Essential no forward peaks, where no known particles can be exchanged (i. e. the cross-channel is exotic).

In all cases the correlation with known particles is perfect and for classes B, C, D suggests that trajectories have roughly parallel extrapolations, with slope $\sim 1\ (\text{GeV})^2$, from the lowest state to $t=0$.

For class A the situation is more confused. The obvious candidate for a particle with vacuum quantum numbers is the $f^0, 2^+$ at $m^2 = 1.57\ \text{GeV}^2$. However with a trajectory of the above slope this would have an $\alpha(0)$ too small to explain the constancy of total cross-section, though as required it would be the highest trajectory. The slope needed is in fact $0.64\ (\text{GeV})^2$, which is not incompatible with the slope required for the Pomeranchon in the $t<0$ region. It should be noted that class A contains, in addition to elastic processes, which are coherent, certain incoherent inelastic processes and it is a, physically somewhat surprising, prediction of Regge theory that these processes have similar high energy behaviour; this prediction is certainly confirmed in some cases. The fact that the Pomeranchen apparently does not contribute to the s-channel spin flip amplitude (see lectures of *G. Höhler*) is somewhat surprising from this point of view.

It is in class E processes that one would first expect to find evidence for other singularities. As an example consider the process $\pi^- p \to K^+ +$ Missing Mass, which would require a $Q=2, S=1$ meson trajectory, for which there are no known particles (i. e. it is 'exotic'). All models of cuts suggest that there should be a branch-point at

$$\alpha_c(0) = \alpha_{K^*}(0) + \alpha_\varrho(0) - 1$$

corresponding to the 'exchange' of a K^* and a ϱ trajectory (see Ref. [1], Chapter V). This occurs at about -0.4, so the energy dependence should be similar to that of class D above. However experiments on this process at 6 and 10 GeV/c reveal no peak at all (except for a small effect when a Y_1^* is produced), in striking contrast to baryon exchange (see footnote on page 153 of Ref. [7]). Another interesting case is $K^- p \to pK^-$ which requires an exotic $Q=2, S=1$ baryon trajectory. At very low energies there is a peak which behaves as $d\sigma/dt \sim s^{-9}$ and could be fitted by an exotic baryon of mass about 2 GeV and spin 1/2. A cut contribution should behave as s^{-3} and this behaviour is not seen. *Michael* [8] has calculated the magnitude of the cut contribution to this process in a particular model. His estimate is smaller than the data at low energies and within the upper limits given by higher energy experiments (see Fig. 2). It will be interesting to see at what energy this cut contribution becomes apparent.

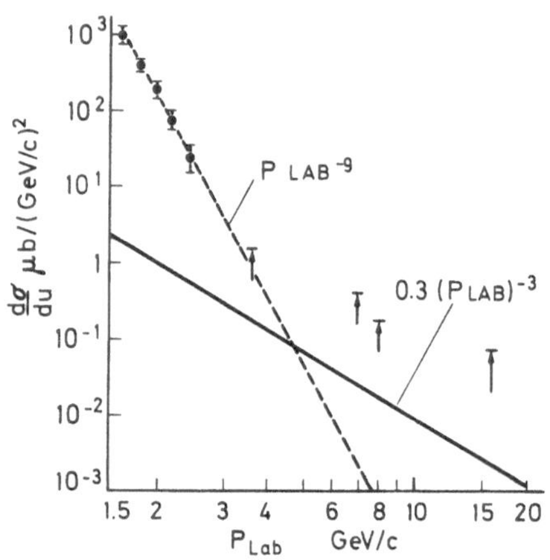

Fig. 2. $K^- p \to pK^-$ data and upper limits

Thus at present our (inadequate) theory confirms the smallness of cut contributions compared to pole contributions, even in the same region of the J-plane. One should qualify this by noting that the so-called 'absorptive correction' cuts due to Pomeranchon + other trajectory and having therefore the same quantum numbers as the trajectory do not contribute to class E (if we exclude exotic trajectories), so such cuts are not limited in this way. Further, it is generally to be expected that $\alpha_c'(t)$ is smaller than typical pole trajectory slopes, so cuts should become more in evidence at high t.

Detailed fits to experimental data are available (see, for example, Ref. [9], Chapter VII of Ref. [1] and the review of *Hite* [10]) and in general confirm the successes above. The most striking feature of these fits, essentially because of the non-shrinking $\pi - p$ differential cross-section, is that the Pomeranchon trajectory is required to have a small slope [11–13] $\lesssim 0.5\,(\mathrm{GeV})^2$. (*Barger* and *Cline* [14]) have recently obtained confirmatory evidence of this slope from a study of $\gamma p \to \phi p$). In the few places where the phase can be checked it also appears to be given correctly by the leading Regge poles under the assumption that α and β are real below threshold as suggested by theoretical arguments (see Chapter II of Ref. [1].

These detailed fits, however, reveal several problems, the most significant of which we now discuss.

Problem 1. Polarisation in $\pi^- p \to \pi^0 n$

Single ϱ-exchange gives zero polarisation whereas polarisations of the order of 15% are observed [15] at 6 and 11 GeV/c. The errors on the data are large but it is clear that if we are to fit it with a new trajectory it must have $\alpha(0) \gtrsim 0$, so it is hard to see why particles lying on it have not been observed.

Problem 2. π-contribution to Forward Direction

Experimental data (up to 8 GeV/c) on $pn \to np$ has a sharp forward peak and behaves with energy as though dominated by the π trajectory, with $\alpha_\pi(0) \approx 0$. However the cross-channel pion contributes to only two of the s-channel helicity amplitudes, $\langle + + | - - \rangle$ and $\langle + - | - + \rangle$, to which its contributions are equal. In the forward direction, $t = 0$, $\langle + - | - + \rangle$ must vanish since it involves a change of total helicity. So, either the π contribution to $\langle + - | - + \rangle$ vanishes at $t = 0$, in which case its contribution to $\langle + - | - + \rangle$ also vanishes so that the π cannot give the observed forward peak, or the π contribution to $\langle + - | - + \rangle$ cancels exactly with another contribution to the same amplitude. The first (and simplest) possibility is a disaster for the pure Regge pole model. In the other possibility, which we call a 'conspiracy' it is necessary that the cancellation at $t = 0$ must hold for all s, so the conspiracy trajectory must coincide with the π trajectory at $t = 0$. The simplest such conspiracy requires a trajectory with the same quantum numbers as the π but opposite parity; it then contributes with opposite sign to $\langle + - | - + \rangle$ and $\langle + + | - - \rangle$ so that its contribution can cancel with that of the π for $\langle + - | - + \rangle$ and double the π contribution to $\langle + + | - - \rangle$. There is of course no observed parity doublet of the π, so the new trajectory must either not pass through $\alpha = 0$ or must have a zero residue there. Either possibility looks unnatural, but models can be constructed, e. g. *Phillips* [16] whose fit requires a rapidly varying residue for the π-conspirator. Some additional theoretical understanding of the conspiracy problem has been given group theoretically, using the extra symmetry which occurs at $t = 0$, but this does not appear to help in phenomenological fitting (see Chapter IV, Ref. [1]).

Problem 3. New Serpukov Data

Data on total cross-sections for $\pi^- p$, $\pi^- n\,(=\pi^+ p)$, $\bar{p}p$, $\bar{p}n$, $K^- p$ and $K^- n$, for laboratory momenta between 20 and 65 GeV/c have recently been obtained from the IHEP machine at Serpukov, and have not followed the expected pattern [17]. This can be seen from Fig. 3 where the new data is compared to the extrapolated fit to lower energy data of *Barger*

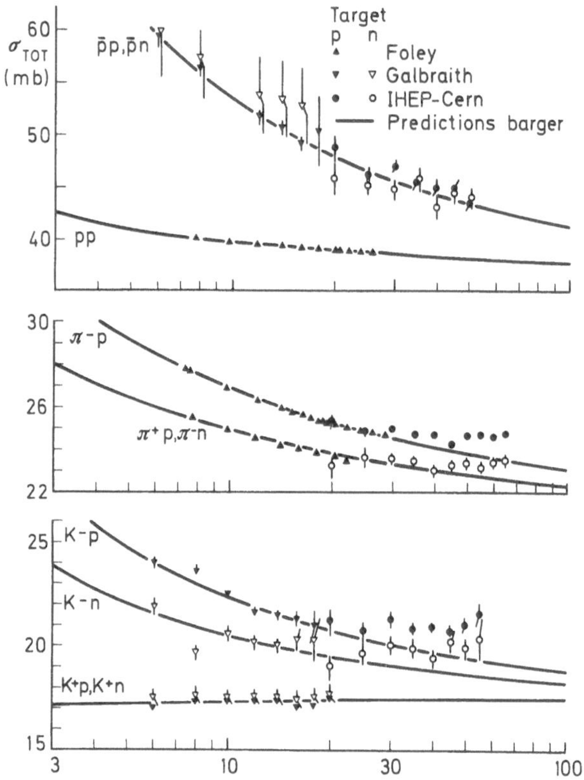

Fig. 3. Serpukov results and cf. with a fit

et al. [18]. The $\bar{p}p$ data is as expected, but the $\pi^- p$ and $K^- p$ cross-sections, instead of continuing to decrease (roughly as $p_{\mathrm{LAB}}^{-\frac{1}{2}}$) and to approach the corresponding limits of π^+ and K^+, as required by the 'Pomeranchuk theorem', remain approximately constant. This phenomenon is especially marked if we plot the data against $p_{\mathrm{LAB}}^{-\frac{1}{2}}$ – a graph which certainly enables us to include asymptotia – as shown in Fig. 4 (taken from *R. J. N. Phillips* lectures at the Schladming Winter School, 1970).

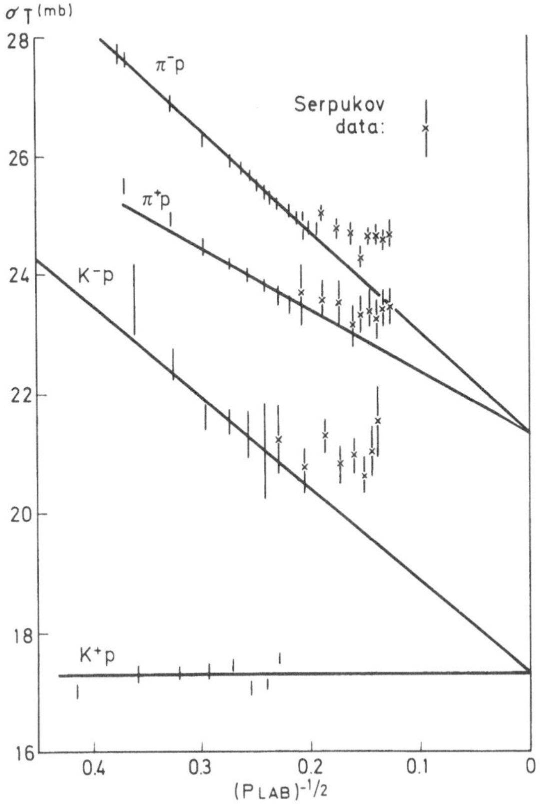

Fig. 4. High energy data against $(P_{\mathrm{LAB}})^{-\frac{1}{2}}$

Explanations fall into two classes. In the first one accepts the violation of the Pomeranchuk theorem, recalling that it is based on certain assumptions about the amplitudes which may not be true in practice (see *Martin* [19] and *Eden* [20] for recent discussions). It is not easy to accommodate such behaviour in Regge theory, since a difference between, say, $\pi^- p$ and $\pi^+ p$ total cross-sections requires an odd signature trajectory. If this is still to contribute in the infinite energy limit it must have $\alpha(0)=1$, but then its contribution is purely real and it cannot contribute to the total cross-section; so it is irrelevant in fitting the Serpukov data. *Barger* and *Phillips* [21] have shown that an odd signature dipole trajectory has the necessary properties; such a trajectory leads to $\mathrm{Re}A/\mathrm{Im}A \to \infty$, which is a general feature of a class of Pomeranchuk theorem violating theories [19, 20]. *Arnowitt* and *Rottelli* [22] have a similar model except that they obtain the unconventional behaviour by having two intersecting complex odd signature trajectories, $\alpha(t)$ and $\alpha^*(t)$.

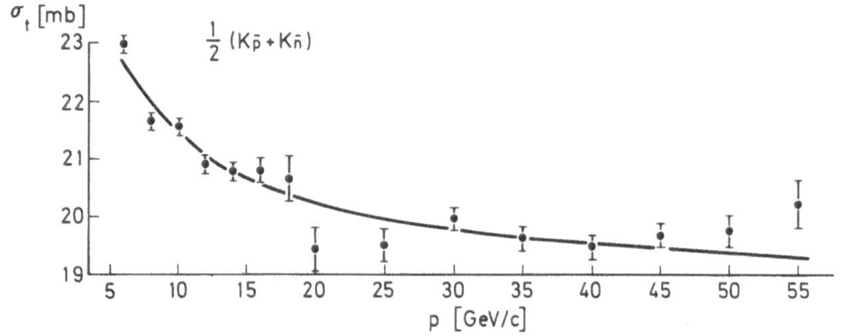

Fig. 5. Restignoli and Violini fit to $K^-p + K^-n$ total cross-section (data points are scaled)

Alternative explanations do not involve any violation of the Pomeranchuk theorem. At this stage there is certainly no positive evidence to permit one to abandon it (e. g. *Wit* [23]). *Restignoli* and *Violini* [24] fit the KN data with just the conventional Regge poles. In order to obtain this fit they scale the new data points by a factor 1.037, and the resulting fit is shown in Fig. 5. It is not good but is sufficiently close to make one cautious about accepting anything revolutionary at this stage.

Ass. 5. Straight Parallel Trajectories

In the early days of Regge theory when one was lucky to have two points on a single trajectory one tended to draw a straight line connecting them and add an apology for such an obvious oversimplification (e. g. the remark at bottom of p. 71 of Ref. [25]). However, it now appears that, at least for positive t, trajectories are over several GeV approximately linear. The evidence is best for baryons where in several cases we have trajectories with 4 particles with the correct quantum numbers lying on trajectories which are straight within the errors. For the mesons the spins of the higher mesons have not been measured but narrow peaks of the correct mass to lie on the straight trajectories (ϱ and A_2) are observed (see review of *Maglic* [26]).

In addition all observed trajectories (with the possible exception of the Pomeranchon trajectory) are approximately parallel. The best fits are:

	$I=1$ Mesons	$\alpha = 0.45 + 1.05t$
Baryons	Even P Even S (α)	$\alpha = -0.39 + 1.01t$
	Odd P Odd S (γ)	$\alpha = -0.46 + 0.88t$
	Even P Odd S (δ)	$\alpha = -0.15 + 0.90t$
	Odd P Even S (β)	$\alpha = -0.70 + 0.95t$

There is some evidence, particularly for the ϱ and N, that the linear trajectories continue smoothly into the negative t region.

Straight line trajectories for baryons lead us to

Problem 4. Baryon Parity Doublets

Because of MacDowell symmetry, a trajectory passing through a physical J-value for $\sqrt{t}$ negative corresponds to a particle of opposite parity to the corresponding particle with $\sqrt{t}$ positive. Since a linear trajectory in t is a parabola in $\sqrt{t}$ such a trajectory should lead to parity doublets. Some of these are certainly not observed (e.g. the nucleon is not a parity doublet). Since the residue will in general have a branch point at $t=0$, it is quite possible to remove some of the lowest states of the opposite parity by giving the residue a zero at the appropriate points, and there does not seem to be any reason why all parity doublets cannot be removed in this way but this looks very unnatural and artificial (note however that there are some possible candidates for parity pairs and the sometimes quoted argument that since these couple differently they cannot be on the same trajectory is not necessarily valid). *Lyth* [27] (see also *Jones* [28]) suggested a mechanism for changing the form of baryon trajectories at thresholds and so moving the parity doublets.

An amusing alternative has recently been suggested by *Carlitz* and *Kislinger* [29] who, using a particular model, construct an amplitude with fixed J-plane cuts at $J=\alpha(0)$, such that the parity doublets lie on the unphysical sheet through these cuts. An immediate difficulty with this is that the presence of cuts at $J=\alpha(0)$ should be apparent in high energy data. For example $\pi^+ p$ backward scattering should behave like $s^{\alpha(0)} \sim s^{-0.39}$ (taking $\alpha(t)$ as the nucleon trajectory) rather than $s^{\alpha(t)}$ and the nonsense dips should not be present (see next section). However *Halzen et al.* [30] show that it is possible to get a reasonable fit to the data, with no more parameters than in the more conventional model. However their fit extrapolates to a nucleon coupling constant about 100 times too large (this appears to be because of the need to suppress the cut in the region of the dip) so the situation is not satisfactory.

Theoretical considerations do not predict and indeed find it hard to explain straight line trajectories. In potential models and conventional bootstrap models trajectories turn over quickly [31]. Possibly they can be kept straight by the inclusion of higher threshold channels but there is no obvious mechanism for this. In quark models, where the scale of mass is much larger one would not be surprised to see apparently straight trajectories, but they should not be so steep unless the 'potential' between the quarks is required to satisfy a large number of moment conditions [6].

Many recent theories essentially use straight line trajectories as input, the slope being a free parameter. The approximate linearity is then compatible with the narrow widths of the observed resonances [6].

Ass. 6. Residues are Simple

Regge theory does not predict the residue function but it is usual in fitting data to assume that when Mandelstam symmetry [32] has been imposed (see Chapter 2 of Ref. [1]), then the resulting residue function is smoothly varying and has no poles or zeros. Here, e.g. for spinless particle scattering, the residue $\beta(t)$ is defined by writing for the Regge asymptotic form

$$\frac{\beta(t)}{\Gamma(\alpha(t)+n)}\,\frac{(e^{-i\pi\alpha}+1)}{\sin\pi\alpha}\left(\frac{s}{s_0}\right)^{\alpha(t)},$$

where s_0 is chosen suitably to minimise the t variation of $\beta(t)$ and n is equal to 1 if $\alpha(0)<0$ and to zero if $\alpha(0)>0$. For the spinless case one also includes sense-nonsense decoupling zeros (Chapter IV of Ref. [1]).

In general detailed fits to the data are in agreement with this assumption. In particular the zeros that come from the Γ function give rise to dips in differential cross-sections. Such dips occur, for example, in $\pi^- p \to \pi^0 n$ at $t=-0.6\,(\mathrm{GeV}/c)^2$ which is exactly where the linear ϱ trajectory passes through $\alpha(t)=0$, and in $\pi^+ p \to p\pi^+$ at $t=-0.2\,(\mathrm{GeV}/c)^2$ where the nucleon trajectory passes through $-1/2$.

There are two problems:

Problem 5. Cross-over Zeros

Near the forward direction $\frac{d\sigma}{dt}(\bar{p}p) > \frac{d\sigma}{dt}(pp)$ but it falls off more rapidly with t and the difference changes sign at $t \simeq -0.1$. According to the detailed fits these cross-sections are dominated by the P, P' and ω trajectories; the first two of which contribute the same to both, whilst the ω contribution to the non-spin-flip amplitude, which dominates near the forward direction, must pass through zero. This is a zero which has to be inserted artificially as it does not correspond to any special point on the ω-trajectory.

Problem 6. Nonsense Dips

There are cases where the wrong signature nonsense dips do not appear, e.g. $K^+ p \to pK^+$ does not have a dip where the Λ_α trajectory passes through $\alpha=-1/2$. This can be explained by the presence of the Λ_γ

trajectory, but then one finds it hard to predict when dips will occur (see *Harrari* [33]).

A final theoretical remark is that it is now known [34] that amplitudes must have fixed poles at nonsense wrong signature J-values. These do not violate unitarity because of moving J-plane cuts, and do not contribute asymptotically because of the signature factor. However there is no reason why residues of Regge poles should not have these fixed poles which would then remove the wrong-signature nonsense zeros and their associated dips. The evidence from $\pi^- p \to \pi^0 n$, etc., is that the dips and hence presumably the zeros are present in at least some cases. However, there are alternative views as mentioned below (see Section III).

Ass. 7. Duality and Exchange Degeneracy

It is a consequence of Regge asymptotic behaviour and analyticity that amplitudes satisfy sum rules of the form

$$\int^{N} d\upsilon\, \upsilon^n \operatorname{Im} A(\upsilon, t) \simeq \operatorname{Im} \sum_i \frac{\beta_i(t)\, N^{\alpha_i + 1 + n}}{\alpha_i + 1 + n}$$

where n is integral and where we have incorporated signature factors, etc., into the $\beta_i(t)$. The sum on the right is over all Regge poles that are significant in the region $\upsilon > N$, and the error in this equation is due to the neglect of lower lying poles, background terms, etc. in $\upsilon > N$. If one includes only poles with $\alpha_i + n > -1$, then the right-hand-side of this equation is the Regge pole contribution integrated from threshold to $\upsilon = N$, so in this sense the leading Regge pole contribution averages the imaginary part of the amplitude. As this is approximately true for all moments one might expect it to be true 'locally', i. e. averaging over a smaller region of υ. Of course it is a poor approximation in the strictly local sense, i. e. without any averaging, since the Regge poles contributions do not have resonance bumps.

Relations of the above type, known as finite energy sum rules, are well satisfied in practice but are often used as constraints on Regge pole parameters determined just from high energy data. These matters will be dealt with elsewhere at this school.

If, however, one does a partial wave analysis of the Regge asymptotic behaviour, including the leading Regge trajectories, then one does obtain phase shifts reminiscent of resonant behaviour, and *Schmid* [35] suggested that perhaps the Regge behaviour does indeed 'contain' in this way the direct channel resonance poles. We shall not here enter the controversy as to how seriously one should take this suggestion but,

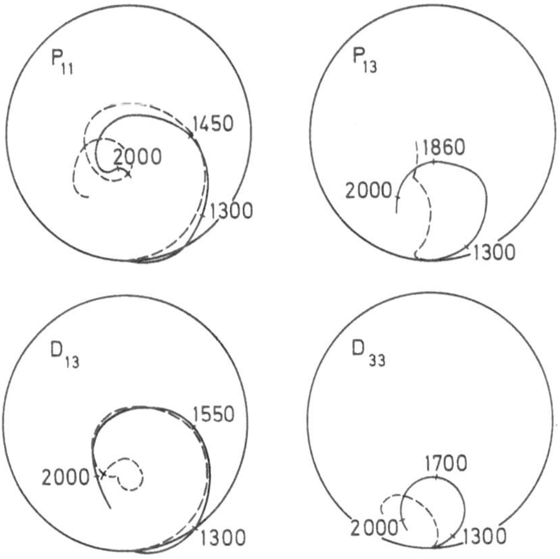

Fig. 6. Comparison of Regge form for $\pi - N$ partial waves with experiment

to illustrate this extreme example of the possibility of the Regge pole model explaining almost everything we show [36] plots of some of the $\pi - N$ partial-wave amplitudes beginning at threshold obtained from a Regge pole fit to high energy data. One should note that one cannot show energy dependence on these Argand plots very easily – if we did the agreement with the experimental data would not be so good.

Since, with the assumption above one is relating cross-channel and direct channel resonances one has the possibility of a linear bootstrap theory. Such bootstraps are however normally obtained from the FESR by approximating the L. H. S. by a sum of resonances; in this way one again obtains a linear relation between direct and cross-channel resonances. An interesting special case is when the direct channel is exotic in which case one predicts cancellation between the cross-channel trajectories – this implies 'exchange degeneracy' (essentially that there is either a purely 'direct' or purely 'exchange' force). Thus one predicts from $\pi\pi \to \pi\pi$ that the ϱ and f are exchange degenerate and from $\pi K \to \pi K$ that the ϱ and A_2 are likewise. The experimental evidence is at least roughly in support of this.

The idea of duality we have discussed is usually predicated on the assumption of a flat Pomeranchon trajectory [37, 38] if the Pomeranchon has a slope of about one-half as suggested above then it does not fit readily into the scheme. However it is probably sufficiently flexible to permit this.

Ass. 8. Factorisation

The simple experimental prediction of factorisation cannot be checked and it turns out that there are no good tests except for the Pomeranchon. Here one can test directly the ratios

$$\frac{\dfrac{d\sigma}{dt}(NN\to NN)}{\dfrac{d\sigma}{dt}(\pi N\to\pi N)}=\frac{\dfrac{d\sigma}{dt}(NN\to NN^*)}{\dfrac{d\sigma}{dt}(\pi N\to\pi N^*)}$$

for those N^*'s which have the same internal quantum numbers as the N, and can therefore be produced by P exchange – with a constant total cross-section. *Freund* [39] found values 2.7, 3.2, 2.9 for these ratios for $N, N^*(1400)$ and $N^*(1688)$ respectively, with errors about ± 0.6 for the last two. Thus factorisation holds to within the errors. An alternative way of studying this is due to *Bari* and *Razmi* [40] who use the fact that at high energy, Pomeranchon exchange should dominate the whole of the production amplitude, i. e. not only the resonance production. In this way they avoid the uncertainties of background subtraction needed to calculate the resonance production cross-section and are able to cover a wide range of πN energies; on the other hand there are uncertainties due to the fact that there is contamination from the $I=3/2$ πN states to which the Pomeranchon does not contribute. They compare the ratio $\frac{d\sigma}{dt}(\pi N\to\pi\pi)\Big/\frac{d\sigma}{dt}(NN\to\pi NN)$ for various

Table 1a. *Ratio of the inelastic double differential cross-sections of pp and $\pi^+ p$*

W(MeV)	Ratio	W(MeV)	Ratio	W(MeV)	Ratio	W(MeV)	Ratio
2113	2.28	1926	2.71	1719	2.34	1338	2.45
2091	2.31	1901	2.70	1691	2.42	1315	2.37
2068	2.31	1876	2.61	1663	2.58	1278	2.35
2045	2.35	1851	2.44	1634	2.50	1241	2.43
2022	2.44	1825	2.28	1483	2.17	1202	2.43
1998	2.50	1799	2.44	1451	2.32	1162	2.27
1974	2.61	1773	2.46	1418	2.40	1121	2.27
1950	2.71	1744	2.46	1384	2.33	1033	2.32

Table 1b. *Ratio of the elastic differential cross-sections of pp and $\pi^+ p$.*

t	Ratio	t	Ratio	t	Ratio
0.058	2.48	0.157	2.16	0.256	2.31
0.084	2.39	0.203	2.29	0.268	2.10
0.116	2.33	0.250	2.20		

values of the πN final state energy (1926 MeV to 1033 MeV), with the corresponding value for the elastic process, and find good agreement throughout the range (see Table 1).

In other cases factorisation tests invariably fail. We list four specific failures.

Problem 7. π-conspiracy and Factorisation

If one accepts the conspiracy explanation of problem 2, above, then one runs into trouble with factorisation. For example [41], the model predicts a forward dip in $\pi^+ p \to \varrho^0 \Delta^{++}$ whereas the data gives a peak. *Dass* and *Froggott* [42] give a detailed discussion of difficulties of this type.

Problem 8. Cross-over Zeros and Factorisation

If the ω residue has a zero, as apparently required by the cross-over mentioned in problem 5, then this zero should show up in other processes dominated by ω exchange. It is seen in the difference between $\frac{d\sigma}{dt}(K^\pm p)$, which has a cross-over at the same value of t, but not in the $I_t=0$ part of $\pi N \to \varrho N$ which should be dominated by ω exchange and can be isolated experimentally by using the I-spin crossing matrix as in

$$2\frac{d\sigma}{dt}(I_t=0)=\frac{d\sigma}{dt}(\pi^+ p\to\varrho^+ p)+\frac{d\sigma}{dt}(\pi^- p\to\varrho^- p)-\frac{d\sigma}{dt}(\pi^- p\to\varrho^0 n).$$

Instead of a minimum at the cross-over point one gets a maximum (at 4 and 8 GeV/c, see *A. P. Contogouris* [43]).

Problem 9. Nonsense Choosing ϱ

If the ϱ, f and A_2 are exchange degenerate then the ϱ must choose nonsense at $\alpha_\varrho=0$ (the A_2 has to choose nonsense otherwise there would be a dip in $\pi^- p\to\eta n$), so the dip in $\pi^- p\to\pi^0 n$ would be a *zero*, whereas it is fitted well with an s^{α_ϱ} dependence.

Problem 10. Exotic Mesons and $B\bar{B}$ Scattering

The absence of exotic mesons is incompatible with duality and factorisation, when applied to $B\bar{B}$ scattering. That it is factorisation that causes the trouble has recently been shown by *Kugler* [44].

III. Summary and Discussion

In general the situation is very satisfactory. In Table 2 we list the problems we have found, in an 'order of significance' (clearly this is highly subjective!).

Table 2. *Problems*

1	Polarisation in $\pi^- p \to \pi^0 n$
2 or 7	π in forward direction or conspiracy and factorisation
8	Cross-over zeros and factorisation
9	Nonsense choosing ϱ
5	Cross-over zeros
6	Dips
3	Serpukov data
4	Baryon parity doublets
10	Exotic mesons

The conventional solution to most of the above problems lies through the introduction of Regge cuts. If one allows these arbitrarily then clearly we lose all predictive power, and we have no problems because we have no theory. All models of cuts are based on the absorptive model in which one takes the Regge pole as the Born term and then introduces absorptive corrections (see *Fox* [45] for a review). Attempts to justify this procedure are not convincing; it is hard to see why the procedure does not involve 'double-counting' and, alternatively, the cuts obtained look similar to the AFS cuts [46] which are known to cancel exactly with other contributions [47]. Nevertheless the method works well. It gives a natural explanation of problems 1 (in particular it predicts the correct sign) and 2 (an absorbed π does have the forward peak); problems 7, 8, 9, 10 are removed because cut contributions do not factorise; problem 5 is solved because interference between cut and pole contributions can give zeros; there is a ready explanation [48] of the Serpukov data, in particular it is expected that where cuts are important elastic cross-sections should approach their asymptotic values from below [49] – this comes from the $P+P'$ cut. The problem of dips has been extensively studied in the presence of cuts and here one has to distinguish two models – the 'weak-cut' model in which the Regge contributions have nonsense zeros, the resulting oscillations being partially responsible for the weakness of the cut, and the strong cut model in which there are no nonsense zeros and hence much stronger cuts (which are enhanced over the absorption model prediction by an arbitrary factor), which produce the usual dips by interference with the poles. For a recent comparison and references we refer to *Drago et al.* [50].

Finally we ask whether it is possible to survive without cuts, i. e. does the data force one inevitably to accept cuts of the magnitude given by the absorption model in one of its forms or are there alternative solutions to the difficulties with the pure Regge-pole model?

Two possibilities have recently been discussed. The first involves having some fixed poles at nonsense right-signature points. These contribute to the asymptotic behaviour (in contrast to fixed poles at wrong signature points discussed previously), and if they exist can solve many of the problems of Regge theory as has recently been shown by *Finkler* [51]. They are also particularly useful in fitting photoproduction data. In order that these poles do not violate unitarity Finkler makes use of the fixed kinematic cuts in the J-plane which are conveniently taken along the real axis between $-\sigma_T$ and $\sigma_T - 1$, where σ_T is the maximum total spin in either the initial or final channel (see Chapter V, Section 4 of Ref. [1]). These cuts allow fixed poles at the nonsense points in their *interior*, i. e. at the nonsense values in the interval $-\sigma_T + 1 \leqq J \leqq \sigma_T - 2$, which, when first considered by *Calogero et al.* [52], were dismissed as being too low to be significant in phenomenological analyses. Finkler, however, uses only physical unitarity, so that, for example, in $N\bar{N} \to N\bar{N}$ scattering the multipion states having effectively infinite spin must be included in the unitarity equation, and he is thus able to allow fixed poles even at the end points of the nonsense region, e. g. at $J=0$ in $N\bar{N} \to N\bar{N}$. Unfortunately it is necessary also to include the constraints of generalized unitarity which makes the presence of lower threshold channels irrelevant [53], so the original conclusion that possible fixed poles associated with kinematic fixed cuts are too low to be important remains valid.

Another possibility [54] begins with the observation that the factorisation assumption is the one that causes most problems. Now factorisation is a consequence of the poles of the many-channel S-matrix being simple poles, but there is no theoretical reason why poles should be simple poles (in conventional bootstrap theories of course it would be surprising if they were not), so since the experiments clearly do not want them to be simple perhaps we should abandon factorisation and, for example, allow doubled trajectories. If one takes the view that the degeneracy is broken to a small degree, e. g. by unitarity corrections, then one would in fact expect the doubled trajectories to be separated by the order of the width of particles on them. The split A_2 would then be good evidence in support of this. Alternatively if one started from the observation of the split A_2 and used exchange degeneracy and $SU(3)$ then one would rapidly conclude that all meson trajectories should be split. There is an immediate objection to this, namely, that other mesons (e. g. ϱ, π) are clearly *not* observed as doublets. This however can be

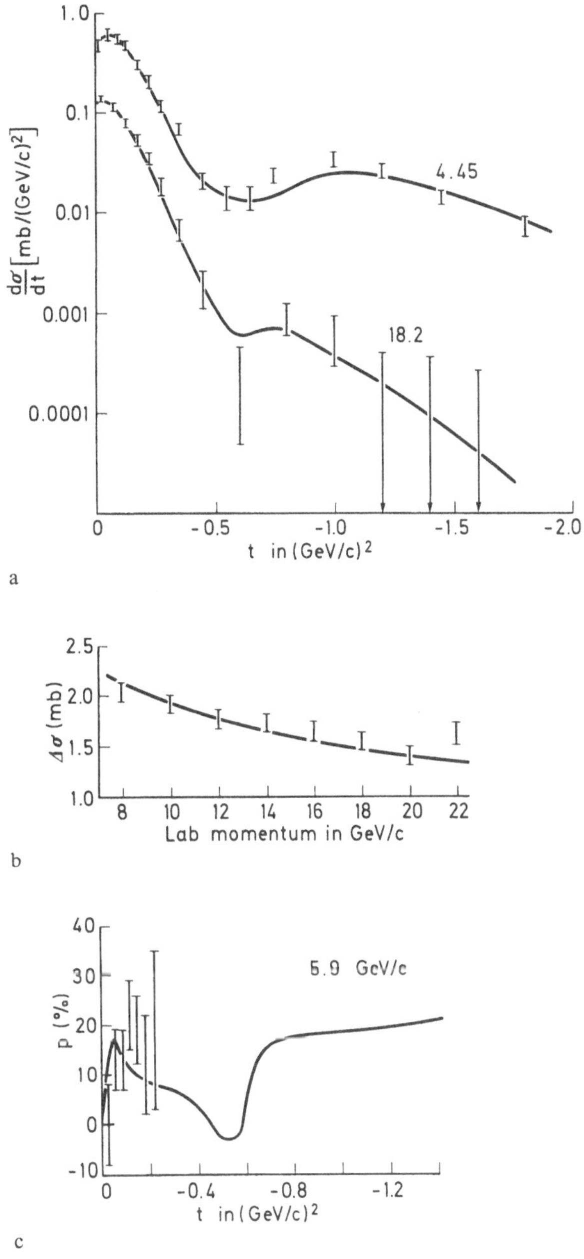

Fig. 7a–c. Typical examples of the doubled trajectory fit to $\pi^- p \rightarrow \pi^0 n$ data. (a) Differential cross-sections at 4.45 and 18.2 GeV/c. (b) Difference between $\pi^- p$ and $\pi^+ p$ cross sections. (c) Polarisation fit at 5.9 GeV/c and prediction for larger $|t|$ values

explained in a natural way in the framework of dual models. The 6π amplitude of *Rubinstein* and *Rittenberg* [55] has a doubled π trajectory*, but of necessity one of the trajectories does not have $J=0, 1, 2$ particles on it, i. e. it starts with the $J=3$ state. Attempts to construct a split A_2 in this manner using the 6π amplitude [56] have not been successful but this may well turn up in other processes and certainly the mechanism is there (see also later work of *Dorren*, *Rittenberg*, and *Rubinstein* [57]).

It is clear that problems 2, 8, 9 and 10, all of which depend on factorisation are removed by doubled trajectories. Also we find a natural mechanism for explaining cross-over zeros (see below), and, provided we permit a small breaking if the degeneracy as explained above we can solve problem 1. To illustrate this all $\pi^- p \to \pi^0 n$ data has been fitted in Ref. [54] with a model involving a doubled ϱ trajectory. The quality of the fit is shown in Fig. 7. In this fit the trajectories were taken as parallel; the separation was a free parameter which turns out to be $\delta\alpha = 0.1$.

The doubling of trajectories does not say anything new with regard to problems 6, 3 and 4, except that it does allow some extra freedom.

We conclude that while cuts do offer a means of explaining the major difficulties with the Regge pole model, the doubling of trajectories seems to remove these problems and therefore to offer a possible alternative suggestion which is in many ways more satisfactory.

References

1. *Collins, P. D. B., Squires, E. J.:* Springer Tracts in Modern Physics **45**, "Regge Poles in Particle Physics". Berlin-Heidelberg-New York: Springer 1968.
2. *Höhler, G., Schaile, H., Sonderegger, P.:* Phys. Letters **20**, 79 (1966).
3. *Freund, P. G. O.:* Phys. Letters **3**, 123 (1963).
4. *Warburton, A. E. A.:* Phys. Rev. **137**, B 993 (1964), and *Barut, A. O., Zwanziger, O. E.:* Phys. Rev. **127**, 974 (1962).
5. *Chiu, C. B., Stack, J. D.:* Phys. Rev. **153**, 1575 (1967).
6. *Collins, P. D. B., Johnson, R. C., Squires, E. J.:* Phys. Letters **26** B, 223 (1967).
7. Proceedings of the 14th International Conference on High-Energy Physics, Vienna 1968.
8. *Michael, C.:* Phys. Letters **29** B, 4, (1969).
9. *Barger, V., Olsson, M.:* Phys. Rev. **146**, 1080 (1966).
10. *Hite, G.:* Rev. Mod. Phys. (to be published) (1970).
11. *Barger, V., Phillips, R. J. N.:* Phys. Rev. **187**, 2210 (1969).
12. *Daum, C., Michael, C., Schmid, C.:* CERN Th. 1124 (1970).
13. *Ter-Martirosyan, K. A.:* JETP Letters **10**, 285 (1969).
14. *Barger, V., Cline, D.:* Phys. Rev. Letters **24**, 1313 (1970).
15. *Bonamy, P., et al.:* Phys. Letters **23**, 501 (1966).
16. *Phillips, R. J. N.:* Nucl. Phys. B **2**, 657 (1967b).
17. *Allaby, J. V., et al.:* Phys. Letters **30** B, 500 (1969).
18. *Barger, V., Olsson, M., Reader, D. D.:* Nucl. Phys. B **5**, 411 (1968).
19. *Martin, A.:* CERN Th. 1075 (1970).

* Unfortunately it also has ghosts; this problem is solved however in Ref. [57].

20. *Eden, R. J.*: to be published in Phys. Rev. (1970).
21. *Barger, V., Phillips, R. J. N.*: Phys. Letters **31** B, 643 (1970).
22. *Arnowitt, R.,Rottelli, P.*: Imperial College preprint, ICTP/69/14 (1970).
23. *Wit, R.*: CERN Th. 1137 (1970).
24. *Restignoli, M., Violini, G.*: Phys. Letters **31** B, 533 (1970).
25. *Squires, E. J.*: "Complex Angular Momentum and Particle Physics. New York: Benjamin 1963.
26. *Maglic, B.*: Proceedings of the 1969 Lund Conference, p. 269 (1969).
27. *Lyth, D. H.*: Phys. Rev. Letters **20**, 641 (1968).
28. *Jones, H. F.*: Nuovo Cimento **55**, 354 (1968).
29. *Carlitz, R., Kislinger, M.*: Phys. Rev. Letters **24**, 186 (1970).
30. *Halzen, F., Kumar, A., Martin, A. D., Michael, C.*: Phys. Letters **32** B, 111 (1970).
31. *Collins, P. D. B., Johnson, R. C.*: Phys. Rev. **182**, 1755 (1969).
32. *Mandelstam, S.*: Ann. Phys. **19**, 254 (1962).
33. *Harrari, H.*: Proceedings of the 4th International Conference on Electron and Photon Interactions at High Energies p. 107 (1969).
34. *Mandelstam, S., Wong, L. L.*: Phys. Rev. **160**, 1490 (1967).
35. *Schmid, C.*: Phys. Rev. Letters, **20**, 689 (1968).
36. *Collins, P. D. B., Johnson, R. C., Squires, E. J.*: Phys. Letters **27** B, 23 (1968).
37. *Freund, P. G. O.*: Phys. Rev. Letters **20**, 235 (1968).
38. *Harrari, H.*: Phys. Rev. Letters **20**, 1395 (1968).
39. *Freund, P. G. O.*: Phys. Rev. Letters **21**, 1375 (1968).
40. *Bari, M. B., Razmi, M. S. K.*: Islamabad preprint (1970).
41. *Le Bellac, M.*: Phys. Letters **25** B, 524 (1967).
42. *Dass, G. V., Frogatt, C. P.*: Nucl. Phys. B **8**, 661 (1968).
43. *Contogouris, A. P.*: Phys. Rev. Letters **19**, 1352 (1967).
44. *Kugler, M.*: Phys. Letters **32** B, 107 (1970).
45. *Fox, G. C.*: "Skeletons in the Regge Cupboard" Talk given at the Stony Brook Conference, September 1969.
46. *Amati, D., Fubini, S., Stanghellini, A.*: Phys. Letters **1**, 29, (1962).
47. *Mandelstam, S.*: Ann Phys. **21**, 302 (1963).
48. *Barger, V., Phillips, R. J. N.*: Phys. Rev. Letters **24**, 291 (1970).
49. *Frautschi, S., Margolis, B.*: Nuovo Cimento **56** A, 1155 and **57** A, 427 (1968).
50. *Drago, F., Love, A., Phillips, R. J. N., Ringland, G. R.*: Phys. Letters **31** B, 647 (1970).
51. *Finkler, P.*: Phys. Rev. D **1**, 1172 (1970).
52. *Calogero, F., Charap, J. N., Squires, E. J.*: "Proc. Sienna Conference" Vol. **1**, 416 (1964).
53. *Squires, E. J.*: "Generalised unitarity and fixed *J*-plane pules", Durham preprint to be published in Phys. Rev. (1970).
54. *Johnson, R. C., Squires, E. J.*: Durham preprint (1970).
55. *Rittenberg, V., Rubinstein, R.*: Phys. Rev. Letters **25**, 191 (1970).
56. *Dorren, J. D., Rittenberg, V., Rubinstein, H. R., Chaichian, M., Squires, E. J.*: Weizmann/Durham preprint (1970).
57. *Dorren, J. D., Rittenberg, V., Rubinstein, H. R.*: CERN preprint Th. 1192 (1970).

Prof. Dr. *Euan J. Squires*
University of Durham
Dept. of Mathematics
Science Laboratories
Durham/England

Certain Problems of Two-Body Reactions with Spin

A. P. Contogouris

Contents

Introduction

It is now generally accepted that in two-body reactions, in particular of the quasi-elastic type, the spin is an important complication. For theoretical as well as phenomenological purposes, particularly powerful is the approach through crossed- (t-) channel helicity amplitudes (THA): Introduction of nonsense factors, which offer today the most plausible and only theoretically self-consistent explanation of the dips observed in quasi-elastic reactions, can be easily done only via THA; the same holds for the treatment of certain types of conspiracy (Section 4 of Part I), when THA offer also the only transparent approach; last but not least, the study of the properties of THA under well established analyticity and factorization ($\approx$ unitarity) requirements leads to a powerful classification of Regge trajectories along with a number of predictions concerning their asymptotic behaviour for forward scattering.

The Part I of these lectures is devoted to the general features of the THA approach and to a derivation of the Toller classification [1] of Regge trajectories by non group-theoretical methods [2, 3]. Section 1 reviews briefly the analyticity properties of THA and the extraction of their kinematical singularities. Section 2 gives a solution to the problem of determining the behaviour of the residue functions near the forward direction subject to analyticity and factorization conditions for general external masses and helicities; in a Regge pole approach this solution is very useful for phenomenological purposes as well. Section 3, after a brief explanation of the origin of conspiracy relations (≡ kinematical or analyticity constraints), makes full use of the solution of Section 2 to derive some of the essential conclusions of the Toller classification. Section 4, by means of well-known analyticity conditions of individual THA, deduces the existence of daughter Regge trajectories; as explained in Ref. [3], via the solution of Section 2, these conditions lead to an alternative non group-theoretical derivation of the Toller classification (without use of conspiracy relations); Section 4 uses this approach to deduce additional aspects of *Toller's* work. Finally, Section 5 discusses certain physical implications as well as certain problems related to the classification of the pion trajectory and also outlines the solution to these problems offered by the Regge cut approach.

The Part II of these lectures is, to a large extent, an application of the foregoing approach to a problem of current interest: the frame dependence of certain vector dominance predictions. Here a simple parameter-free (and satellite-free) Veneziano model is used as a specific tool to study the forward structure of $\pi^- + p \to \varrho^0_{\text{transv}} + n$ and the ratio $\varrho_{1\,-1}/\varrho_{11}$ of related density matrix elements. Our conclusion is that vector dominance should be formulated and tested in the helicity frame. Most important, it follows that through our approach (which is shown to be more general than the Veneziano model), the presence of a forward peak in $\pi^- p \to \varrho^0_{\text{transv}} n$ in the helicity frame offers a new test of the ϱ-meson universality: For values of the ratio $g_{\varrho NN}/g_{\varrho\pi\pi}$ away from the universality value 1 the forward peak disappears completely and/or the shape of the forward $\pi^- p \to \varrho^0_{\text{transv}} n$ cross-section is completely distorted. The presence of such a peak is supported by recent experimental analysis; thus our approach offers a new verification of ϱ universality.

Part I: Forward Structure of Helicity Amplitudes, Kinematical Constraints and the Toller Classification of Regge Trajectories

1. Analyticity Properties of Helicity Amplitudes

Consider the t-channel two-body reaction

$$1 + \bar{2} \to \bar{3} + 4\,; \tag{1.1}$$

we are interested in the asymptotic behaviour of $1+3\to 2+4$ and eventually shall let $s\to\infty$. Denote by $f_{\lambda_3\lambda_4,\lambda_1\lambda_2}(t,\cos\theta_t)$ the corresponding helicity amplitudes. For Reggeization the first step is to factor out the kinematic zeros which occur at forward and backward t-channel scattering because of angular momentum conservation. Thus introduce

$$\bar{f}_{\lambda_3\lambda_4,\lambda_1\lambda_2}=\left(\sqrt{2}\cos\frac{\theta_t}{2}\right)^{-|\lambda+\mu|}\left(\sqrt{2}\sin\frac{\theta_t}{2}\right)^{-|\lambda-\mu|}f_{\lambda_3\lambda_4,\lambda_1\lambda_2} \quad (1.2)$$

where

$$\lambda=\lambda_1-\lambda_2 \qquad \mu=\lambda_3-\lambda_4.$$

Consider forward scattering. In the center-of-mass (cm) system λ is the projection of the total angular momentum of the initial state in the direction of motion; and μ of the final state. If $\lambda\neq\mu$ the amplitude $f_{\lambda_3\lambda_4,\lambda_1\lambda_2}\to 0$ as $\theta_t\to 0$; however, the kinematical factor $(\sin\theta_t/2)^{|\lambda-\mu|}\to 0$, as well. Likewise, for backward scattering ($\theta_t\to\pi$), if $\lambda\neq-\mu$: $f_{\lambda_3\lambda_4,\lambda_1\lambda_2}\to 0$, but also $(\cos\theta_t/2)^{|\lambda+\mu|}\to 0$. Thus $\bar{f}_{\lambda_3\lambda_4,\lambda_1\lambda_2}$ is free of the kinematical zeros imposed by angular momentum conservation.

In terms of $\bar{f}_{\lambda_3\lambda_4,\lambda_1\lambda_2}$ we can construct helicity amplitudes asymptotically dominated by exchange of definite *normality* σ; for boson exchanges

$$\sigma\equiv P(-)^J=\pm 1\,; \quad (1.4)$$

$\sigma=+1(-1)$ is also referred to as exchange of natural (unnatural) parity. The construction is based on the following transformation under parity P of the helicity state $|JM_J;\lambda_3\lambda_4\rangle^{(4)}$:

$$P|JM_J;\lambda_3\lambda_4\rangle=\eta_3\eta_4(-)^{J-S_3-S_4}|JM_J;-\lambda_3-\lambda_4\rangle$$

where s_3 and η_3 the spin and intrinsic parity of particle 3. We define

$$\lambda_m=\max\{|\lambda|,|\mu|\} \quad (1.5)$$

and construct the following amplitudes:

$$\bar{f}^{\sigma}_{\lambda\mu}=\bar{f}^{\pm}_{\lambda\mu}=\bar{f}_{\lambda_3\lambda_4,\lambda_1\lambda_2}\pm\eta_3\eta_4(-)^{s_3+s_4+\lambda+\lambda_m}\bar{f}_{-\lambda_3-\lambda_4,\lambda_1\lambda_2}\,.$$

Now it can be shown that $\bar{f}^{\sigma}_{\lambda\mu}$ have partial wave expansions of the form [5]:

$$\bar{f}^{\sigma}_{\lambda\mu}(t,z)=\sum_J(2J+1)\,\{e^{J+}_{\lambda\mu}(z)\,F^{J,\sigma}_{\lambda\mu}(t)+e^{J-}_{\lambda\mu}(z)\,F^{J,-\sigma}_{\lambda\mu}(t)\} \quad (1.6)$$

with $z=\cos\theta_t$. The functions $e^{J\pm}_{\lambda\mu}$ are proportional to Jacobi polynomials $P^{|\lambda-\mu|,|\lambda+\mu|}_{J-\lambda_m}(z)$ and $F^{J,\pm\sigma}_{\lambda\mu}(t)$ are generalized partial wave amplitudes [5].

Since $P_{J-\lambda_m}$ are polynomials of finite order in z, $\bar{f}^{\sigma}_{\lambda\mu}(t, z)$ is a sum over terms containing no kinematical singularities in z. On the other hand z is a linear function of s:

$$z = \cos\theta_t$$
$$= \frac{2ts + t^2 - t\sum_i m_i^2 + (m_1^2 - m_2^2)(m_3^2 - m_4^2)}{\{[t-(m_1+m_2)^2]\,[t-(m_1-m_2)^2]\,[t-(m_3+m_4)^2]\,[t-(m_3-m_4)^2]\}^{\frac{1}{2}}}\,; \tag{1.7}$$

thus the terms in the expansion of $\bar{f}^{\sigma}_{\lambda\mu}$ have no kinematic singularities in s.

After a Sommerfeld-Watson transform a Regge pole $J = \alpha(t)$ of normality σ contributes to $\bar{f}^{\sigma}_{\lambda\mu}$ through $e^{J+}_{\lambda\mu}(z)$ a contribution of order $s^{\alpha(t)-\lambda_m}$. If $\lambda \neq 0$ and $\mu \neq 0$, then $e^{J-}_{\lambda\mu}(z) \neq 0$ and $\bar{f}^{\sigma}_{\lambda\mu}$ may also receive a contribution from a Regge pole $J = \bar{\alpha}(t)$ of normality $-\sigma$ but of order $s^{\bar{\alpha}(t)-\lambda_m-1}$. Thus $e^{J+}_{\lambda\mu}$ always provides the asymptotically dominant contribution.

Consider now the residue function corresponding to the asymptotically dominant contribution to $\bar{f}^{\sigma}_{\lambda\mu}(s, t)$. In general, this residue contains kinematic singularities in t because the particles in the reaction (1.1) may have : (i) unequal masses and (ii) nonzero spin.

In connection with (i) notice that e.g. the c.m. momentum $p_{1\bar{2}}$ of the initial state is

$$p_{1\bar{2}} = \frac{[t-(m_1+m_2)^2]^{\frac{1}{2}}\,[t-(m_1-m_2)^2]^{\frac{1}{2}}}{2(-t)^{\frac{1}{2}}}\,, \tag{1.8}$$

i.e. for $m_1 \neq m_2$ singular at $t = 0$; hence, simple threshold considerations imply that the partial helicity amplitudes have singularities in t. As follows from $e^{J}_{\lambda\mu}(\sim P_{J-\lambda_m})$ the effective angular momentum of the state is $J - \lambda_m$ ($\to \alpha(t) - \lambda_m$ after Reggeization). Thus this set of singularities is easily removed by extracting from the residue the threshold factor

$$(p_{1\bar{2}}\,p_{\bar{3}4})^{\alpha(t)-\lambda_m}\,. \tag{1.9}$$

In connection with (ii), the spin of the particles implies additional singularities at

$$t = 0,\ (m_1 \pm m_2)^2,\ (m_3 \pm m_4)^2\,.$$

For a given mass and helicity configuration these singularities can be determined by either of the following two methods:

(A) Use of the crossing matrix $X^{\{\lambda_s\}}_{\{\lambda_t\}}(s, t)$ which relates the THA $f_{\{\lambda_t\}}$ to the s-channel HA $M_{\{\lambda_s\}}$:

$$f_{\{\lambda_t\}} = \sum_{\{\lambda_s\}} X^{\{\lambda_s\}}_{\{\lambda_t\}}\, M_{\{\lambda_s\}}$$

and is explicitly given in Ref. [6]. We can further write the corresponding relations between reduced HA:

$$\bar{f}_{\{\lambda_t\}} = \sum_{\{\lambda_s\}} \bar{X}^{\{\lambda_s\}}_{\{\lambda_t\}} \bar{M}_{\{\lambda_s\}}$$

with

$$\bar{X}^{\{\lambda_s\}}_{\{\lambda_t\}} = X^{\{\lambda_s\}}_{\{\lambda_t\}} \frac{(\sin\theta_s/2)^{|\lambda_s-\mu_s|}(\cos\theta_s/2)^{|\lambda_s+\mu_s|}}{(\sin\theta_t/2)^{|\lambda-\mu|}(\cos\theta_t/2)^{|\lambda+\mu|}}$$

where λ, μ as in Eq. (1.3) and λ_s, μ_s the corresponding differences of s-channel helicities. It is expected* from the discussion above that $\bar{M}_{\{\lambda_s\}}$ are free of kinematical singularities in t, hence all kinematical singularities in t of $\bar{f}_{[\lambda_t]}$ are given by $\bar{X}^{[\lambda_s]}_{[\lambda_t]}$.

(B) For a given reaction use the relations between THA and Mandelstam amplitudes in its invariant decomposition. As an example to be subsequently discussed in detail consider the invariant decomposition of the reaction $\pi N \to VN$ ($V \equiv$ vector meson):

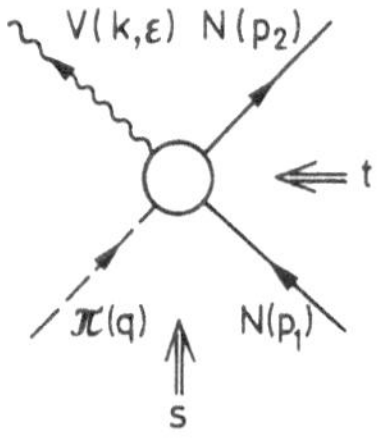

$$T = \bar{u}(p_2)\gamma_5[-B_1 \not{\varepsilon}\not{k} + 2B_2\,\varepsilon\cdot P - 2B_3\,\varepsilon\cdot q + B_4 \not{k}\not{\varepsilon}\cdot q - B_5\not{\varepsilon} - B_6\not{k}\not{\varepsilon}\cdot P + M\not{\varepsilon}\not{k})]\,u(p_1) \tag{1.10}$$

where $P = \frac{1}{2}(p_1 + p_2)$, $M =$ nucleon mass and $B_i = B_i(s, t, u)$. Decomposition of T into THA gives e.g. [7]:

$$\bar{f}^{\,+}_{11} = K^{+}_{11}(t)\left(M B_1 + \frac{t}{4} B_6\right); \qquad K^{+}_{11}(t) = \frac{[t-(m+\mu)^2]^{\frac{1}{2}}\,[t-(m-\mu)^2]^{\frac{1}{2}}}{\sqrt{2}\,M\,t^{\frac{1}{2}}}$$

with $m(\mu)$ the vector meson (pion) mass. Since the invariant amplitudes $B_i(s, t, u)$ are known to satisfy Mandelstam analyticity, it is clear that $\bar{f}^{\,+}_{11}$ has kinematical singularities at $t = 0$ and $t = (m \pm \mu)^2$.

However, the last example also shows that we can introduce a singularity-free PCTHA $\tilde{f}^{\,+}_{11}$ by simply factoring out $K^{+}_{11}(t)$. In general

$$\bar{f}^{\,\sigma}_{\lambda\mu}(s,t) = K^{\sigma}_{\lambda\mu}(t)\,\tilde{f}^{\,\sigma}_{\lambda\mu}(s,t) \tag{1.11}$$

where $\tilde{f}^{\,\sigma}_{\lambda\mu}(s,t)$ is free of kinematical singularities and $K^{\sigma}_{\lambda\mu}(t)$ is a known kinematical factor ($\equiv$ Wang factor) [8, 9].

---

* Provided that the summation in (1.6) introduces only the dynamical singularities in s (no kinematical ones). This can be checked e.g. by method (B).

If $\beta_{\lambda\mu}(t)$ denotes the residue of $F^{J,\sigma}_{\lambda\mu}(t)$ and $J=\alpha(t)$, then as $z\to\infty$:

$$\bar{f}^{\sigma}_{\lambda\mu}(z,t)\simeq(2\alpha+1)\frac{1\pm e^{-i\pi\alpha}}{\sin\pi\alpha}\,\beta_{\lambda\mu}(t)\,[e^{J+}_{\lambda\mu}(-z)]_{J=\alpha(t)}\,. \tag{1.12}$$

In view of the discussion above we shall write

$$\beta_{\lambda\mu}(t)=(p_{1\bar{2}}\,p_{\bar{3}4})^{\alpha(t)-\lambda_m}\,K_{\lambda\mu}(t)\,\gamma'_{\lambda\mu}(t)$$

where $\gamma'_{\lambda\mu}$ contains only dynamic singularities in t (existing at $t\geqq t_{\text{thr}}>0$); in particular $\gamma'_{\lambda\mu}(t)$ is regular at $t=0$. Near $t=0$ we can further write

$$(p_{1\bar{2}}\,p_{\bar{3}4})^{\alpha(t)-\lambda_m}\sim t^{T(\lambda\mu)}\qquad K_{\lambda\mu}(t)\sim t^{W(\lambda\mu)}$$

with $T(\lambda\mu)$ and $W(\lambda\mu)$ known for each configuration λ,μ. Hence

$$\beta_{\lambda\mu}(t)=t^{T(\lambda\mu)+W(\lambda\mu)}\,\gamma'_{\lambda\mu}(t)\qquad(t\approx 0)\,.$$

2. Factorization Requirements at $t=0$

We further accept that the residues $\beta_{\lambda\mu}(t)$ satisfy the factorization condition:

$$\beta_{\lambda\mu}(12\to 34)\,\beta_{\lambda'\mu'}(1'2'\to 3'4')=\beta_{\lambda\lambda'}(12\to 1'2')\,\beta_{\mu\mu'}(34\to 3'4')\,. \tag{2.1}$$

Now the point is that if we require for all λ,μ

$$\gamma'_{\lambda\mu}(0)\neq 0$$

then the (known) functions $T(\lambda\mu)$ and $W(\lambda\mu)$ do not satisfy factorization; the condition (2.1) demands that for some λ,μ: $\gamma'_{\lambda\mu}(t)\underset{t\to 0}{\sim}t^n$, $n>0$. Thus we set

$$\gamma'_{\lambda\mu}(t)=t^{X(\lambda\mu)}\,\gamma_{\lambda\mu}(t)\qquad\gamma_{\lambda\mu}(0)\neq 0$$

and try to determine the *smallest* $X(\lambda\mu)$ consistent with the known T and W (i.e. with analyticity) and factorization. This is a problem important in itself, for the exact value of $X(\lambda\mu)$ determines which amplitudes vanish as $s\to\infty$ and $t\to 0$. In this and the next section we will consider leading (parent) trajectories, when the regularity of $\gamma'_{\lambda\mu}(t)$ is easily seen to imply

$$X(\lambda\mu)=\text{integer}\geqq 0\,. \tag{2.3}$$

Start from the case $m_1\neq m_2$, $m_3\neq m_4$ (called unequal-unequal mass case and denoted by UU) when

$$T_{UU}(\lambda\mu)=\lambda_m-\alpha(0)\qquad W_{UU}(\lambda\mu)=-\frac{|\lambda|+|\mu|}{2}\,. \tag{2.4}$$

Replacing (1.12) and (2.2) in (2.1) we get

$$X(\lambda\lambda)+X(\mu\mu)=2X(\lambda\mu)+\big||\lambda|-|\mu|\big|\,. \tag{2.5}$$

In the UU case, $X(\lambda\mu)>0$ implies a contribution to the s-channel which vanishes in the forward direction. Now, suppose that there is at least one pair λ, μ which gives a nonvanishing contribution. Because of angular momentum conservation the only t-channel amplitudes that may give a finite contribution to the s-channel have either $\lambda=\mu$ or $\lambda=-\mu$ [10]. Hence, if for $\lambda=\lambda_0$ we get a nonvanishing contribution, either $X(\lambda_0, \lambda_0)=0$ or $X(\lambda_0, -\lambda_0)=0$. Then define

$$|\lambda_0|=M \tag{2.6}$$

assume $X(\lambda_0, \lambda_0)=0$ and apply Eq. (2.5) for $\lambda=\lambda_0$:

$$X(\mu\mu)=2X(\lambda_0\mu)+\big|M-|\mu|\big|\,.$$

Since $X(\lambda_0\mu)\geqq 0$ we conclude

$$X(\mu\mu)\geqq\big|M-|\mu|\big|$$

and since we are interested in the smallest $X(\lambda\mu)$ ("minimal" solution) take the equality sign. Then we obtain

$$X_{UU}(\lambda\mu)=\tfrac{1}{2}\{\big|M-|\lambda|\big|+\big|M-|\mu|\big|-\big||\lambda|-|\mu|\big|\} \tag{2.7}$$

which is the required solution for the dominant trajectory in the UU case.

The foregoing discussion shows that in UU

$$|\lambda|=|\mu|=M$$

offers the only possibility for a nonzero forward contribution in the s-channel. Hence M can be defined as the helicity flip at both vertices which gives an asymptotically finite amplitude in the forward direction of the s-channel. Clearly M should be identified with the Toller quantum number [1] and, as will be seen, is a number characterizing the exchanged trajectory. In view of (2.7):

$$0\leqq M\leqq \min(s_1+s_2, s_3+s_4)\,.$$

We shall outline the way to obtain the solution in the other mass configurations. Take $m_1=m_2$ and $m_3=m_4$ (EE case) when:

$$T_{EE}(\lambda\mu)=0\,, \qquad W_{EE}(\lambda\mu)=\tfrac{1}{4}(1-(-)^{\lambda+\mu})\,. \tag{2.8}$$

The forms (1.12) and (2.2) replaced in (2.1) give:

$$2X(\lambda\mu)+2W(\lambda\mu)=X(\lambda\lambda)+X(\mu\mu)$$

since for boson trajectories λ and μ are integers so that $W(\lambda\lambda)=W(\mu\mu)=0$.

Now it is clear that $2W(\lambda\mu)=0$ or 1. When $2W(\lambda\mu)=0$ the minimal solution is

$$X(\lambda\mu)=X(\lambda\lambda)=X(\mu\mu)=0\,.$$

When $2W(\lambda\mu)=1$ the minimal solution follows by taking $X(\lambda\mu)=0$, one of the $X(\lambda\lambda)$, $X(\mu\mu)$ equal to 0 and the other equal to 1. We conclude that in all cases $X(\lambda\lambda)=1$ or 0, so that write

$$X_{EE}(\lambda\lambda)=\tfrac{1}{2}(1-(-)^k) \tag{2.9}$$

where k is an integer function of λ and of the quantum numbers of the trajectory.

To determine k apply factorization for $m_1=m_2$ and $m_3\neq m_4$ (EU case) in the form

$$(\beta^{EU}_{\lambda\mu}(t))^2=\beta^{EE}_{\lambda\lambda}(t)\,\beta^{UU}_{\mu\mu}(t)\,.$$

The threshold and Wong factors are known to be

$$T_{EU}=\tfrac{1}{2}(\lambda_m-\alpha(0))\qquad W_{EU}(\lambda\mu)=-s_1+\tfrac{1}{4}[1-\sigma(-)^{2s_1+\lambda+\mu+\lambda_m}] \tag{2.10}$$

and with the expressions (2.7) and (2.9) for $X_{UU}(\mu\mu)$ and $X_{EE}(\lambda\lambda)$ we get

$$X_{EU}(\lambda\mu)=s_1-\tfrac{1}{4}[(-)^k-\sigma(-)^{2s_1+\lambda+\mu+\lambda_m}]+\tfrac{1}{2}[|M-|\mu||-\lambda_m]\,.$$

It can be shown that the condition $X_{EU}(\lambda\mu)=$ integer ($\geqq 0$) implies

$$(-)^k=\sigma(-)^{\lambda+M}$$

which determines completely the minimal solution for the dominant trajectory:

$$X_{EE}(\lambda\mu)=\tfrac{1}{4}[1-\sigma(-)^{\lambda+M}]\,[1-\sigma(-)^{\mu+M}]\,, \tag{2.11}$$

$$X_{EU}(\lambda\mu)=s_1-\frac{\sigma}{4}(-)^{\lambda}[(-)^M-(-)^{2s_1+\mu+\lambda_m}]+\tfrac{1}{2}(|M-|\mu||-\lambda_m)\,. \tag{2.12}$$

3. Conspiracy Relations and Toller Classification: Families $M=1$ and $M=0$, $\sigma=+$

For processes with spin Mandelstam analyticity implies certain constraints between PCTHA at $t=0$, $(m_1\pm m_2)^2$ and $(m_3\pm m_4)^2$ (= "conspiracy" relations). A general method for their derivation is given in Ref. [9]. Here it sufficies to give an example from $\pi N\to VN$. With the decomposition (1.10) straightforward calculation gives:

$$B_2(s,t)=\frac{2}{t}\left\{\tilde f^{\,-}_{01}+\frac{t+m^2-\mu^2}{t-4M^2}\left(M\tilde f^{\,+}_{11}+\frac{t}{4}\tilde f^{\,+}_{01}\right)\right\}.$$

Consider the point $t=0$. Mandelstam analyticity demands B_2 to be analytic there; but this is possible only if for $t\to 0$:

$$\tilde{f}_{01}^{-} - \frac{m^2-\mu^2}{4M}\tilde{f}_{11}^{+} \sim t \quad \text{or} \quad i\bar{f}_{01}^{-} + \bar{f}_{11}^{+} \sim t^{\frac{1}{2}}. \tag{3.1}$$

The important point is that analyticity of B_2 implies a constraint between helicity amplitudes dominated by exchange of different quantum numbers ($\sigma=-$ and $\sigma=+$). For EE reactions (only) the $t=0$ constraints can also be derived from conservation of angular momentum in the s-channel.

We shall outline now the steps leading to the Toller classification of Regge trajectories by means of conspiracy relations. Because of (1.12) and (2.2) the limit of $\bar{f}_{\lambda\mu}^{\sigma}(s,t)$ for $s\to\infty$ and $t=$ nonzero but small is

$$\bar{f}_{\lambda\mu}^{\sigma}(s,t) \simeq \gamma_{\lambda\mu}\, t^{T(\lambda\mu)+W(\lambda\mu)+X(\lambda\mu)} z^{\alpha(t)-\lambda_m}$$

leaving out unimportant factors $(2\alpha+1)(1\pm e^{-i\pi\alpha})(\sin\pi\alpha)^{-1}$ etc. However, (1.7), (1.8) give

$$z^{\alpha(t)-\lambda_m} \sim \left(\frac{s}{p_{1\bar{2}}\,p_{\bar{3}4}}\right)^{\alpha(t)-\lambda_m} \sim \frac{s^{\alpha(t)-\lambda_m}}{t^{T(\lambda\mu)}}$$

so that

$$\bar{f}_{\lambda\mu}^{\sigma}(s,t) \simeq \gamma_{\lambda\mu}\, t^{W(\lambda\mu)+X(\lambda\mu)}\, s^{\alpha(t)-\lambda_m}. \tag{3.2}$$

As our derivation of the Toller classification will be based on (3.1) we consider the EU case, when (2.10) and (2.12) give

$$W_{EU}(\lambda\mu) + X_{EU}(\lambda\mu) = \tfrac{1}{4}[1-\sigma(-)^{\lambda+M}] + \tfrac{1}{2}(|M-|\mu|| - \lambda_m). \tag{3.3}$$

We start with the most impressive case $M=1$. We shall assign $M=1$ to the pion trajectory $\alpha^{\sigma}(t)=\alpha^{-}(t)$, which contributes to $\bar{f}_{01}$ and gives

$$W_{EU}(01) + X_{EU}(01) = -\tfrac{1}{2}$$

Hence, from (3.2), as $t\to 0$:

$$\bar{f}_{01}^{-}(s,t) = \gamma_{01}\, t^{-\frac{1}{2}}\, s^{\alpha^{-}(0)-1}, \qquad \gamma_{01} = \text{const} \neq 0\,.$$

Superficially, such a t behaviour seems to contradict (3.1). However, consider also the exchange of a trajectory $\alpha^{+}(t)$ with all quantum numbers of the pion except $\sigma=+$; this contributes to $\bar{f}_{11}^{+}$ and (3.3) gives

$$W_{EU}(11) + X_{EU}(11) = -\tfrac{1}{2}$$

Hence, as $t\to 0$

$$\bar{f}_{11}^{+}(s,t) = \gamma_{11}\, t^{-\frac{1}{2}}\, s^{\alpha^{+}(0)-1}, \qquad \gamma_{11} = \text{const} \neq 0\,.$$

Then, the only way to satisfy (3.1) for all s is

$$\alpha^{-}(0) = \alpha^{+}(0) \qquad i\gamma_{01} = -\gamma_{11}\,.$$

This is the famous conspiracy with parity doubling; the trajectory $\alpha^+(t)$ is called pion conspirator. All trajectories classified with Toller quantum number $M=1$ should be accompanied by parity doubling (Toller class III) [12].

We consider now a trajectory with $M=0$ and $\sigma=+$. Typical case is the ω, which also contributes to $\pi N \to VN$ (e.g. $\pi^\pm p \to \varrho^\pm p$). As shown in all phenomenological analysis of KN elastic scattering, ω is strongly coupled to $K\bar{K}$ near the forward direction; and since

$$0 \leqq M \leqq s_K + s_{\bar{K}} = 0$$

we conclude that $M_\omega=0$; also $\sigma_\omega=+$. Now, ω contributes to $\bar{f}_{11}^{\,+}$ and (3.3) gives

$$W_{EU}(11) + X_{EU}(11) = \tfrac{1}{2}$$

so that, as $t \to 0$

$$\bar{f}_{11}^{\,+}(s,t) = \gamma_{11}\, t^{\frac{1}{2}}\, s^{\alpha_\omega(0)-1}\,.$$

Here $\bar{f}_{01}^{\,-}$ receives no contribution. Hence, the conspiracy relation (3.1) is satisfied again, but this time by having the ω residue vanishing in the forward direction. Then it is usually said that (3.1) is satisfied by evasion and this is the typical situation for trajectories belonging to Toller class I [12].

4. Daughter Trajectories and Toller Classification — Family $M=0$, $\sigma=-$

In the asymptotically dominant contribution (1.12) consider an expansion of $e_{\lambda\mu}^{J+}$ in powers of z. With

$$\hat{\alpha} \equiv \alpha(t) - \lambda_m$$

we get

$$\bar{f}_{\lambda\mu}^{\,\sigma}(t,z) \sim \beta_{\lambda\mu}(t)\,(c_0 z^{\hat{\alpha}} + c_1 z^{\hat{\alpha}-2} + \cdots) \tag{4.1}$$

where the c_i behave like constants at $t=0$. In view of (1.7), (1.8):

$$z = \frac{s}{2p_{1\bar{2}}\,p_{\bar{3}4}}\left(1 + \frac{\Delta}{s}\right)$$

where

$$\Delta \equiv \frac{1}{2t}\,(t^2 - t\,\Sigma m_i^2 + (m_1^2 - m_2^2)(m_3^2 - m_4^2))\,. \tag{4.2}$$

Expanding further in powers of s:

$$\bar{f}_{\lambda\mu}^{\,\sigma}(t,z) \sim \beta_{\lambda\mu}(t)\,(2p_{1\bar{2}}\,p_{\bar{3}4})^{-\alpha} \cdot\left\{c_0 s^{\hat{\alpha}} + c_0 \hat{\alpha}\,\Delta s^{\hat{\alpha}-1} + \left[c_0 \frac{\hat{\alpha}(\hat{\alpha}-1)}{2}\,\Delta^2 + c_2 (2p_{1\bar{2}}\,p_{\bar{3}4})^2\right] s^{\hat{\alpha}-2} + \cdots\right\}. \tag{4.3}$$

Consider first the UU case. For $t \to 0$ (4.2) and (1.8) give:

$$\Delta \sim t^{-1} \qquad p_{1\bar{2}} p_{\bar{3}4} \sim t^{-1} .$$

Hence, the term of order $s^{\hat{\alpha}-n}$ has for $t \to 0$ an extra singularity t^{-n} as compared to the first. But $\bar{f}^{\sigma}_{\lambda\mu}(t, z)$ has a maximal singularity given by the Wang factor $(= -\frac{1}{2}(|\lambda| + |\mu|) = \text{finite})$. Thus if the Regge pole $J = \alpha(t)$ is the only singularity in complex J, the last expansion will, in general, violate analyticity. In this way we conclude the existence of lower-lying trajectories (daughters) which cancel the extra singularities [13], the daughter of order n must have an intercept

$$\alpha^n(0) = \alpha(0) - n , \qquad n = 1, 2, 3, \ldots .$$

Next, consider the EE case, when as $t \to 0$

$$\Delta \sim \text{const} \qquad p_{1\bar{2}} p_{\bar{3}4} \sim \text{const}$$

and all the terms of the expansion (4.3) behave alike. Here, analyticity on individual PCTHA does not imply the existence of daughters; however, these having being established in a UU process we conclude that the nth order EE daughter residue cannot be more singular than the EE mother.

Finally, consider the EU case. Here as $t \to 0$

$$\Delta \sim \text{const} \;\; p_{1\bar{2}} \sim \text{const} \;\; p_{\bar{3}4} \sim t^{-\frac{1}{2}}$$

so that in (4.3) the terms of order $2n$ and $2n+1$ have both an extra singularity t^{-n} in comparison to the first. Thus, in contrast to UU when daughters of all orders are necessary, the EU case requires only even order ones:

$$\alpha^{\sigma, 2n}(0) = \alpha^{\sigma}(0) - 2n .$$

Before we proceed notice that, as a consequence of the generalized Pauli principle, when $m_1 = m_2$ and $\lambda_1 = \pm \lambda_2$ the dominant contribution to $\bar{f}^{\sigma}_{\lambda\mu}$ has, apart from σ, a well defined value of the quantum number

$$\sigma_c = C(-)^J \qquad (C = \text{charge conjugt. quantum no}) .$$

In particular, it follows that $\lambda_1 = \lambda_2$ leads to $\sigma_c = +$ and $\lambda_1 = -\lambda_2$ to $\sigma\sigma_c = +$.

Now, for the residue of the n daughter near $t = 0$ write again

$$\beta^n_{\lambda\mu}(t) \simeq t^{T(\lambda\mu, n) + W(\lambda\mu) + X(\lambda\mu, n)} \gamma^n_{\lambda\mu}(t)$$

where

$$T_{UU}(\lambda\mu, n) = \lambda_m - \alpha(0) + n , \qquad T_{EE}(\lambda\mu) = 0 ,$$
$$T_{EU}(\lambda\mu, n) = \tfrac{1}{2}(\lambda_m - \alpha(0) + n)$$

and, of course, the Wang factor as before. The requirement that the PCTHA $\bar{f}^{\sigma}_{\lambda\mu}$ have a maximal singularity $t^{W(\lambda\mu)}$ plus the discussion above imply the following conditions on the exponents $X(\lambda\mu, n)$ (barring accidental cancellations between terms in the expansions of $e^{J+}_{\lambda\mu}$ and $e^{J-}_{\lambda\mu}$:

$$X_{UU}(\lambda\mu, n) \geqq -n\,, \qquad X_{EE}(\lambda\mu, n) \geqq 0\,, \tag{4.4–5}$$

$$X_{EU}(\lambda\mu, 2n) \geqq -n\,, \qquad X_{EU}(\lambda\mu, 2n+1) \geqq -n\,; \tag{4.6a–b}$$

of course, always $X(\lambda\mu, n) =$ integer.

In this way, use of factorization leads, as in Section 2, to the following minimal solution:

$$\begin{aligned} X_{UU}(\lambda\mu, n) &= \tfrac{1}{2}\{|M-|\lambda|| + |M-|\mu|| - ||\lambda| - |\mu||\} - n\,,\\ X_{EE}(\lambda\mu, n) &= \tfrac{1}{4}\{1-\sigma(-)^{\lambda+M+n}\}\,\{1-\sigma(-)^{\mu+M+n}\}\,,\\ X_{EU}(\lambda\mu, n) &= s_1 - \tfrac{1}{4}\sigma(-)^{\lambda}\{(-)^{M+n} - (-)^{2s_1+\mu+\lambda_m}\} + \tfrac{1}{2}(|M-|\mu|| - \lambda_m - n). \end{aligned} \tag{4.7}$$

We shall concentrate in the EU case. It is easily seen that the condition (4.6a) places no restrictions, but (4.6b) does. To find them write (4.7) with $n \to 2n+1$ as follows:

$$\begin{aligned} &X_{EU}(\lambda\mu, 2n+1)\\ &= s_1 + \frac{\sigma}{4}(-)^{\lambda}\{(-)^{M} + (-)^{2s_1+\mu+\lambda_m}\} + \tfrac{1}{2}(|M-|\mu|| - \lambda_m) - n - \tfrac{1}{2}\,. \end{aligned} \tag{4.8}$$

We distinguish the following cases:

(a) $|\mu| \leqq |\lambda|$. Since

$$\tfrac{1}{4}\sigma(-)^{\lambda}[(-)^{M} + (-)^{2s_1+\mu+\lambda_m}] - \tfrac{1}{2} \geqq -1 \tag{4.9}$$

we deduce from (4.8)

$$X_{EU}(\lambda\mu, 2n+1) \geqq s_1 + \tfrac{1}{2}|M-|\mu|| - \frac{|\lambda|}{2} - 1 - n\,.$$

Since $X_{EU} =$ integer and $|\lambda| \leqq 2s_1$, the condition (4.6b) could be violated only if

$$|\lambda| = 2s_1 \quad \text{and} \quad |\mu| = M\,.$$

In that case (4.8) becomes

$$X_{EU}(\lambda\mu, 2n+1) = \tfrac{1}{2}\sigma(-)^{\lambda+M} - \tfrac{1}{2} - n$$

which shows that violation of (4.6b) will occur for

$$\sigma(-)^{\lambda+M} = -1 \tag{4.10}$$

(b) $|\lambda| < |\mu|$ and $M \leqq |\mu|$. Here (4.8) becomes

$$X_{EU}(\lambda\mu, 2n+1) = s_1 + \frac{\sigma}{4}(-)^\lambda[(-)^M + (-)^{2s_1}] - \frac{M}{2} - \tfrac{1}{2} - n$$

which is

$$\geqq s_1 - \frac{M}{2} - 1 - n$$

Since $s_1 \geqq M/2$ we conclude that (4.6b) will be violated only if

$$s_1 = \frac{M}{2}$$

together with (4.10).

(c) $|\lambda| < |\mu| < M$. Here

$$X_{EU}(\lambda\mu, 2n+1) = s_1 + \frac{\sigma(-)^\lambda}{4}[(-)^M + (-)^{2s_1}] + \frac{M}{2} - |\mu| - \tfrac{1}{2} - n\,.$$

Violation requires $s_1 + \frac{M}{2} - |\mu| \leqq 0$. But since always $s_1 \geqq \frac{M}{2}$:

$$s_1 + \frac{M}{2} - |\mu| \geqq M - |\mu| > 0\,. \tag{4.11}$$

Hence, case (c) never leads to violation.

Specialize now to the case $M=0$. Then case (b) cannot lead to violation. For, condition (4.11) demands $s_1 = 0$ (i.e. E vertex with spin 0 particles); but then $\lambda = 0$ and condition (4.10) demands $\sigma = -$ which contradicts angular momentum conservation (a system of two scalar or pseudoscalar particles has $\sigma = +$). Thus for $M=0$ violation is possible only in case (a) with

$$|\lambda| = 2s_1 \quad \text{and} \quad \mu = 0$$

(notice then $e^{J-}_{\lambda\mu} = 0$ in (1.6)). But $|\lambda| = 2s_1$ clearly implies

$$\lambda_1 = -\lambda_2$$

which, as said, corresponds to $\sigma\sigma_c = +1$, i.e. to trajectories with

$$(\sigma\sigma_c) = (+\,+) \quad \text{or} \quad (-\,-)\,. \tag{4.12}$$

In all this we have been considering the exponent $X_{EU}(\lambda\mu, 2n+1)$, i.e. the residue of a daughter of odd order. Our conclusion is that if trajectories $(+\,+)$ or $(-\,-)$ appear in that order, analyticity will be violated. Hence the trajectories (4.12) must necessarily be classified as daughters of *even* order only. It is easy to see that while σ is the same for all trajectories of a family, σ_c alternates with n. Hence for $M=0$ trajectories with $(+\,-)$ or $(-\,+)$ correspond to $n=$ odd.

The family of A_1 ($J^{PG} = 1^{+-}$, $I=1$ so that $\sigma = -$, $\sigma_c = -$) is usually classified with $M=0$ and offers an example of the Toller class II.

5. Physical Implications. — Certain Difficulties

A. According to (4.12), the ω trajectory $(\sigma, \sigma_c) = (+\,+)$, which must be assigned $M=0$ (Section 3), has to be necessarily taken as a parent trajectory ($n=0$) and thus should be considered of class I. Direct evidence for this can be obtained from forward $\gamma p \to \pi^0 p$. Vector dominance (and experiment) suggests that this is dominated by ω-exchange. The main feature of class I is the satisfaction of the conspiracy relations by evasion. Then, if ω belongs to class I, the high energy differential cross-section for $\gamma p \to \pi^0 p$ is expected to show a forward dip; this is the case in all h.e. data [14].

Even better, this can be tested by isolating the ω-exchange contribution to $\pi N \to \varrho N$ which, through isospin invariance, can be directly done with a linear combination of cross-sections for $\pi^\pm p \to \varrho^\pm p$ and $\pi^- p \to \varrho^0 n$. The analysis clearly supports $M_\omega = 0$ [15].

B. Suppose we assign $M=0$ to the pion trajectory $(-\,+)$; then (4.12) implies that it has to be taken as a daughter ($n=1$). The parent trajectory will have $(-\,-)$, i.e. the quantum numbers of A_1; and if we accept usual trajectory slope for α_π:

$$\alpha_{A_1}(0) = \alpha_\pi(0) + 1 \approx 1\,.$$

Such a trajectory seems to be ruled out by all relevant phenomenological analyses and this seems to suggest $M_\pi = 1$.

This assignment has been used as a natural solution of the following problem: Consider $\gamma p \to \pi^+ n$ or $np \to pn$ near the forward direction; due to the proximity of its pole, pion exchange is expected to be important. However, e.g. in the former reaction a conspiracy relation of the form (3.1) holds again:

$$\tilde{f}_{01}^{-}(s,0) \sim \tilde{f}_{11}^{+}(s,0)\,. \tag{5.1}$$

In the absence of conspiracy (i.e. disregarding the Toller classification) $\tilde{f}_{11}^{+}(s,t) - 0$ (for, it receives no contribution) π contributes to $\tilde{f}_{01}^{-}$ and (5.1) implies $\tilde{f}_{01}^{-}(s,0) = 0$. The resulting contribution is shown in Fig. 3 ("Regge conventional") and is in clear contradiction with experiment.

By assigning $M=1$ to the pion we conclude, as in Section 3, the existence of a parity doublet which allows finite forward contributions to the PCTHA's entering in (5.1). The experimental data can be accounted very successfully (Fig. 3).

There are, however, certain difficulties of the Toller classification and in particular of assigning $M=1$ to the pion. The main are the following:

(i) *The Forward Structure of* $\pi^+ p \to \varrho^0 \Delta^{++}$ [16]. Here again pion exchange is expected to be important and this is verified by detailed experimental information on the density matrix elements, suggesting that at small $|t|$ the dominant contribution comes from exchange of $\sigma = -$ and, in particular, from the PCTHA $f_{00}^{-}(s,t)$. But from (2.4) and (2.7):

$$W_{UU}(\lambda\mu) + X_{UU}(\lambda\mu) = -\tfrac{1}{2}(|\lambda| + |\mu|) + \tfrac{1}{2}\{|M - |\lambda|| + |M - |\mu|| - ||\lambda| - |\mu||\}$$

With the assignment $M_\pi = 1$ we get

$$W_{UU}(00) + X_{UU}(00) = 1\,.$$

With (3.2) this implies that the residue of $f_{00}^{-}(s,t)$ vanishes at $t=0$. In contrast, experiment gives a very neat forward peak (of width $\sim \mu_\pi^2$, as in $\gamma + p \to \pi^+ + n$).

It is still possible to fit the data with an $M=1$ pion if, in addition, A_1 (which has also $\sigma = -$) gives a significant forward contribution. However, the picture is complicated and contains free parameters.

(ii) *The Soft-Pion Limit* [17]. The difficulty is most easily seen if we write the residues of a given Regge pole $\alpha(t)$ of normality σ in the following form:

$$\beta_{\lambda\mu}(t) = \beta_\lambda(t)\,\beta_\mu(t)$$

This satisfies factorization (2.1) and our solution (2.7), (2.11) and (2.12) is equivalent to

$$\beta_\lambda^E(t) \sim t^{\frac{1}{4}(1-\sigma(-)^\lambda + M)}\;\beta_\mu^U(t) \sim t^{\frac{1}{2}(M - |\mu| - \alpha(0))} \qquad (t \to 0)$$

where E stands for the vertex $m_1 = m_2$ and U for $m_3 \neq m_4$. Assigning $M=1$ to the pion we get for a helicity nonflip U vertex:

$$\beta_{\mu=0}^U(t) \sim t^{\frac{1}{2} - \alpha(0)}\,.$$

In the soft pion limit

$$\mu_\pi \to 0 \qquad \alpha(0) \to 0\,,$$

the function $\beta_\mu{}^U(t)$ becomes the actual amplitude for the process $3 + \pi \to 4$; then (5.4) implies that this amplitude vanishes. Taking $3 \equiv \pi$ and $4 \equiv \sigma$, ϱ or f_0 we conclude that for $M_\pi = 1$ there is no zero-mass $\pi - \pi$ scattering (and in fact at any energy, since m_4 is variable!) [18].

Because of these difficulties recent phenomenological analyses of quasi-elastic reactions tend to associate the pion trajectory with no definite Toller class in the above sense [19]. Pion exchange is taken to satisfy (5.1) by evasion and the forward peak in $\gamma p \to \pi^+ n$ is generated by a pair of Regge cuts satisfying (5.1) by conspiracy [14]. Notice that, in general, factorization does not hold for moving branch points; then, the solution (2.7), (2.11). (2.12) is not relevant. In this way, $\pi^+ p \to \varrho^0 \Delta^{++}$ has

$$f_{00}^{-} \approx \frac{b}{t - \mu^2}\, s^{\alpha(t)} \qquad b \neq 0$$

and is easily seen to produce a narrow forward peak [14].

Part II: Vector Dominance Relations in the Veneziano Model

1. Frame Dependence of Vector Dominance

There are two important statements of vector dominance which have created much controversy during the past two years:

(A) If γ_1 denotes isovector photon and ϱ_T transverse ϱ meson, the amplitudes for $\gamma_1 p \to \pi^+ n$ and $\pi^- p \to \varrho_T^0 n$ are related by a known proportionality factor.

(B) If $\sigma_\perp(\sigma_{//})$ denotes the differential cross-section for $\gamma_1 p \to \pi^+ n$ (or $\gamma_1 n \to \pi^- p$) by photons polarized perpendicular (parallel) to the production plane and if ϱ_{mn} denote the density matrix elements for $\pi^- p \to \varrho^0 n$, then the following relation holds:

$$\frac{\sigma_\perp - \sigma_{//}}{\sigma_\perp + \sigma_{//}} = \frac{\varrho_{1-1}}{\varrho_{11}}. \tag{1.1}$$

However, both statements (a), (b) are ambiguous unless the quantization axis for the spin of the ϱ meson is chosen. Most important choices, so far, are the following: (i) The direction of the ϱ^0 in the cm system of the s-channel ($\equiv$ helicity frame). Then if $M_{0\lambda_N,\lambda_\varrho\lambda'_N}$, are the s-channel helicity amplitudes for $\pi^- p \to \varrho^0 n$:

$$\varrho_{mn}^{(H)} = \sum_{\lambda_N \lambda'_N} M_{0\lambda_N, m\lambda'_N} \overset{*}{M}_{0\lambda_N, n\lambda'_N}. \tag{1.2}$$

(ii) The direction of incident pion in the ϱ^0 rest system (Jackson frame). Here with $f_{\lambda_N\lambda_{\bar{N}},\lambda_\varrho 0}$ the t-channel helicity amplitudes:

$$\varrho_{mn}^{(J)} = \sum_{\lambda_N \lambda_{\bar{N}}} f_{\lambda_N\lambda_{\bar{N}}, m0} \overset{*}{f}_{\lambda_N\lambda_{\bar{N}}, n0}. \tag{1.3}$$

(iii) A frame defined by the condition (Donohue-Hogaasen) [2]:

$$\mathrm{Re}\,\varrho_{10} = 0 \tag{1.4}$$

and obtained from the helicity frame via definite rotation (by angle φ about the production normal, where

$$\tan 2\varphi = 2\sqrt{2}\,\mathrm{Re}\,\varrho_{10}^{(H)} / (\varrho_{11}^{(H)} - \varrho_{00}^{(H)} - \varrho_{1-1}^{(H)}).$$

Notice that there are two solutions of (1.4) corresponding to $\varrho_{1-1}/\varrho_{11}$ = maximum/minimum; we shall be interested in the first.

With respect to statement (A), the most prominent feature of $\gamma p \to \pi^+ n$ is a forward narrow peak (of width $\approx$ (pion mass)^2); hence it is essential to determine the frame in which $\pi^- p \to \varrho_T^0 n$ has the same shape. With respect to the statement (B), it is important to determine the frame in which (1.1) is expected to be valid, at least in some limiting sense. Of course, the frames for (A) and (B) must be identical if vector dominance is to make sense.

It should be stressed that the choice of proper frame is a *dynamical* question, not to be solved by purely kinematical analysis. An answer can be obtained by using appropriate dynamical models and here we shall proceed with simple Veneziano models containing no free parameters and no satellite terms [3]. Our conclusion is that vector dominance should be defined and tested in the helicity frame only. Most important, it follows that through our procedure (which, as it will become clear, is more general than the Veneziano model) the presence of a sharp forward peak in $\pi^- p \to \varrho_T^0 n$ offers a new test and verification of ϱ-meson universality.

2. Basic Kinematical Relations-Structure in Jackson Frame

As in Part I, we shall denote by $\bar{f}^{\sigma,\sigma_c}_{\lambda_N - \lambda_{\bar{N}}, \lambda_\varrho}$ the PCTHA ($\sigma = P(-)^J$, $\sigma_c = C(-)^J$) and by $\tilde{f}^{\sigma,\sigma_c}_{\lambda_N - \lambda_{\bar{N}}, \lambda_\varrho}$ the singularity-free PCTHA (with the Wang kinem. factor extracted). Then the conspiracy relation derived in Part I Eq. (3.1) will hold:

$$\tilde{f}_{01}^{-+}(s,t) \to \frac{m^2 - \mu^2}{4M} \tilde{f}_{11}^{++}(s,t) \quad (t \to 0) \tag{2.1}$$

and will subsequently be of particular interest.

For a possible peak of width $\approx \mu^2$ in $\pi^- p \to \varrho_T^0 n$ pion exchange is expected to be responsible. Apart from π, A_2 and A_1 are also exchanged. A_2 is believed to belong to the Toller class I and satisfy (2.1) by evasion; then its overall contribution near $t = 0$ vanishes. Finally, A_1 is expected to be of no particular importance in the generation of a sharp forward peak.

With these remarks and the fact that A_1 dominates the PCTHA $\tilde{f}_{10}^{--}$ and $\tilde{f}_{11}^{--}$ it turns out that it is sufficient to consider the invariant amplitudes B_1, B_2, and B_3 only (in the notation of Part I); in terms of them the PCTHA are [4]:

$$\bar{f}_{00}^{-+} = \frac{\sqrt{-t}}{2Mm\tau}(2m^2 s B_1 + (t + m^2 - \mu^2) s B_2 + \tau^2 B_3), \tag{2.2}$$

$$\tilde{f}_{01}^{-+} = \tfrac{1}{4}(t + m^2 - \mu^2) B_1 + \frac{t}{2} B_2, \tag{2.3}$$

$$\tilde{f}_{01}^{++} = -B_1, \tag{2.4}$$

$$\tilde{f}_{11}^{++} = M B_1 \tag{2.5}$$

and, for our purpose:

$$\tilde{f}_{10}^{--} = \tilde{f}_{11}^{--} = 0 . \tag{2.6}$$

In the expression of $\tilde{f}_{00}^{-+}$ it is:

$$\tau^2 = [t - (m+\mu)^2]\,[t - (m-\mu)^2] .$$

Clearly, the pion contributes to the first two PCTHA.

It can be seen now very easily that the transverse cross-section for $\pi^- p \to \varrho_T^0 n$ in Jackson frame ($\sigma_T^{(J)}$) has no forward peak: This transverse cross-section clearly receives contributions only from helicity states with $\lambda_\varrho = \pm 1$; then, in view of (1.3), from the PCTHA containing π exchange only $\tilde{f}_{01}^{-+}$ contributes to $\sigma_T^{(J)}$. Clearly, a sharp forward peak will arise in $\sigma_T^{(J)}$ if and only if $\tilde{f}_{01}^{-+}$ has a term $\sim (t-\mu^2)^{-1}$, i.e. with a pion pole. However, the amplitude $\tilde{f}_{01}^{-+}$ is sense-nonsense at $\alpha_\pi(t) = 0$ and its residue must contain a nonsense factor $\alpha_\pi(t)$; this cancels the pole at $t = \mu^2$. Thus we conclude, and in fact without appeal to any special model, that $\sigma_T^{(J)}$ has no forward peak [5]; hence the Jackson frame is not proper for a test of vector dominance.

3. A Veneziano Model. — Structure in Helicity Frame

Each of the invariant amplitudes B_i receives the isospin decomposition:

$$B_i = B_i^{(+)} \delta_{\alpha\beta} + B_i^{(-)} \tfrac{1}{2} [t_\alpha, t_\beta] .$$

The amplitudes $B_i^{(+)}$ correspond to t-channel isospin exchange $I_t = 0$ and do not contribute to $\pi^- p \to \varrho^0 n$. Thus, we proceed with the construction of simple Veneziano models for $B_1^{(-)} - B_3^{(-)}$.

In accord with duality arguments for $pp \to \pi^+ d$ and $\gamma N \to N\pi$ [6] and phenomenological fits to backward $\gamma N \to \pi N$ we shall use an $N_\alpha - N_\gamma$ exchange degenerate baryon trajectory $\alpha_N(s)$ but a nondegenerate Δ_δ trajectory $\alpha_\Delta(s)$.

In Eqs. (2.2)–(2.5) $\tilde{f}_{00}^{-+}$ is the only PCTHA containing B_3; it is also the only amplitude containing a pion pole. Under $s \leftrightarrow u$ crossing $B_3^{(-)}$ is even: Hence we take the following Veneziano representation:

$$\begin{aligned} B_3^{(-)}(s,t,u) &= c_1 \{B(-\alpha_\pi(t), \tfrac{1}{2} - \alpha_N(s)) + s \leftrightarrow u\} \\ &+ c_2 \{B(-\alpha_\pi(t), \tfrac{1}{2} - \alpha_\Delta(s)) + s \leftrightarrow u\} + c_3 \{B(\tfrac{1}{2} - \alpha_N(s), \tfrac{1}{2} - \alpha_\Delta(u)) + s \leftrightarrow u\} \end{aligned} \tag{3.1}$$

where $B(x,y) =$ Euler B-function. In the limit $s \to \infty$ $t =$ fixed: $B_3^{(-)} \sim s^{\alpha_\pi(t)}$ as (2.2) demands. To make Δ exchange nondegenerate it is sufficient to impose

$$c_2 = -c_3 .$$

Near the pion pole ($t \to \mu^2$) and assuming a pion trajectory $\alpha_\pi(t) = \lambda(t-\mu^2)$ we have:

$$B_3^{(-)} \approx -\frac{2(c_1+c_2)}{\lambda(t-\mu^2)}.$$

Comparison with the contribution of an elementary pion exchange (Fig. 1 (a)) gives

$$c_1 + c_2 = g g_{\varrho\pi\pi} \qquad (g \equiv g_{\pi NN}) \tag{3.2}$$

thus determining $c_1 + c_2$ in terms of the $\pi - N$ and $\varrho - N$ couplings.

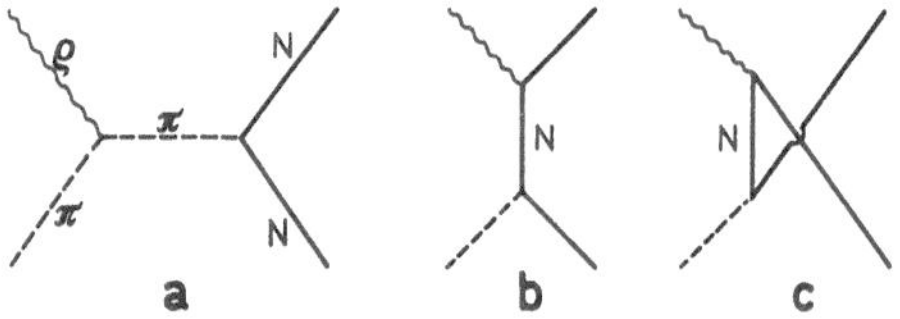

Fig. 1. The Born model

$B_2^{(-)}$, which appears in $\tilde{f}_{00}^{-+}$ and $\tilde{f}_{01}^{-+}$, is expected to contain a contribution from pion exchange; however, $\tilde{f}_{01}^{-+}$ has no pole at $t = \mu^2$. Taking into account that $B_2^{(-)}$ is odd under $s \leftrightarrow u$ crossing the simplest form (with nondegenerate Δ) is:

$$\begin{aligned} B_2^{(-)}(s,t,u) &= b_1\{B(1-\alpha_\pi(t), \tfrac{1}{2}-\alpha_N(s)) - s \leftrightarrow u\} \\ &+ b_2\{B(1-\alpha_\pi(t), \tfrac{1}{2}-\alpha_\Delta(s)) - s \leftrightarrow u + B(\tfrac{1}{2}-\alpha_N(s), \tfrac{1}{2}-\alpha_\Delta(u)) - s \leftrightarrow u\}. \end{aligned} \tag{3.3}$$

Near the nucleon pole ($s \to M^2$):

$$B_2 \approx -\frac{b_1+b_2}{\lambda(s-M^2)}.$$

In our model we assume that the ϱ is coupled to the nucleon via "electric" coupling only [7]; then with a $\varrho N \bar{N}$ vertex of the form

$$g_{\varrho N\bar{N}} \bar{u}(p_2)\, \gamma_\mu u(p_1)\, \varepsilon_\varrho^\mu$$

the one elementary nucleon graph of Fig. 1(b) gives

$$b_1 + b_2 = -\lambda g g_{\varrho NN}. \tag{3.4}$$

Finally, $B_1^{(-)}$ is also odd under $s \leftrightarrow u$ and its simplest representation is the same as for $B_2^{(-)}$:

$$\begin{aligned} B_1^{(-)}(s,t,u) &= a_1\{B(1-\alpha_\pi(t), \tfrac{1}{2}-\alpha_N(s)) - s \leftrightarrow u\} \\ &+ a_2\{B(1-\alpha_\pi(t), \tfrac{1}{2}-\alpha_\Delta(s)) - s \leftrightarrow u + B(\tfrac{1}{2}-\alpha_N(s), \tfrac{1}{2}-\alpha_\Delta(u)) - s \leftrightarrow u\}; \end{aligned} \tag{3.5}$$

the exact implications of this form will be discussed at the end of this section. Comparison with the elementary nucleon pole contribution of Fig. 1 (b) gives

$$a_1 + a_2 = \lambda g g_{\varrho NN} \,. \tag{3.6}$$

Notice, in passing, that Fig. 1 (b) also gives $c_1 - c_3 = \lambda g g_{\varrho\pi\pi}$ so that all the parameters of $B_3^{(-)}$ are completely determined; however, $B_2^{(-)}$ and $B_1^{(-)}$ need additional information, e.g. at the Δ pole.

It is straightforward now to calculate the limits of the above Veneziano amplitudes for $s \to \infty$ $t =$ fixed. In view of (3.2), (3.4), and (3.6) the result is:

$$-B_1^{(-)}(s,t) \to B_2^{(-)}(s,t) \to g g_{\varrho NN} s^{-1} K(s,t) \,, \tag{3.7}$$

$$B_3^{(-)}(s,t) \to g g_{\varrho\pi\pi} (\mu^2 - t)^{-1} K(s,t) \tag{3.8}$$

where

$$K(s,t) \equiv \Gamma(1 - \alpha_\pi(t)) \, (1 + e^{-i\pi\alpha_\pi(t)}) \, (\lambda s)^{\alpha_\pi(t)} \,.$$

The expressions of the PCTHA (2.2–5) easily give the longitudinal (σ_{L}) and transverse (σ_{T}) *differential* cross-sections in the Jackson frame:

$$\begin{aligned} \sigma_{\mathrm{L}}^{(J)} &= \sum_{\lambda_N, \lambda_{\bar N}} |f_{\lambda_N \lambda_{\bar N}, 00}|^2 \\ &= \frac{(-t)}{4M^2 m^2 \tau^2} |2m^2 s B_1^{(-)} + (t + m^2 - \mu^2) s B_2^{(-)} + \tau^2 B_3^{(-)}|^2 \,, \end{aligned}$$

$$\begin{aligned} \sigma_{\mathrm{T}}^{(J)} &= \sum_{\lambda_N, \lambda_{\bar N}, \lambda_\varrho = \pm 1} |f_{\lambda_N \lambda_{\bar N}, \lambda_\varrho 0}|^2 \\ &= \frac{s^2}{4M^2 \tau^2} \{\tau^2 |B_1^{(-)}|^2 + |(t + m^2 - \mu^2) B_1^{(-)} + 2t B_2^{(-)}|^2\} \end{aligned}$$

and with (3.7) and (3.8):

$$\sigma_{\mathrm{T}}^{(J)} = \frac{|K(s,t)|^2}{4M^2 \tau^2} g^2 g_{\varrho NN}^2 \{\tau^2 + (t - m^2 + \mu^2)^2\} \,. \tag{3.9}$$

Thus, we verify that $\sigma_{\mathrm{T}}^{(J)}$ is a smooth function of t and has no forward peak.

We turn now to a calculation of the longitudinal and transverse differential cross-sections in the helicity frame. In view of (1.2) the simplest way proceeds in terms of SHA as follows: In our model with $B_5 = B_6 = B_8 = 0$ the SHA corresponding to $\lambda_\varrho = 0$ have the following expressions:

$$\begin{aligned} M_{0\frac{1}{2},0\frac{1}{2}} &= 0 \,, \\ M_{0-\frac{1}{2},0\frac{1}{2}} &= \frac{\sqrt{-t}}{2Mm} (s B_2 + (t + m^2 - \mu^2) B_3) \,. \end{aligned} \tag{3.9^1}$$

Hence in our model:

$$\sigma_{\mathrm{L}}^{(H)} = \sum_{\lambda_N \lambda'_N} |M_{0\lambda_N, 0\lambda'_N}|^2 = \frac{-t}{4M^2 m^2} |s B_2^{(-)} + (t + m^2 - \mu^2) B_3^{(-)}|^2 .$$

The transverse helicity cross-section is most easily obtained by using

$$\sigma_{\mathrm{L}}^{(J)} + \sigma_{\mathrm{T}}^{(J)} = \sigma_{\mathrm{L}}^{(H)} + \sigma_{\mathrm{T}}^{(H)}$$

which follows from the orthogonality of the crossing matrix $X_{\{\lambda t\}}^{\{\lambda s\}}$ of part I. Thus we get

$$\sigma_{\mathrm{T}}^{(H)} = \frac{1}{2M^2} \{|-s B_1^{(-)} + t B_3^{(-)}|^2 + |t B_3^{(-)}|^2\}$$

and with the asymptotic expressions (3.7) and (3.8):

$$\sigma_{\mathrm{T}}^{(H)} = g^2 g_{\varrho\pi\pi} \frac{1}{2M^2} |K(s,t)|^2 \left\{ \left| \frac{g_{\varrho NN}}{g_{\varrho\pi\pi}} + \frac{t}{\mu^2 - t} \right|^2 + \frac{t^2}{(\mu^2 - t)^2} \right\}. \quad (3.10)$$

So far we have proceeded without imposing any relation on the ϱN and $\varrho\pi$ couplings. The ϱ-meson universality requires that the ratio

$$R \equiv g_{\varrho NN} / g_{\varrho\pi\pi} \quad (3.11)$$

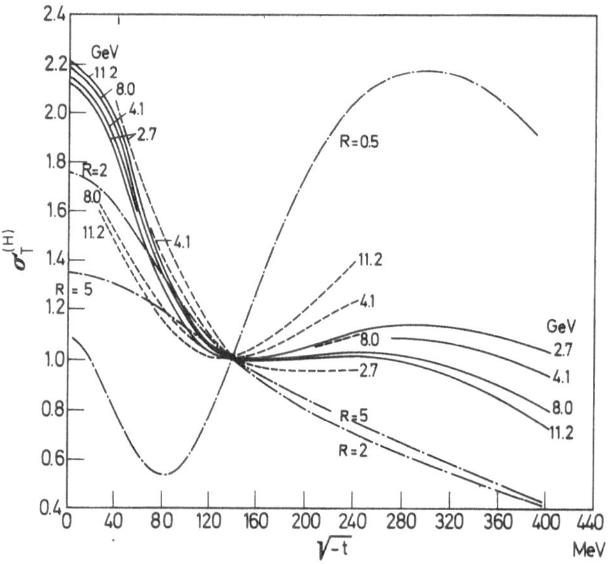

Fig. 2. Transverse cross-sections for $\pi^- p \to \varrho^0 n$. Solid curves: Differential cross-sections $\sigma_{\mathrm{T}}^{(H)}$ at pion energies $p_0 = 2.7, 4.1, 8.0$ and 11.2 GeV predicted by our (parameter-free) Veneziano model, Eq. (3.12) (part II). — Dot-dashed curves: the same at $p_0 = 8.0$ GeV for values of the ratio $R = g_{\varrho NN}/g_{\varrho\pi\pi} = 0.5, 2,$ and 5. — Dashed curves: The Rutherford-Purdue fits to experimental data (Ref. 8). The cross-sections are always normalized at $t = -\mu^2$

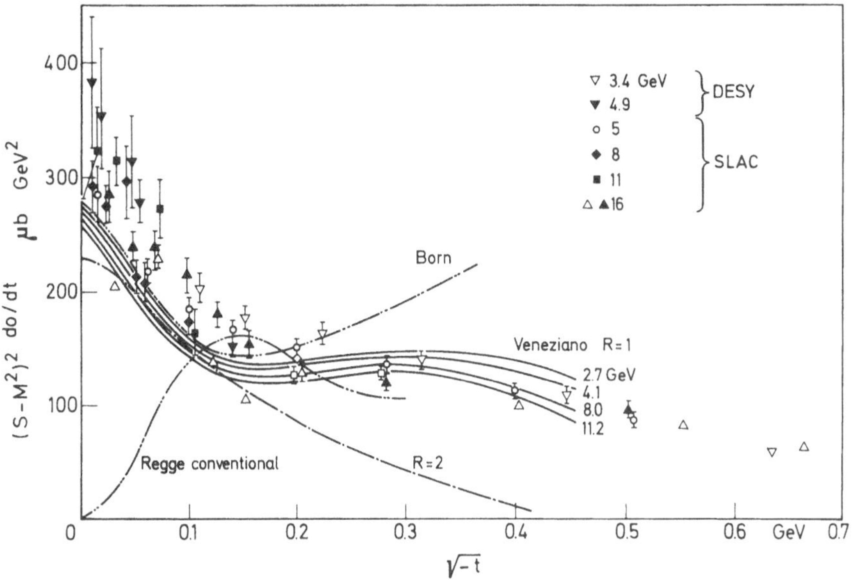

Fig. 3. Comparison with photoproduction data. Solid curves: Properly normalized differential cross-sections $\sigma_{\mathrm{T}}^{(H)}$ as in Fig. 2 (= the $\gamma p \to \pi^+ n$ diff. cross-section of Eq. (5.10)) plotted against the $\gamma p \to \pi^+ n$ data of *R. Diebold*, SLAC-Pub-673 (1969). — Dot-dashed curve: $\sigma_{\mathrm{T}}^{(H)}$ calculated from Eq. (3.10) with $R=2$ at 8 GeV (properly normalized). — Dot-dot-dashed curve: The gauge invariant Born model and the Regge conventional (= non-conspiring) pion exchange

be equal to unity. Then

$$\sigma_{\mathrm{T}}^{(H)} = g^2 g_{\varrho\pi\pi} \frac{1}{2M^2} |K(s,t)|^2 \left\{ \frac{\mu^4}{(t-\mu^2)^2} + \frac{t^2}{(t-\mu^2)^2} \right\}. \tag{3.12}$$

The term $\mu^4/(t-\mu^2)^2$ produces now a forward peak of width $\approx \mu^2$. The results of a calculation at pion lab. energies 2.7, 4.1, 8.0 and 11.2 GeV are given in Fig. 2 (solid curves). These are compared to the fits to data from a recent experiment of the *Rutherford-Purdue* collaboration [8] (dashed curves of Fig. 2); the agreement is good. The same cross-sections are plotted against experimental data for $\gamma p \to \pi^+ n$ (Fig. 3); clearly, the model works fairly well.

4. A Test of ϱ-Universality

Let us see, however, what happens if we vary the ratio (3.11) away from the universality value $R=1$. From Fig. 2 (dot-dashed curves), it is clear that $R=0.5$ or $R=5$ are strongly contradicted by the Rutherford-Purdue results; for $R=5$ the forward peak disappears; for $R=0.5$ (and all $R<0.5$ as we find) the resulting shape does not resemble at all the experimental curves. It appears that $R=2$ is not excluded, but its trend at $-t \gtrsim \mu^2$ is not supported. This is best seen when the corresponding cross-section $\sigma_T^{(H)}$ is plotted against the photoproduction data (Fig. 3); at $-t \gtrsim \mu^2$ $R=2$ falls definitely below the experimental points. It is clear that the precise shape of $\sigma_T^{(H)}$ is very sensitive to the exact value of the ratio (3.11). We conclude that, within the present approach, the existence of a sharp forward peak in $\sigma_T^{(H)}$ is a new test (and verification) of the ϱ universality, in addition to other well-known tests.

Apart from the Regge factor $K(s,t)$, the form (3.10) of $\sigma_T^{(H)}$ is the same with the Born approximation model of Fig. 1 (elementary pion in the t-channel plus elementary nucleon in the s- and u-channel with electric ϱN coupling) [9]. Near $t=0$ it is $K(s,t) \approx 1$, so that the same test of ϱ-universality can also be carried in the Born model. However, for $\sqrt{-t} \gtrsim \mu$ the latter model fails completely; this is clear in Fig. 3 (dot-dot-dashed curve) when we plot the gauge invariant Born model of $\gamma p \to \pi^+ n$ (i.e. the equivalent to the elementary particle exchanges of Fig. 1 with $R = g_{\varrho NN}/g_{\varrho\pi\pi} = 1$). In contrast, the Regge form (3.12) leads to cross-sections of reasonable shape up to much larger $-t$.

From (2.2)–(2.5) it is seen that B_1 contributes to PCTHA with $\sigma=+$ as well as with $\sigma=-$. Then the form (3.5), which contains exchange of $\alpha_\pi(t)$, appears at first sight to introduce leading contributions to $\tilde{f}_{01}^{++}$ and $\tilde{f}_{11}^{++}$ of normality $\sigma=-$, an untenable situation. However, this is not so. What really happens is that $B_1^{(-)}$ contains two parts, say $B_{1\pm}^{(-)}$, corresponding to $\sigma=\pm$. Then the conspiracy relation (2.1) is equivalent to:

$$B_{1+}^{(-)}(s,t=0) = B_{1-}^{(-)}(s,t=0)\,.$$

Clearly, to produce a forward peak this relation (or (2.2)) requires a conspiring pair of either Regge poles or Regge cuts. In simple Veneziano models (known to contain no cuts) the only possible conspiracy is between Regge poles. Thus, in our model, while $B_{1-}^{(-)}$ contains pion exchange, the part $B_{1+}^{(-)}$ contains the exchange of the conspirator π'; this has a trajectory

$$\alpha_{\pi'}(t) \equiv \alpha_\pi(t)$$

so that it is $B_{1+}^{(-)}(s,t) \equiv B_{1-}^{(-)}(s,t)$ for all t (not just $t=0$). Thus we need only one Veneziano representation for $B_1^{(-)}(s,t)$, Eq. (3.5).

This remark, however, indicates that another model leading to equivalent results can be made out of a nonconspiring Regge pion plus a conspiring pair of Regge cuts [3]. It is also known that in forward photoproduction of charged pions duality (via finite energy or continuous moment sum rules) does not discriminate between conspiring poles and conspiring cuts. Thus the same procedure and test of universality can be carried through a Regge cut model in which the magnitude of the cuts is determined in terms of the dual s- and u-channel poles. Thus, our procedure is more general than the simple Veneziano model of Section 3.

5. The Ratio $\varrho_{1-1}^{(H)}/\varrho_{11}^{(H)}$. — Veneziano Model for $\gamma p \to \pi^+ n$

It is straightforward to calculate the ratio $\varrho_{1-1}^{(H)}/\varrho_{11}^{(H)}$ in the model of Section 3. In addition to the expressions (3.9^1) we have:

$$\begin{aligned} M_{0\frac{1}{2},\,\pm 1\frac{1}{2}} &= 0\,, \\ M_{0-\frac{1}{2},\,1\frac{1}{2}} &= \frac{1}{\sqrt{2}M}(-sB_1+tB_3)\,, \\ M_{0-\frac{1}{2},\,-1\frac{1}{2}} &= -\frac{1}{\sqrt{2}M}\,tB_3 \end{aligned} \tag{5.1}$$

so that with (3.7), (3.8):

$$\varrho_{1-1}^{(H)} = \frac{1}{2M^2}\, g^2 g_{\varrho NN}^2 |K(s,t)|^2 \frac{(-t)\,\mu^2}{(t-\mu^2)^2}\,.$$

With the expressions (5.1) it is $\varrho_{-1-1} \neq \varrho_{11}$ so that we should take the average:

$$\tfrac{1}{2}(\varrho_{11}^{(H)}+\varrho_{-1-1}^{(H)}) = \frac{1}{2M^2}\, g^2 g_{\varrho NN}^2 |K(s,t)|^2 \frac{t^2+\mu^4}{(t-\mu^2)^2}\,.$$

Hence

$$\frac{\varrho_{1-1}^{(H)}}{\varrho_{11}^{(H)}} \left(\Rightarrow \frac{\varrho_{1-1}^{(H)}}{\frac{1}{2}(\varrho_{11}^{(H)}+\varrho_{-1-1}^{(H)})} \right) = \frac{2(-t)\,\mu^2}{t^2+\mu^4}\,. \tag{5.2}$$

To make a theoretical prediction about the asymmetry ratio $(\sigma_\perp - \sigma_{//})/(\sigma_\perp + \sigma_{//})$ we should construct an equivalent model for photoproduction of charged pions. In the notation of Ref. [10] the invariant amplitudes $A^{(-)}, B^{(-)}, D^{(-)}$ are related to PCTHA for $\gamma N \to \pi N$ by

$$A^{(-)} + tB^{(-)} = \tilde{f}_{01}^{-+} \qquad (\pi \text{ exchange})\,, \tag{5.3}$$

$$2MA^{(-)} - tD^{(-)} = 2\tilde{f}_{11}^{++} \qquad (\pi', A_2 \text{ exchange}) \tag{5.4}$$

and the conspiracy relation (3.1) of part I becomes

$$\tilde{f}_{11}^{++}(s,t) \to M \tilde{f}_{01}^{-+}(s,t) \qquad t \to 0\,. \tag{5.5}$$

We shall again neglect the A_2. Moreover, the contribution of $D^{(-)}$ in (5.4) is multiplied by t and does not affect the forward direction; thus we take

$$D^{(-)}(s,t,u) = 0\,.$$

Hence, the simplest Veneziano model for near-forward $\gamma p \to \pi^+ n$ is [11]

$$A^{(-)} + tB^{(-)} = \beta_1 \left\{ \frac{B(-\alpha_\pi(t), \frac{1}{2} - \alpha_N(s))}{\frac{1}{2} - \alpha_\pi(t) - \alpha_N(s)} - s \leftrightarrow u \right\} \tag{5.6}$$
$$+ \beta_2 \{ B(1 - \alpha_\pi(t), \tfrac{1}{2} - \alpha_\Delta(s)) - s \leftrightarrow u + B(\tfrac{1}{2} - \alpha_N(s), \tfrac{1}{2} - \alpha_\Delta(u)) - s \leftrightarrow u \} .$$

Straightforward comparison to the gauge invariant Born model gives

$$\beta_1 = -\lambda^2 \mu^2 eg \qquad \beta_2 = \tfrac{1}{2} \lambda eg \qquad \left(\frac{e^2}{4\pi} = \frac{1}{137} \right).$$

Hence, for $s \to \infty$, $t =$ fixed:

$$A^{(-)} + tB^{(-)} \to \tfrac{1}{2} \lambda eg \frac{\mu^2 + t}{\mu^2 - t} K(s, t) . \tag{5.7}$$

The amplitude $A^{(-)}$ containing the pion conspirator π' has the following Veneziano representation

$$A^{(-)} = a_1 \{ B(1 - \alpha_{\pi'}(t), \tfrac{1}{2} - \alpha_N(s)) - s \leftrightarrow u \} \tag{5.8}$$
$$+ a_2 \{ B(1 - \alpha_{\pi'}(t), \tfrac{1}{2} - \alpha_\Delta(s)) - s \leftrightarrow u + B(\tfrac{1}{2} - \alpha_N(s), \tfrac{1}{2} - \alpha_\Delta(u)) - s \leftrightarrow u \} .$$

Taking, as in Section 3, $\alpha_{\pi'}(t) \equiv \alpha_\pi(t)$ and using (5.5) we find for $s \to \infty$ $t =$ fixed:

$$A^{(-)} \to \tfrac{1}{2} \lambda eg \, K(s, t) . \tag{5.9}$$

Finally, the invariant amplitude $C^{(-)}$ is related to A_1 exchange and shall be neglected.

In this model the differential cross section for $\gamma p \to \pi^+ n$ is

$$\frac{d\sigma}{dt} = \frac{1}{32\pi} \{ |A^{(-)}|^2 + |A^{(-)} + tB^{(-)}|^2 \} = \frac{\lambda^2 e^2 g^2}{64\pi} |K(s, t)|^2 \frac{t^2 + \mu^4}{(t - \mu^2)^2} \tag{5.10}$$

i. e. exactly proportional to $\sigma_T^{(H)}$ of (3.10). In Section 2 and Eq. (3.9) we found that $\sigma_T^{(J)}$ has a completely different form, without forward peak. The same is true in the Donohue-Hogäasen frame, where $\sigma_T^{(D-H)}$ is readily found to be smooth, without any structure at all near $t = 0$.

The well-known Stichel theorem states that $\sigma_\perp (\sigma_{//})$ receives asymptotic contributions only from $+(-)$ normality exchange. Hence, the asymmetry ratio:

$$\Sigma \equiv \frac{\sigma_\perp - \sigma_{//}}{\sigma_\perp + \sigma_{//}} = \frac{|A^{(-)}|^2 - |A^{(-)} + tB^{(-)}|^2}{|A^{(-)}|^2 + |A^{(-)} + tB^{(-)}|^2} = \frac{2(-t)\mu^2}{t^2 + \mu^4} \tag{5.11}$$

identical to the ratio of (5.2). Since our models for both $\pi^- p \to \varrho_T^0 n$ and $\gamma p \to \pi^+ n$ are of the Regge type and thus asymptotic, the conclusion is that, at least at small $|t|$, with increasing energy the ratios $\varrho_{1-1}^{(H)}/\varrho_{11}^{(H)}$ and Σ are expected to tend to the same value.

The same conclusion is obtained in the Born model [9]. Moreover, straight-forward calculation shows that the ratio $\varrho_{1-1}/\varrho_{11}$ in the Jackson or Donohue-Hogäasen frame is completely different from (5.11).

Thus, in agreement with other considerations [9, 12], our final conclusion is that the helicity frame offers the only meaningful choice for the definition and test of vector dominance relations.

At present we have experimental information for $\varrho_{1-1}^{(H)}/\varrho_{11}^{(H)}$ at 4 GeV and for Σ at 3.4 GeV showing a significant discrepancy. This may be due to the following reasons: (i) Finite ϱ-mass corrections important at relatively low energies, (ii) Nonleading contributions in s left out in the above asymptotic models and (iii) Significant experimental error in the present data for $\varrho_{1-1}^{(H)}/\varrho_{11}^{(H)}$.

References to Introduction and Part I

1. *Toller, M.*: Nuovo Cimento **37**, 631 (1965); **53** A, 671 (1968); **54** A, 295 (1968). *Consenza, G., Sciarrino, A., Toller, M.*: Phys. Letters **27** B, 398 (1968); Nuovo Cimento **62** A, 999 (1969).
2. *Capella, A., Contogouris, A. P., Tran Thanh Van, J.*: Phys. Rev. **175**, 1892 (1968).
3. *Capella, A., Contogouris, A. P., Tran Thanh Van, J.*: Nucl. Phys. B **12**, 167 (1969).
4. *Jacob, M., Wick, G. C.*: Ann. Phys. (N. Y.) **7**, 404 (1959).
5. *Gell-Mann, M., et al.*: Phys. Rev. **133**, B 145 (1964).
6. *Trueman, T. L., Wick, G. C.*: Ann. Phys. (N. Y.) **26**, 322 (1964).
7. *Diu, B., LeBellac, M.*: Nuovo Cimento **53** A, 158 (1968).
8. *Ling-Lie Wang*: Phys. Rev. **142**, 1187 (1966).
9. *Cohen-Tannoudji, G., Morel, A., Navelet, H.*: Ann. Phys. (N. Y.) **46**, 239 (1968).
10. Using the $s-t$ crossing relations, it can be seen that as $t\to 0$ the SHA are

$$M_{\{\lambda^s\}} \equiv M_{\lambda_3^s\lambda_4^s,\lambda_1^s\lambda_2^s} \to \sum_{\{\lambda_i\}} t^{\frac{1}{2}(|\lambda_1-\lambda_1^s|+|\lambda_2-\lambda_2^s|+|\lambda_3-\lambda_3^s|+|\lambda_4-\lambda_4^s|)} f_{\lambda_3\lambda_4,\lambda_1\lambda_2}(t,\cos\theta_t)\,.$$

For UU and $t<0$, $s\to\infty$: $\cos^2\theta_t/2 \sim \sin^2\theta_t/2 \sim st$, so that (1.2) and (3.2) give:

$$f_{\lambda_3\lambda_4,\lambda_1\lambda_2} = \gamma_{\lambda\mu} t^{\frac{1}{2}||\lambda|-|\mu||+X(\lambda\mu)} s^{\alpha(t)}\,.$$

Hence, for an amplitude finite at $t=0$ we must have:

$$X(\lambda,\mu)=0 \qquad \text{and} \qquad |\lambda|=|\mu|\,.$$

Notice that in UU scattering $\theta_s=0$ corresponds to $\theta_t=0$ or π, where angular momentum conservation demands $\lambda=\mu$ or $\lambda=-\mu$.
11. With the $X_{UU}(\lambda\mu)$ of (2.7) the last equation of Ref. [10] gives:

$$M_{\{\lambda^s\}} \equiv M_{\lambda_3^s\lambda_4^s,\lambda_1^s\lambda_2^s} \to \gamma_{\lambda^s\mu^s} t^{\frac{1}{2}(|M-|\lambda^s||+|M-|\mu^s||)}\ \lambda^s\equiv\lambda_1^s-\lambda_2^s,\ \mu^s\equiv\lambda_3^s-\lambda_4^s$$

and with all $\lambda_i^s=\lambda_i$. Hence $M_{\{\lambda^s\}}\to$ finite demands

$$M=|\lambda^s|=|\lambda| \qquad \text{and} \qquad M=|\mu^s|=|\mu|\,.$$

12. *Freedman, D. Z., Wang, J. M.*: Phys. Rev. **160**, 1560 (1967).
13. — — Phys. Rev. **153**, 1596 (1967).

14. *Contogouris, A. P., Lebrun, J. P.*: Nuovo Cimento **64** A, 627 (1969) and Nucl. Phys. B **13**, 246 (1969); see also *Blackmon, M., et al.*: Nucl. Phys. B **12**, 495 (1969) and Phys. Rev. **183**, 1452 (1969) and *Capella, A., Tran Thanh Van, J.*: Nuovo Cimento Letters **1**, 321 (1969).
15. *Contogouris, A. P., Lubatti, H., Tran Thanh Van, J.*: Phys. Rev. Letters **19**, 1352 (1967).
16. *LeBellac, M.*: Phys. Letters **25** B, 524 (1967).
    See also *Squires, E. J.*: Lectures at the 1970 Heidelberg-Karlsruhe Summer Institute.
17. *Mandlestam, S.*: Phys. Rev. **168**, 1884 (1968). – *Sawyer, R. F.*: Phys. Rev. Letters **21**, 764 (1968).
18. It is possible that the limit (5.5) for the Regge residues is singular. However, this appears to contradict the smooth extrapolation assumption of the usual Current Algebra approach and so far has received no support from any concrete model.
19. In this approach the pion is taken as a mother trajectory ($n=0$). Then it can be shown that the $t=0$ behaviour of its residue functions follows from Eqs. (4.7) by taking the pion Toller quantum number as $M=0$ in at least the following two cases: (i) UU scattering (ii) EU and EE scattering with the equal-mass vertex made out of $N\bar{N}$; clearly, (i) and (ii) cover all reactions of present physical interest. However, in general, the E vertex introduces additional zeros. See *DiVecchia, P., Drago, F., Paciello, M.*: Nuovo Cimento **61** A, 421 (1969) and references therein. Work along the Toller classification via analyticity and factorization has independently been carried also by *Jones, L., Shepard, H.*: Phys. Rev. (1969), *Bronzan, J. B., et al.*: Phys. Rev. **181**, 2111 (1969) and references therein, *Wang, L. L., Wang, J. M.*: Phys. Rev. (1969) and *Frampton, P. H.*: Nucl. Phys. B **7**, 507 (1968).

References and Footnotes to Part II

1. *Diebold, R., Poirier, J.*: Phys. Rev. Letters **22**, 255 (1969).
2. *Donohue, J., Hogaasen, H.*: Phys. Letters **25** B, 554 (1967).
3. *Contogouris, A. P., Gaskell, R., Sveč, M.*: Nucl. Phys. (to appear).
4. *Diu, B., LeBellac, M.*: Nuovo Cimento **53** A, 158 (1968).
5. In agreement with *Avni, Y., Harrari, H.*: Phys. Rev. Letters **23**, 262 (1969).
6. *Roy, D. P.*: Phys. Rev. Letters **23**, 1417 (1969).
7. The model can be extended to include "magnetic" ϱN coupling, as well. The magnetic coupling contributes only to B_5, B_6 and B_8 which in the t-channel are controlled by A_2 and A_1 exchange and thus do not affect the structure of $\pi^- p \to \varrho_T^0 n$ at small $-t$.
8. *Scharenguivel, J., et al.*: Phys. Rev. Letters **24**, 332 (1960).
9. *Cho, C., Sakurai, J.*: Phys. Letters **30** B, 119 (1969). See also: *Horn, D., Jacob, M.*: Nuovo Cimento **56** A, 63 (1968).
10. *Chew, G., Goldberger, M., Low, F., Nambu, Y.*: Phys. Rev. **106**, 1345 (1957).
11. *Argyres, E., Contogouris, A. P., Lam, C. S., Ray, S.*: Nuovo Cimento (to appear). See also: *Ahmad, M., Fayyazuddin, Riazuddin*: Phys. Rev. Letters **23**, 504 (1969).
12. *LeBellac, M., Plaut, G.*: Nuovo Cimento **54** A, 97 (1969).

Dr. *A. P. Contogouris*
Department of Physics
McGill University
Montréal, Canada

Duality and Regge Theory

REINHARD OEHME

Contents

I. Introduction

Some seven years ago, at the Villa Monastero [1] and at Newbattle Abbey [2], I gave a set of lectures on complex angular momentum in elementary particle scattering. It was a time when Regge theory was considered to be on the way out, mainly because pion-nucleon scattering did not show the characteristic shrinkage of the diffraction peak. This shrinkage had been previously observed in proton-proton scattering, and was considered as the main, initial success of the model. In these lectures in 1963, I tried to show that complex angular momentum interpolations, Regge poles and branch points are an intimate and natural part of dispersion theory. This fact is quite independent of the singularity structure which eventually will be required in order to explain the experimental situation, and which may perhaps be derived from a more comprehensive theory.

Dispersion theory is a general scheme rather than a theory. It is a framework based upon a number of basic assumptions; it does not provide a really constructive method to calculate the Green's functions which are required for the description of experimental results. So far, it has not been possible to write down an explicit and practical S-matrix which satisfies simultaneously all the constraints of the dispersion scheme. There are, of course, approximate methods which generally start by neglecting one or more of the fundamental postulates in the hope of obtaining a first approximation which is sufficiently near the general solution so that the constraints, which *a priori* have been left out, can be recovered by a perturbation method. In previous years, certain analyticity and unitarity conditions in a given channel were mainly used as a starting point. Recently, the emphasis has changed, and one tries to make Ansätze which are fully crossing symmetric, leaving the implementation of the complete unitarity condition as a perturbative correction. In particular, dual meromorphic models have lately been considered extensively. Because meromorphic functions are so much simpler and more constrained than general non-schlicht ones, these models have been rather successful in providing interesting, explicit forms for many-particle amplitudes which, of course, are rather oversimplified.

My lectures are divided into two parts. In this first part I discuss duality and Regge behaviour for meromorphic amplitudes and for amplitudes with cuts. I consider only binary reactions because I am interested here in the concept of duality rather than its applications, which will be covered by other lecturers.

II. High Energy Limits

Let us begin by repeating [1, 2] briefly some heuristic considerations which indicate the relevance of Regge interpolations in relativistic dispersion theory. In order to have a simple model, we consider the invariant amplitude $F(s, t)$ for the elastic scattering of spinless particles with the same mass. In terms of the scattering amplitude in the centre-of-mass system, we have

$$F(s,t)=\frac{\sqrt{s}}{2}\,T_{\mathrm{c.m.}}(s,\cos\vartheta)\,, \tag{1}$$

where

$$\cos\vartheta=1+\frac{t}{2q^2}\,,\quad \text{and}\quad 4q^2(s)=s-4\mu^2\,.$$

For fixed values of t, the function $F(s, t)$ is analytic in the cut s-plane. We assume that our system has several stable single-particle states and/or resonances in the s-channel, which are described by poles on the real axis or in secondary Riemann sheets respectively. Poles describing resonances of spin l have residua which are proportional to $P_l(\cos\vartheta)$. We have terms of the form

$$\frac{b_l\, q^{2l} P_l\left(1+\dfrac{t}{2q^2}\right)}{m^2(l)-s}\,, \tag{2}$$

and similar contributions from actual resonances. For the purpose of our argument, we can ignore the branch cuts and consider only pole terms. We write $F(s, t)$ in the form

$$F(s,t)=\sum_{\text{families}}\sum_{l}\frac{b_l\, q^{2l} P_l(\cos\vartheta)}{m^2(l)-s}+R(s,t)\,, \tag{3}$$

where we have indicated several families of resonance sequences with increasing spin, and where $R(s, t)$ describes the rest of the amplitude which has no resonance poles.

Let us now consider the high-energy limit of $F(s, t)$ in the t-channel, where F corresponds to the physical amplitude for negative values of the momentum transfer s. We obtain

$$\lim_{t\to\infty} F(s, t) = \sum_{\text{families}} \sum_{l} \frac{b_l t^l}{m^2(l) - s} + \lim_{t\to\infty} R(s, t), \tag{4}$$

and we see that individual pole terms with $l \geqq 2$ violate the Froissart bound [3]

$$|F(s, t)| < O(t(\lg t)^2) \tag{5}$$

for $s \leqq 0$, which is a direct consequence of the general assumptions of dispersion theory (or even quantum field theory), with unitarity playing a major part. Consequently, all powers t^l with $l \geqq 2$ must be cancelled by contributions from $R(s, t)$ as we continue in s from the neighbourhood of the resonances $s = m^2(l)$ to real $s \leqq 0$.

A natural way to achieve a compensation is to introduce an interpolating function $\alpha(s)$ such that $\alpha(m^2(l)) = l$ and $\alpha(s) \leqq 1$ for real $s \leqq 0$. Considering l as a continuous variable, we can take $\alpha(s)$ as the inverse function of $s(l) \equiv m^2(l)$, which describes the dependence of the mass upon the strength of the centrifugal forces. We may then write in place of Eq. (4)

$$\lim_{t\to\infty} F(s, t) = \sum_{\text{families}} b(s)\, t^{\alpha(s)} \sum_{l} \frac{1}{m^2(l) - s} + \lim_{t\to\infty} \bar{R}(s, t). \tag{6}$$

Although there is no direct need to do so, we have also included poles with $l = 0, 1$ in the interpolation. They could in principle remain "elementary poles" which do not belong to a particular trajectory $\alpha(s)$ [4].

In modifying the meromorphic part of $F(s, t)$ in Eq. (4), we have added non-pole contributions which were, *a priori*, contained in $R(s, t)$. For our later considerations, it is of interest to see that a Regge-type asymptotic behaviour as in Eq. (6) can be obtained directly from an infinite sequence of resonance poles [5, 6].

Suppose we consider the most divergent terms in the limit $t \to \infty$ of an infinite sequence of resonance poles:

$$F(s, t) \to \sum_{l=0}^{\infty} \frac{b_l t^l}{m^2(l) - s}. \tag{7}$$

For simplicity, we assume that b_l is independent of l and that $m^2(l)$ is linear, i.e.,

$$m^2(l) = \frac{l-\alpha(0)}{\alpha'(0)}, \quad \text{or} \quad \alpha(s) = \alpha(0) + \alpha'(0)\,s\,. \tag{8}$$

Then the series

$$\sum_{l=0}^{\infty} \frac{t^l}{l-\alpha(s)}$$

can be summed for $|t|<1$ with the result

$$\begin{aligned}\sum_{l=0}^{\infty} \frac{t^l}{l-\alpha(s)} &= \int_1^{\infty} \mathrm{d}t' \frac{t'^{\alpha(s)}}{t'-t} \\ &= -\frac{1}{\alpha(s)}\, {}_2F_1(1, -\alpha(s); 1-\alpha(s), t)\,.\end{aligned} \tag{9}$$

The right-hand side of Eq. (9) can now be continued to $|t| \geqq 1$, and in particular for $t \to \infty$, we find the asymptotic form

$$-\frac{\pi}{\sin \pi \alpha(s)} (-t)^{\alpha(s)}\,. \tag{10}$$

This form can, of course, also be obtained directly from the sum (7) by using the Sommerfeld-Watson transformation [5]. But as a contrast to our later discussion of dual amplitudes, it is of interest to see the explicit relation in Eq. (9) between a meromorphic s-channel function and a non-schlicht, Regge-behaved t-channel expression. We note that the sum (9) is not Regge behaved for $|s| \to \infty$, $|\arg s| > \varepsilon$.

III. Mathematical Duality [7]

In the previous section we saw that, in certain cases, an infinite sequence of resonances can be summed in order to obtain an amplitude with Regge-assymptotic behaviour in the crossed channel. However, the simple example (9) given there was not crossing symmetric. In this section we want to consider amplitudes which are meromorphic functions in *both* variables s and t. In particular, we take first the simple case of an amplitude $F(s,t)$ which has resonance poles only in the s- and t-channels, the u-channel being "empty" like, for example, the $\pi^+\pi^+ \to \pi^+\pi^+$ channel. Of course, empty elastic channels also show diffraction peaks, which are therefore not contained in the meromorphic approximation and must be considered separately [8].

On the real s-axis, the function $F(s,t)$ has poles at $s = s_n$, $n = 1, 2, \ldots,$ and for reasons of simplicity we assume equidistant poles corresponding to a linear trajectory $\alpha(s)=\alpha(0)+\alpha'(0)s$. Furthermore, we require that the amplitude has an asymptotic expansion in s with the leading term given by

$$F(s,t)\sim -\beta(t)(-s)^{\alpha(t)}, \tag{11}$$

for

$$|s|\to\infty, \quad |\arg s|\geqq\varepsilon.$$

This asymptotic form corresponds to a pair of exchange-degenerate positive and negative signature trajectories in the t-channel.

For fixed values of $t\in D$, where D is some domain with $\mathrm{Re}\,\alpha(t)<0$, we can then write $F(s,t)$ in the form

$$F(s,t)=\sum_{n=0}^{\infty}\frac{R_n(t)}{n-\alpha(s)}+E(s,t), \tag{12}$$

where $E(s,t)$ is entire in s. The resonance content of the pole at $s=s_n$, $\alpha(s_n)=n$ is restricted by the assumption that the residua $R_n(t)$ are polynomials of maximal order n

$$R_n(t)=\sum_{k=0}^{n}c_k(n)\,t^n. \tag{13}$$

In our simple example, the amplitude $F(s,t)$ is a symmetric function of s and t, and therefore it has in the *t-channel* a representation analogous to Eq. (12).

The important question now is whether in Eq. (12) the t-channel singularities are contained in $E(s,t)$ or whether they are generated by controlled divergence of the infinite sum of resonances. The assumptions we have made so far are not quite sufficient to decide this question. However, we can prove that $E(s,t)$ must vanish identically if we impose the following additional restrictions [9]:

1) $E(s,t)$ is separately bounded by the Regge term (11) for $|s|\to\infty$, $|\arg s|>\varepsilon$.

2) For $|s|\to\infty$, $|\arg s|<\varepsilon$, the growth of the entire function is restricted by

$$|E(s,t)|<O(e^{\tau|s|^{\varrho}}), \tag{14}$$

with $\tau>0$ and $\varrho<\pi/2\varepsilon$. From the physical point of view, this restriction to functions of finite order and type is presumably quite reasonable*. In the narrow-width approximation, the angular region $|\arg s|<\varepsilon$ corresponds in a certain sense to the secondary sheets in the more general amplitudes with unitarity branch cuts.

We give a brief sketch of the proof that $E\equiv 0$ [7, 9]. For $t\in D$, we have $\mathrm{Re}\,\alpha(t)<0$, and hence $E(s,t)\to 0$ for $|s|\to\infty$, $|\arg s|\geqq\varepsilon$. Because the

* It is relevant to realize that the entire functions excluded by this condition are essentially those which grow at least as fast as $\exp(\exp|s|)$ for $|s|\to\infty$, $|\arg s|<\varepsilon$, and which are bounded by a polynomial for $|\arg s|\geqq\varepsilon$.

the function E is of finite order and type, it follows from the Phragmén-Lindelöf theorem [10] that we have also $E \to 0$ for $|\arg s| < \varepsilon$. Since E is entire in s, we conclude that $E(s, t) = 0$ for $t \in D$. But $E(s, t)$ is an analytic function of both variables s and t, and hence it follows that $E(s, t) \equiv 0$.

Granting the rather general restriction [2], we have shown that the meromorphic amplitude cannot be written in the form of an interference model [11]

$$F(s, t) = G_t(s, t) + G_s(s, t) \tag{15}$$

where G_t is an entire function in s which is Regge bounded for $|s| \to \infty$, $|\arg s| > \varepsilon$, and meromorphic in t with resonance poles at $\alpha(t) = n$. The function G_s has analogous properties with s and t interchanged. However, it should be noted that a splitting of F corresponding to Eq. (15) becomes possible if we allow the function G_t to exceed a polynomial bound for $|s| \to \infty$, $|\arg s| > \varepsilon$. For example, there are many entire functions which increase exponentially for $|s| \to \infty$, $|\arg s| > \pi/2$, i.e., in the left half plane, but which satisfy a bound like (11) for $|\arg s| < \pi/2$. Since we want to maintain the Regge asymptotic form (11) for the complete amplitude $F(s, t)$, an exponential increase of $G_t(s, t)$ for $\operatorname{Re} s \to -\infty$ would have to be cancelled by a corresponding term in $G_s(s, t)$. This would imply a strong correlation between s- and t-channel pole parts of the amplitude, which is not in the spirit of the interference model.

We conclude that complete Regge boundedness in the sense described above implies essentially that the amplitude is given by its meromorphic part in one variable:

$$F(s, t) = \sum_{n=0}^{\infty} \frac{R_n(t)}{n - \alpha(s)} . \tag{16}$$

We call such amplitudes "dual" [12]. In particular, we have introduced the notion of "mathematical duality" [7] in order to describe the situation where an amplitude is completely given in terms of its meromorphic part in one independent variable*.

We have discussed the amplitude $F(s, t)$ so far only for fixed values of s or t. As has been assumed in the beginning, the u-channel contains no resonances. Let us now consider F for fixed values of u, denoting it by

* If, instead of the bound (11), the function $E(s, t)$ satisfies only

$$|E(s, t)| < O(s^N)$$

for $t \in D$, with $N \geqq 0$, then our arguments can be generalized to show that E must be a polynomial in s of maximal degree N

$$E(s, t) = \sum_{n=0}^{N} a_n(t) s^n .$$

If the function E is required to contain all t-channel resonance poles, then these resonances would be limited to spins $J \leqq N$, which is contrary to our assumptions.

$F(s, t(s, u))$. We can write a meromorphic expansion of the form

$$F(s, t(s, u)) = \sum_{n=0}^{\infty} r_n(u) \left\{ \frac{1}{n-\alpha(s)} + \frac{1}{n-\alpha(t)} \right\}, \tag{17}$$

with

$$\alpha(t) = c - \alpha(u) - \alpha(s), \qquad c = 3\alpha(0) + \alpha'(0)\,\Sigma .$$

The residua in Eq. (17) can be expressed in terms of the $R_n(t)$ given in Eq. (13):

$$\begin{aligned} r_n(u) &= R_n(t(s_n, u)) = R_n(s(t_n, u)) \\ &= R_n\left(\Sigma - u - \frac{n-\alpha(0)}{\alpha'(0)}\right). \end{aligned} \tag{18}$$

As we will see later, the Regge behaviour of $F(s, t)$ for $s \to \infty$ and fixed t implies that $R_n(t) \sim n^{\alpha(t)}$ for $n \to \infty$. For fixed values of u, we may expect that the amplitude decreases faster than any inverse power for $s \to \infty$, $\pi - \varepsilon > |\arg s| > \varepsilon$. This can be achieved if the polynomials $r_n(u)$ have oscillatory behaviour for $n \to \infty$.

If the amplitude $F(s, t)$ has resonances in all three channels, the notion of mathematical duality implies, as before, that the amplitude is completely determined by its meromorphic part in one invariant variable. For example, if $F(s, t)$ is symmetric under interchange of s, t and u, we have for fixed values of t

$$F(s, t) = \sum_{n=0}^{\infty} \sigma_n(t) \left\{ \frac{1}{n-\alpha(s)} + \frac{1}{n-\alpha(u)} \right\}, \tag{19}$$

with

$$\alpha(u) = c - \alpha(t) - \alpha(s)$$

and

$$\sigma_n(t) = \sum_{k=0}^{n} c_k(n)\, t^k . \tag{20}$$

In order to determine the amplitude, we need the sequence of all resonance poles in two channels. The t-channel resonances are then obtained *via* the asymptotic expansion of the residua $\sigma_n(t)$ for $n \to \infty$.

Meromorphic amplitudes with resonances in all three channels are often written as a sum of three terms which contain only resonance poles in two channels. For example, in the symmetric case considered above, we may write

$$F(s, t) = H(s, t) + H(s, u) + H(u, t) \tag{21}$$

where the function H is symmetric and mathematically dual as discussed before. Writing

$$H(s, t) = \sum_{n=0}^{\infty} \frac{R_n(t)}{n-\alpha(s)}, \tag{22}$$

we may compare the expression (21) with the meromorphic expansion (19). With the help of an expansion of $H(s, u)$ for fixed t similar to Eq. (17), we find

$$\sigma_n(t) = R_n(t) + R_n(u(s_n, t)), \tag{23}$$

with

$$u(s_n, t) = \Sigma - s_n - t, \qquad s_n = \frac{n - \alpha(0)}{\alpha'(0)}.$$

A special example of the meromorphic amplitudes discussed in this section is given by the Veneziano model [13], which we do not need to write out here.

IV. Phenomenological Duality

At first, we consider again a meromorphic amplitude with resonance poles in the s-and t-channels only. We assume that $F(s, t)$ is Regge behaved for $|s| \to \infty$, $|\arg s| > \varepsilon$, and we write in the form (16), namely,

$$F(s, t) = \sum_{n=0}^{\infty} \frac{R_n(t)}{n - \alpha(s)}. \tag{24}$$

We now want to see explicitly how the singularities in the t-channel are generated by the infinite sum (24). The Regge behaviour of F in the s-plane implies for the sequence of polynomials $R_n(t)$ for $n \to \infty$ an asymptotic expansion of the form

$$R_n(t) \sim \sum_{k=0}^{\infty} b_k(\alpha(t))\, n^{\alpha(t) - k}. \tag{25}$$

We split the series (24) into two terms

$$\sum_{n=n_a}^{\infty} \frac{R_n(t)}{n - \alpha(s)} + \sum_{n=0}^{n_a} \frac{R_n(t)}{n - \alpha(s)}, \tag{26}$$

where n_a is a very large but finite. Substitution of the asymptotic expansion (25) into Eq. (26) gives the contribution

$$\begin{aligned} &\sum_{n=n_a}^{\infty} \sum_{k=0}^{\infty} \frac{b_k(\alpha(t))\, n^{\alpha(t)-k}}{n - \alpha(s)} + \cdots \\ &\qquad = \sum_{n=n_a}^{\infty} \sum_{k=0}^{\infty} \sum_{r=0}^{\infty} b_k(\alpha(t))\, (\alpha(s))^r\, n^{\alpha(t)-k-r-1} + \cdots \end{aligned} \tag{27}$$

for $|\mathrm{Re}\,\alpha(s)| < n_a$. Here and in the following the dots indicate contributions which are entire in t; we are only interested in the meromorphic part. We now interchange the summations in Eq. (27) and use the familiar representation of the ζ-function in the form

$$\sum_{n=n_a}^{\infty} n^{-z} = \frac{1}{z-1} + \text{entire function}\,. \tag{28}$$

The result is

$$\begin{aligned}\sum_{r=0}^{\infty}\sum_{k=0}^{\infty} b_k(r+k)\,(\alpha(s))^r \frac{1}{r+k-\alpha(t)} + \cdots \\ = \sum_{n=0}^{\infty} \frac{1}{n-\alpha(t)}\left(\sum_{k=0}^{n} b_k(n)\,(\alpha(s))^{n-k}\right) + \cdots\,.\end{aligned} \tag{29}$$

For a symmetric amplitude $F(s,t)$ the residuum at the pole $\alpha(t)=n$ must be equal to $R_n(s)$, and hence we can identify this polynomial with the sum

$$\sum_{k=0}^{n} b_k(n)\,(\alpha(s))^{n-k}\,. \tag{30}$$

For the case of an amplitude with resonances in three channels, as the one given in Eq. (19), the t-channel singularities are obtained in a completely analogous manner. Instead of Eq. (29), we obtain the expression

$$\sum_{n=0}^{\infty} \frac{1}{n-\alpha(t)}\left(\sum_{k=0}^{n} b_k(n)[(\alpha(s))^{n-k} + (c-\alpha(s)-n)^{n-k}]\right), \tag{31}$$

where the residuum should be identified with $\sigma_n(t)$ as given in Eq. (23).

Let us again return to the simple mathematically dual function $F(s,t)$ given in Eq. (24). For $|s|\to\infty$, $|\arg s|>\varepsilon$ it has an asymptotic expansion with the leading term

$$-\beta(t)\,(-s)^{\alpha(t)}\,, \tag{32}$$

where

$$\beta(t) = \frac{b(\alpha(t))}{\sin\pi\alpha(t)}\,. \tag{33}$$

If we evaluate Eq. (32) on the real axis and take the imaginary part, we obtain

$$b(\alpha(t))\,s^{\alpha(t)}\,.$$

On the other hand, the absorptive part of $F(s,t)$ in the s-channel is given by

$$\operatorname{Im} F(s+i0,t) = \pi \sum_{n=0}^{\infty} R_n(t)\,\delta(n-\alpha(s))\,. \tag{34}$$

If we average over an appropriate interval round the points $\alpha(s)=n$, we have

$$\langle \operatorname{Im} F(s+i0,t\rangle_{\mathrm{Av.}} = R_n(t)|_{n=\alpha(s)}\,, \tag{35}$$

and we see that the ratio

$$\frac{b(\alpha(t))\, s^{\alpha(t)}}{\langle \operatorname{Im} F(s+i0,t)\rangle_{\text{Av.}}} \approx 1 \tag{36}$$

for $s \to \infty$. The asymptotic relation (36) is a consequence of our assumptions concerning Regge behaviour, and it can be extended to the more complete asymptotic expansion. However, here we are interested in the possible approximate validity of Eq. (36) for non-asymptotic values of s. This depends, of course, upon the specific model considered, and it is well known that the Veneziano model is an example where the relation (36) is reasonably well satisfied, mainly because the Stirling formula is a fair approximation of the Γ-function even at moderate values of the argument.

We use the term "*phenomenological duality*" [7] for the approximate equality of the properly averaged absorptive part of the amplitude and the corresponding imaginary part of the asymptotic Regge expansion evaluated at non-asymptotic energies. We have introduced phenomenological duality here within the framework of meromorphic amplitudes, but the notion can evidently be extended to more general cases.

The generalization of our considerations concerning phenomenological duality to amplitudes with three active channels is straightforward. If the amplitude is written in terms of three functions H as in Eq. (21), the residuum of the pole at $\alpha(s)=n$ is given by Eq. (23), and it has contributions from $H(s,t)$ and $H(s,u)$. On the other hand, the Regge asymptotic form is given by

$$-\frac{b(\alpha(t))}{\sin \pi\alpha(t)} \{(-s)^{\alpha(t)} + (s)^{\alpha(t)}\} \tag{37}$$

for $|s| \to \infty$, $\pi - \varepsilon > |\arg s| > \varepsilon$, with the first term coming from $H(s,t)$ and the second one from $H(u,t)$. With the appropriate oscillatory character of the term $R_n(u(s_n,t))$ in Eq. (23), the function $H(s,u)$ should be exponentially damped for $|s| \to \infty$, $|\arg s| > \varepsilon$ and fixed values of t. This oscillatory character of $R_n(u(s_n,t))$ is also important for the validity of phenomenological duality. Since only $H(s,t)$ contributes to the imaginary part of the Regge term (37), and since

$$\operatorname{Im} F(s+i0,t) = \pi \sum_{n=0}^{\infty} \{R_n(t) + R_n(u(s_n,t))\}\, \delta(n-\alpha(s)), \tag{38}$$

we see that the second term in Eq. (38) must be removed by the averaging procedure leading to a relation corresponding to Eq. (36).

V. Amplitudes with Branch Cuts

In the previous two sections we have considered only meromorphic amplitudes. We saw that a splitting of amplitudes as required by an interference model generally leads to the introduction of functions with exponentially increasing terms which must cancel between the s- and t-channel resonance pole sums. As long as there are no branch cuts, there is no place to hide such unwanted terms, and one is naturally led to mathematically dual amplitudes. Clearly this situation changes as soon as we allow the amplitude to have branch lines, like the usual unitarity cuts along the real axes.

Suppose the scattering amplitude has a dispersion representation of the form

$$F(s,t)=\frac{1}{\pi}\int_{s_0}^{\infty} ds' \frac{\operatorname{Im} F(s'+i0,t)}{s'-s}+\frac{1}{\pi}\int_{u_0}^{\infty} du' \frac{\operatorname{Im} F(u'+i0,t)}{u'-u} \tag{39}$$

for fixed $t \in D$, and the asymptotic Regge expansion has the leading term

$$-\beta(t)\,\{(-s)^{\alpha(t)}+(s)^{\alpha(t)}\}\,. \tag{40}$$

As before, we consider only amplitudes without diffraction contributions. We split the $\operatorname{Im} F(s+i0,t)$ into two parts:

$$\operatorname{Im} F=\operatorname{Im} F_R+\operatorname{Im} F_E\,, \tag{41}$$

where $\operatorname{Im} F_E$ does not contain any resonance poles in s. Of course, such a splitting of the absorptive part is not unique, but this is not important for our present purpose. What we want to emphasize is that the contribution

$$\frac{1}{\pi}\int_{s_0}^{\infty} ds' \frac{\operatorname{Im} F_E(s',t)}{s'-s}+\frac{1}{\pi}\int_{u_0}^{\infty} du' \frac{\operatorname{Im} F_E(u',t)}{u'-u} \tag{42}$$

to the full amplitude $F(s,t)$ can now dominate asymptotically and give rise to the Regge term (40). The presence of branch cuts along the real s-axis makes the theorems used in the meromorphic case inapplicable. We call the branch cuts associated with the discontinuities $\operatorname{Im} F_E$ "*empty cuts*" [7], because there are no resonance poles in the secondary sheets.

Perhaps the simplest way to generate Regge asymptotic behaviour with the help of empty cuts is *via* an essential singularity at infinity. Then the Regge expansion is a divergent asymptotic expansion. The amplitude can have an asymptotic behaviour corresponding to Eq. (40) in the physical sheet in all directions including those parallel to the real axis. Of course, in a secondary sheet, there must then be some angular region where the function grows exponentially.

In one variable, a simple example of such a behaviour is given by the Kummer function $U(-\alpha, c, s_0-s)$ for $c=\frac{1}{2}$ [14]. This function is analytic in the two-sheeted s-plane with a square root cut from s_0 to $+\infty$. It has an essential singularity at infinity, but in the physical sheet, and in right-half plane of the second sheet, its asymptotic expansion is of the form

$$U(-\alpha, \tfrac{1}{2}, s_0 - s) \sim (-s)^{\alpha}(1 + O(s^{-1}))\,,$$
$$|s| \to \infty, \quad -3/2\pi < \arg(s_0 - s) < +3/2\pi\,.$$

For $\mathrm{Re}\, s \to -\infty$ in the unphysical sheet the function U increases exponentially.

There are, of course, many other ways to generate Regge asymptotic behaviour in the presence of branch cuts, particularly in view of the fact that the actual amplitude has an unlimited number of thresholds with an accumulation point at infinity. Here we remark only that we certainly do not want to simply write an empty cut contribution which is proportional to $(s - s_0)^{\alpha(t)}$, because $F(s, t)$ has no such branch point at finite points in the s-plane. In a model where the Regge asymptotic expansion is assumed to have a finite radius of convergence, the function $F(s, t)$ actually has a branch point corresponding to Eq. (40) at infinity. There is a t-dependent and generally infinite Riemann-sheet structure associated with the point infinity, and hence we would need an infinite number of ordinary branch points in the finite plane, or special logarithmic ones with appropriate t-dependent coefficients.

In an interference type model, the Regge behaviour is associated with the empty cuts mentioned above. We must require that the resonance part of the amplitude does not disturb this asymptotic behaviour, and we may assume that this part decreases faster than any inverse power. A behaviour like this would be natural if there should be only a finite number of resonances, or if the poles in the secondary sheets are very far removed from the real axis for higher energies.

Let us finally consider dual amplitudes in the presence of cuts. In the meromorphic case the Regge behaviour was directly related to the infinite sequence of resonance poles with residua having an asymptotic behaviour like $R_n(t) \sim n^{\alpha(t)}$ for $n \to \infty$. In the presence of unitarity branch lines, we may well have an infinite number of resonance poles in secondary sheets close to the real axis, so that the situation is approximately similar to the meromorphic one. We then have "approximate mathematical duality" in the sense that the nearby resonance poles are responsible for the Regge asymptotic form of the discontinuity across the real axis, which in turn gives rise to the Regge behaviour of the complete amplitude. Of course, at high energies, the distinction between the generation of Regge behaviour by resonances or by empty cuts is rather mathematical because overlapping resonances may not be distinguishable from a continuous distribution. But the interesting empirical fact is that at intermediate energies, where individual resonances can be isolated, phenomenological duality appears to be compatible with the data for several amplitudes [12, 15]. It is, of course, an open question whether or not this implies that we have approximate mathematical duality at high energies, but it is perhaps not implausible.

Acknowledgments: It is a pleasure to thank Professors *Abdus Salam* and *P. Budini* as well as the International Atomic Energy Agency and UNESCO for their very kind hospitality at the International Centre for Theoretical Physics in Trieste.

Some parts of these lectures have also been presented at the International Centre for Theoretical Physics in Trieste during the summer of 1969.

This work has been supported in part by the U.S. Atomic Energy Commission.

References

1. *Oehme, R.:* High-energy scattering and dispersion theory. In: Dispersion relations and their connection with causality, pp. 167—256. Ed. *E. P. Wigner*. New York: Academic Press 1964.
2. — Complex angular momentum in elementary particle scattering. In: Strong interactions and high-energy physics, pp. 129—222. Ed. *R. G. Moorhouse*. Edinburgh-London: Oliver and Boyd 1964.
3. *Froissart, M.:* Phys. Rev. **123**, 1053 (1961); *A. Martin:* Phys. Rev. **129**, 1432 (1963).
4. *Oehme, R.:* Phys. Rev. Letters, **9**, 358 (1962); Phys. Rev. **130**, 424 (1963).
5. *Van Hove, L.:* Phys. Letters **24** B, 183 (1967).
6. *Durand, L.:* Phys. Rev. **166**, 1680 (1968).
7. *Oehme, R.:* Nucl. Phys. B **16**, 161 (1970).
8. *Freund, P. G. O.:* Phys. Rev. Letters **20**, 235 (1968); *Harari, H.:* Phys. Rev. Letters **20**, 1395 (1968).
9. *Oehme, R.:* Nuovo Cimento Letters **2**, 53 (1969).
10. *Phragmén, E., Lindelöf, E.:* Acta Math. **31**, 381 (1908); *Bieberbach, L.:* Lehrbuch der Funktionentheorie, Bd. II, S. 270. Leipzig: Teubner 1931.
11. *Barger, V., Cline, D.:* Phys. Rev. Letters **16**, 913 (1966); *Allessandrini, V. A., Amati, D., Squires, E. J.:* Phys. Letters **27** B, 463 (1968).
12. *Dolen, R., Horn, D., Schmid, C.:* Phys. Rev. **166**, 1768 (1968).
13. *Veneziano, G.:* Nuovo Cimento **57** A, 190 (1968).
14. *Magnus, W., Oberhettinger, F., Soni, R. P.:* Formulas and theorems for the special functions of mathematical physics, p. 262. Berlin-Heidelberg-New York: Springer 1966.
15. For reviews and references see, for example: *Van Hove, L.:* Proceedings of the Third Hawaii Topical Conference on Particle Physics, Honolulu 1969 (to be published); *Jackson, J. D.:* Rev. Mod. Phys. **42**, 12 (1970); *Jacob, M.:* Procs of the Lund International Conference on Elementary Particles, p. 125. Ed. *G. von Dardel* (Berlingska, Sweden 1969).

Professor Dr. *Reinhard Oehme*
The Enrico Fermi Institute
and the Department of Physics
The University of Chicago,
Chicago, Illinois, USA*
and
International Centre for Theoretical Physics,
Trieste, Italy

---
* Permanent Address.

Complex Angular Momentum

REINHARD OEHME

Contents

I. Introduction

In this second part of my lectures, I will discuss several problems in Regge theory which are of current interest. As an introduction, I review some of the relevant properties of complex angular momentum interpolations within the framework of relativistic dispersion theory. The method of complex angular momenta has been developed in detail in my 1963 lectures on the subject [1, 2], and in the following I frequently refer to these publications, which also contain extensive references to the earlier literature. From the more recent topics, I consider the question of shielding cuts, which shield fixed poles from the continued unitarity condition. I discuss the general character of branch point trajectories, the problem of fixed poles at nonsense wrong signature points and the implications of these poles for the dip mechanism. Furthermore, I give an introduction to the problems associated with pole-cut relationships in the complex angular momentum plane. These relationships are of interest for many applications, but in particular they are important in connection with diffraction scattering. Here my lectures contain some new work in which I show that, in a rather natural way, pole-cut relationships give rise to models of amplitudes which do not satisfy the Pomeranchuk theorem. I also give a brief discussion of the Pomeranchuk theorem itself and its implications.

II. Complex Angular Momentum

In part I of these lectures* we have already studied Regge interpolations in connection with high energy limits of amplitudes. It is the purpose of this section to approach the same problem on the basis of complex angular momenta. As we have explained in the introduction, we mention here only the most relevant points and refer by (O, n) to page n of our Scottish lecture notes for further details [1].

Considering particles without spin, we introduce continued partial-wave amplitudes $F_{\pm}(s, \lambda)$ which are uniquely determined by the physical partial-wave amplitudes $F_l(s)$ for $l > N$:

$$F_{\pm}(s, \lambda = l) = F_l(s), \qquad l = \text{even/odd}. \tag{1}$$

The *uniqueness* is most important. It is guaranteed by the Carlson theorem (O, 140) provided $F(s, \lambda)$ is sufficiently bounded for $|\lambda| \to \infty$ in directions parallel to the imaginary axis. It prevents the possibility of adding to $F_{\pm}(s, \lambda)$ arbitrary terms of the type

$$(\sin \pi \lambda)\, \phi(s, \lambda), \tag{2}$$

with $\phi(s, \lambda)$ being regular at $\lambda = l$. From the existence of a fixed s dispersion relation for $F(s, t)$, and from the polynomial boundedness of the amplitude for $|t| \to \infty$, we can obtain the representation

$$F_{\pm}(s, \lambda) = \frac{1}{\pi} \int_{v_0}^{\infty} \mathrm{d}v \, \frac{1}{2q^2(s)} \, Q_\lambda\left(1 + \frac{\vartheta}{2q^2(s)}\right) A_{\pm}(s, v) \tag{3}$$

with

$$A_{\pm}(s, v) = A_t(s, v) \pm A_u(s, v). \tag{4}$$

Eq. (3) is valid *a priori* for $\mathrm{Re}\,\lambda > N$, and it gives continued partial-wave amplitudes which are bounded by a polynomial for $\lambda \to \lambda_0 \pm i\infty$, and hence comply with *Carlson*'s theorem.

For s real and negative, the Froissart bound gives

$$|A_{\pm}(s, v)| \leqq O(v(\lg v)^2), \tag{5}$$

and from this condition we find that Eq. (1) must actually hold for $l \geqq 2$ (O, 156). This fact implies that only the physical partial-wave amplitudes $F_0(s)$ and $F_1(s)$ can in principle have resonance poles which are not present in $F_{\pm}(s, \lambda)$. We call possible poles of this kind "*elementary poles*" (O, 155) because they describe resonances which do not lie on Regge trajectories.

Of particular importance for our later discussion is the *continued unitarity condition*. Using *Carlson*'s theorem, we find that the elastic unitarity condition, satisfied by the physical partial-wave amplitudes

* See the preceding article in this volume.

for $4\mu^2 \leqq s < s_i$, can be extended to the whole complex λ-plane. We can write it in the form*

$$F(s+i0, \lambda) - F(s-i0, \lambda) = 2i\varrho(s+i0)\, F(s+i0, \lambda)\, F(s-i0, \lambda) \qquad (6)$$

with $4\mu^2 \leqq s < s_i$ and

$$\varrho(s) = \left(\frac{s-4\mu^2}{s}\right)^{\frac{1}{2}}. \qquad (7)$$

Eq. (6) defines the continuation of the analytic function $F(s, \lambda)$ through the elastic square-root cut into the second sheet. We find that this continuation is of the form

$$F^{\mathrm{II}}(s, \lambda) = \frac{F(s, \lambda)}{1 + 2i\varrho(s)\, F(s, \lambda)}. \qquad (8)$$

From the Q_λ-representation (3), it follows that $F(s, \lambda)$ is a real analytic function

$$F(s, \lambda) = F^*(s^*, \lambda^*). \qquad (9)$$

In the s-plane it has cuts for $s \leqq 4\mu^2$ and, in the case of scalar particles, for $s \leqq -8\mu^2$. There is also a winding point at $s = 4\mu^2$ due to the factor $q^{2\lambda}(s)$ (O, 145, 151).

The singularities in the λ-plane are directly related to the asymptotic expansion of $A(s, v)$ in Eq. (3), with the exception of fixed poles at certain negative integer values of λ which are due to the Legendre functions Q_λ. An asymptotic term like $A \sim v^{\alpha(s)} (\lg v)^{\beta(s)}$ corresponds to a singularity of $F(s, \lambda)$ of the form

$$F(s, \lambda) \sim \frac{1}{(\lambda - \alpha(s))^{1+\beta(s)}}, \qquad (10)$$

or to $\lg(\lambda - \alpha(s))$ for the special case $\beta(s) \equiv -1$. Because of their connection with the asymptotic part of the integral (3), and since the left-hand cuts of $A(s, v)$ recede to $s \to -\infty$ for $v \to +\infty$, we find that singular surfaces like $\alpha(s)$ do not inherit the left-hand branch points of $F(s, \lambda)$ (O, 148).

III. Regge Poles and Cuts

The *Regge pole trajectories* $\lambda = \alpha(s)$ are analytic surfaces defined by

$$F^{-1}(s, \lambda = \alpha(s)) = 0. \qquad (11)$$

We have already seen that the function $\alpha(s)$ does not inherit the left-hand branch points of $F(s, \lambda)$. It must, however, have a branch point at the elastic threshold $s = 4\mu^2$. This follows simply from the unitarity condition

* Here and in the following we often omit the subscript $\pm$ if the relation is valid for positive and negative signature amplitudes.

(6), or, equivalently, from the continuation (8) into the second sheet. If for $\lambda \to \alpha(s)$ we have the pole

$$F \sim \frac{R(s)}{\lambda - \alpha(s)}, \tag{12}$$

it follows from Eq. (8) that

$$F^{\text{II}} \sim \frac{1}{2i\varrho(s)} \tag{13}$$

in the same limit, i.e., the function $F^{\text{II}}(s, \lambda)$ does not have a pole at $\lambda = \alpha(s)$. We know from the continuity theorem of functions with two or more complex variables (O, 149) that an isolated singularity in the λ-plane cannot simply disappear during an analytic continuation in s. The Regge pole surface $\alpha(s)$ avoids this difficulty by having a branch cut for $s \geqq 4\mu^2$ so that

$$\alpha^{\text{II}}(s) \neq \alpha(s). \tag{14}$$

Then we have from Eq. (11)

$$F^{\text{II}-1}(s, \lambda = \alpha^{\text{II}}(s)) = 0 \tag{15}$$

which is now compatible with Eq. (13). A more detailed analysis (O, 153) shows that $\alpha(s)$ has the threshold behaviour

$$\alpha(s) = \alpha(4\mu^2) + \text{const.}\,(4\mu^2 - s)^{\alpha(4\mu^2) + \frac{1}{2}} + \cdots. \tag{16}$$

The inverse of the trajectory function $\lambda = \alpha(s)$ is a λ-dependent pole in the s-plane which we denote by $s = s_0(\lambda)$ with $\alpha(s_0(\lambda)) = \lambda$. There is a corresponding pole at $s = s_0^{\text{II}}(\lambda)$ in the second sheet of the s-plane. It satisfies $\alpha^{\text{II}}(s_0^{\text{II}}(\lambda)) = \lambda$. For real values of $\lambda < \alpha(4\mu^2)$, the poles at $s = s_0$ and $s = s_0^{\text{II}}$ are on the real axis in different sheets and generally at different positions. As λ increases and crosses the point $\lambda = \alpha(4\mu^2)$, the two poles coincide and then move for real $\lambda > (4\mu^2)$ to complex conjugate positions in the *second sheet* of the s-plane. A meromorphic expansion of the partial-wave amplitude near the pole with $\text{Im}\, s_0 < 0$ then gives the familiar Breit-Wigner formula (O, 171, 150).

Let us now consider *branch point trajectories* of $F(s, \lambda)$ which correspond to branch points of fixed character like, for example, square roots or others. In contrast to a pole trajectory, a branch point trajectory $\lambda = \alpha_c(s)$ cannot disappear from the physical sheet through an elastic cut because, if $F^{\text{II}}(s, \lambda)$ has a branch point at $\lambda = \alpha_c(s)$, so does $F(s, \lambda)$ since

$$F^{-1}(s, \lambda) = F^{\text{II}-1}(s, \lambda) - 2i\varrho(s). \tag{17}$$

But the moving branch point $s = s_c(\lambda)$ in the s-plane, which corresponds to $\lambda = \alpha_c(s)$ with $\alpha_c(s_c(\lambda)) = \lambda$, should not remain in the physical sheet for positive real λ. Since we can show that the function $F(s, \lambda)$ has all thresholds as fixed (λ-independent) branch points in the s-plane (O, 185), the moving cuts $s = s_c(\lambda)$ have to be present in addition to the fixed unitarity cuts, and for increasing real values of λ they would eventually interfere with the unitary properties of the amplitude. We know that singular surfaces like $\alpha_c(s)$ do not have the left-hand cuts which are present in $F(s, \lambda)$. Hence the only way for the branch point trajectories to disappear from the physical sheet is through inelastic thresholds*.

The moving branch points indicated by perturbation theory have been discussed in detail in (O, 187). Here we want to mention only the well known fact that the iteration of two Regge poles, using the elastic unitarity condition in the t-channel, gives rise to an asymptotic contribution to the corresponding discontinutiy function which looks like a branch cut contribution. As far as the continued partial-wave amplitude $F(s, \lambda)$ and the physical amplitude $F(s, t)$ are concerned, this branch point is in a secondary sheet. However, its characteristics are essentially the same as the actual branch point obtained from a corresponding non-planar diagram, which is plausible on the basis of the reduced graph (O, 190).

One important exception to this analogy is the sign of the cut term relative to the pole term. From the actual non-planar graph, this sign comes out negative and opposite to the one obtained by simple iteration. But since we know that the iteration cut is in the second sheet relative to the elastic cut, this sign change is just what one would expect** (O, 189).

As far as the character of the branch points obtained in perturbation theory is concerned, the calculations done so far seem to lead mainly to logarithmic cuts. In the next two sections we will show that general considerations indicate that the moving branch points should rather be of square-root character. Of course it is quite possible that further implementation of crossed channel unitarity in these perturbation theory calculations will change the logarithmic character found *a priori* [5].

IV. Forbidden Singularities and Shielding Cuts

At first, we consider in this section the restrictions imposed by the continued elastic unitarity condition on the existence of fixed poles and other singularities, assuming that there are no shielding cuts which interfere with the application of this condition (O, 160).

* In potential scattering there are only elastic cuts and hence we can conclude from our arguments that the corresponding continued partial-wave amplitude does not have moving branch points in the λ-plane (0, 185).

** We do not discuss here the generation of cuts in the absorption model. For discussion and references see Ref. [3]. This method gives a negative relative sign [4] which also seems to be favoured by phenomenological calculations.

Suppose the continued partial-wave amplitude $F(s, \lambda)$ has an s-independent pole at $\lambda = \lambda_0$. From the continued unitarity condition, it then follows that $F^{\mathrm{II}}(s, \lambda)$ is given by Eq. (13) for $\lambda \to \lambda_0$, and hence the pole would have to get lost as we continue $F(s, \lambda)$ through the elastic cut into the second sheet. As we have already pointed out in Section III, such disappearance of an isolated singularity is not compatible with the continuity theorem. We conclude that $F(s, \lambda)$ cannot have a fixed pole unless the continuation into sheet II is somehow blocked for $\lambda \to \lambda_0$.

The theorem we have described above for fixed poles can be generalized directly to all singularities with fixed position (λ_0) and character in the λ-plane for which the amplitude $F(s, \lambda)$ becomes infinite as $\lambda \to \lambda_0$. In fact, it is also valid for certain moving singularities with s-independent character. If $F(s, \lambda) \to \infty$ for $\lambda \to \lambda_0(s)$, and if $\lambda_0(s)$ does not have a branch point in the s-plane at $s = 4\mu^2$ so that $\lambda_0^{\mathrm{II}}(s) \neq \lambda_0(s)$, then the singularity is again not compatible with the unitarity condition. We have seen already in Section III that Regge poles avoid this difficulty by having a cut for $s \geqq 4\mu^2$ in the trajectory. But we have also shown that branch point trajectories $\lambda = \alpha_c(s)$ generally do not have the elastic branch cut. Consequently a logarithmic moving cut, where the amplitude near $\lambda = \alpha_c(s)$ has an expansion of the form

$$F(s, \lambda) \sim C(s) \log(\lambda - \alpha_c(s)) + \text{non-singular terms} \tag{18}$$

is also not allowed (O, 161). On the other hand, a square root branch point or a logarithmic one of the form

$$F(s, \lambda) \sim C(s)\, (\lambda - \alpha_c(s))^n \log(\lambda - \alpha_c(s)) + \cdots \tag{19}$$

with $n \geqq 1$ does not lead to difficulties with elastic unitarity*.

The implications of the unitarity condition (8) can be exhibited very clearly by considering the function

$$\varphi^{-1}(s, \lambda) \equiv F^{-1}(s, \lambda) + i\varrho(s)\,, \tag{20}$$

which has *no* branch point at $s = 4\mu^2$ and hence satisfies

$$\varphi^{\mathrm{II}}(s, \lambda) = \varphi(s, \lambda)\,. \tag{21}$$

If $F(s, \lambda)$ has a singularity at $\lambda = \lambda_0(s)$ such that $F^{-1}(s, \lambda_0(s)) = 0$, we find from Eq. (20) that

$$\varphi^{-1}(s, \lambda_0(s)) = i\varrho(s)\,, \tag{22}$$

* The elastic unitarity condition also imposes severe restrictions on possible isolated essential singularities. For example, an essential singularity of s-independent position (λ_0) and character is not possible since there is at least one sequence $\{\lambda_n\} \to \lambda_0$ such that $F(s, \lambda_0)$ becomes unbounded for $n \to \infty$.

with $\varrho(s)=[(s-4\mu^2)/s]^{\frac{1}{2}}$. Unless there is a shielding cut, which we will discuss later, it is seen from Eq. (22) that the square-root branch point at $s=4\mu^2$ on the right-hand side must come from a corresponding cut in $\lambda_0(s)$ which is not possible if $\lambda_0(s)=\text{const.}$, for example [6].

The constraints imposed by the unitarity condition can be removed at the cost of introducing moving branch points with very specific properties [6–8]. Suppose $F(s,\lambda)$ and hence also $\varphi(s,\lambda)$, has a branch point surface $\lambda=\alpha_c(s)$ or $s=s_c(\lambda)$. We assume $s_c(\lambda_0)=4\mu^2$ so that the moving branch point coincides with the elastic threshold for $\lambda=\lambda_0$. Furthermore, it must have square-root character in order to completely cancel the branch cut on the right-hand side of Eq. (22).

In order to see more explicitly how such a special moving cut removes the contradiction of a fixed pole with the unitarity condition, let us consider a simple model. Suppose $F(s,\lambda)$ near the point $(s,\lambda)=(4\mu^2,\lambda_0)$ is of the form

$$F(s,\lambda)=\frac{(4\mu^2-s)^{\frac{1}{2}}+(s_c(\lambda)-s)^{\frac{1}{2}}}{\lambda-\lambda_0}+\cdots. \tag{23}$$

The continuation through the elastic cut is then obtained by changing the sign of $(4\mu^2-s)^{\frac{1}{2}}$, and, with $s_c(\lambda_0)=4\mu^2$, we see that

$$F(s,\lambda)=\frac{2(4\mu^2-s)^{\frac{1}{2}}}{\lambda-\lambda_0} \tag{24}$$

for $\lambda\to\lambda_0$, whereas $F^{\mathrm{II}}(s,\lambda)$ has *no* pole at this point.

So far, we have considered fixed singularities in the complex λ-plane only in connection with simple elastic unitarity. However, these arguments can be generalized to systems with many coupled channels (O, 161). As long as all thresholds are different we can always find a Riemann sheet where the fixed pole would have to disappear suddenly unless there are specially arranged shielding cuts which prevent the continuation into this sheet. The situation is different for coupled spin channels with the same threshold. Here we have a degeneracy, and complex fixed poles may exist under certain restrictions. Points on the real axis in the λ-plane which lie on a kinematic cut count as complex points. For real points we can see directly from the unitarity condition that the amplitude must be bounded and hence we cannot have a fixed pole (O, 163).

V. Nonsense Wrong Signature Points

In the two previous sections we have discussed the properties of moving branch points in the λ-plane but not the question of their existence. Of course, there are strong indications from perturbation theory that such cuts should be present. But we will see in this section that there are also very general arguments which point to the presence of cuts and their close association with fixed poles at nonsense wrong signature points. These fixed poles are also important in connection with the problem of dips in the momentum transfer distribution.

The continued partial-wave amplitude $F(s,\lambda)$ has fixed poles at negative integer points of λ. This can be seen from the representation (3) and the fact that

$$Q_\lambda(z) \sim \frac{1}{\lambda+n} P_{n-1}(z) \tag{25}$$

for $\lambda \to -n$ and $n=1,2\ldots$. Of course, this representation is generally not valid in the region of the λ-plane considered here. But we can take the discontinuity of $F(s,\lambda)$ across the left-hand cuts in the s-plane. This expression involves only a finite integral and hence it can always be continued into the left half λ-plane (O, 145). Here we find, for $\lambda \to -n$,

$$\operatorname{disc} F_\pm(s+i0,\lambda)\big|_{s \leqq -8\mu^2,\,\lambda\to -n}$$
$$= \frac{-1}{\lambda+n}\frac{1}{\pi}\int\limits_{C_-(s)}^{C_+(s)} \mathrm{d}c\, P_{n-1}(c)\operatorname{disc} A_\pm(s+i0,c)\,, \tag{26}$$

where

$$C_\pm(s) = \pm\frac{[s(s+8\mu^2)]^{\frac{1}{2}}}{s-4\mu^2}\,,$$

with $|C_\pm(s)| \leqq 1$ for $s \leqq -8\mu^2$ and

$$\operatorname{disc} A_\pm(s,c) = 2i\{\varrho_{tu}(s,c) \pm \varrho_{tu}(s,-c)\}\,. \tag{27}$$

We have written the discontinuity of $A_\pm$ simply as a function s and c. In the language of double dispersion relations, this discontinuity is given in terms of the third spectral function $\varrho_{ut}(u,t)$, but we never assume here the actual existence of a Mandelstam representation, only the corresponding analyticity properties in the finite (s,t) space.

From Eq. (26) we infer that $F_\pm(s,\lambda)$ has fixed poles at $\lambda=-n$, $n=$ odd/even, respectively. Hence we have poles at nonsense wrong signature points. According to the theorem of the previous section, such poles are not compatible with the continued unitarity condition unless we have appropriate shielding cuts. In fact, the positions of the branch points obtained from perturbation theory or similar iteration schemes are just right for the shielding mechanism, e.g. (O, 189)

$$\alpha_c(s) = n\alpha\left(\frac{s}{n^2}\right) - n + 1, \qquad n \geqq 2 \tag{28}$$

(with $\alpha(\mu^2)=0$), gives $\alpha_c(s=n^2) = -n+1$. On the other hand, the question of the character of these cuts has not been cleared up. Unless we consider branch points with s-dependent character, the unitarity condition requires square-root cuts.

By themselves, fixed poles at nonsense wrong signature points do not contribute directly to the high-energy limit of the amplitude $F(s,t)$ in the crossed channel. In order to see this, we write the Sommerfeld-Watson transformation (O, 168) for $t\to\infty$ in the form

$$F(s,t)=\frac{1}{2i}\int_C d\lambda\, F(s,\lambda)\, S(\lambda)\, K(s,\lambda)\, t^{\lambda}, \tag{29}$$

where $S(\lambda)$ is the signature factor

$$S_{\pm}(\lambda)=i+\frac{\mp 1-\cos\pi\lambda}{\sin\pi\lambda} \tag{30}$$

and

$$K(s,\lambda)=q^{-2\lambda}(s)\frac{\Gamma(\lambda+\frac{3}{2})}{\Gamma(\lambda+1)\sqrt{\pi}}\,. \tag{31}$$

At a wrong signature point, $S(\lambda)$ is regular, and for nonsense points the factor $K(s,\lambda)$ has a zero which cancels the pole in $F_{\pm}(s,\lambda)$.

The presence of fixed poles can, however, have important effects in connection with Regge pole trajectories which pass through nonsense wrong signature points. If the Regge pole and the fixed pole appear in $F(s,\lambda)$ in an *additive form*, then we see from Eqs. (29)–(31) that the Regge pole contribution has a zero at the $s=s_0$ where $\alpha(s_0)=-n=$ nonsense wrong signature point. As is well known, these zeros give rise to dips in the momentum transfer distribution. However, it is also possible that fixed pole and Regge pole are present in $F(s,\lambda)$ in a *multiplicative form*:

$$F(s,\lambda)\sim\frac{R(s)}{(\lambda-\alpha(s))(\lambda+n)}, \tag{32}$$

in which case the actual Regge residuum has a pole at $s=s_0$. In this case there is no zero in the differential cross-sections. The dip has been filled in, or partially filled in, depending upon the strength of the fixed pole term. In the following section we will show that the question of additive or multiplicative fixed poles is a dynamical one and cannot be decided on the basis of the s-channel unitarity condition [9].

In concluding this section we want to mention briefly the generalization of our considerations to amplitudes with spin without going into the details of the helicity formalism*. If we denote the helicities in the initial state of the s-channel by λ_1,λ_2, we have a "sense" transition to the exchanged Regge pole if $|\lambda_1-\lambda_2|\leqq n_0$ and a "nonsense" transition for $|\lambda_1-\lambda_2|>n_0$, where n_0 is an integer value attained by $\alpha(s)$ for $s=s_0$. Correspondingly, for the amplitude we have sense-sense, sense-nonsense and nonsense-nonsense components which we denote by F_{ss}, F_{sn} and F_{nn}.

* The literature is quoted in Refs. [3] and [10].

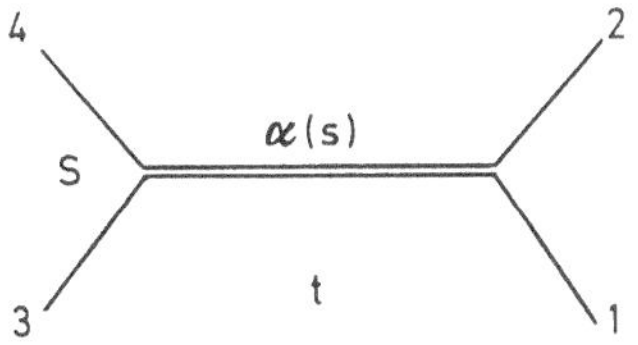

Fig. 1. Regge pole exchange. $|\lambda_1 - \lambda_2| \leqq n_0$ corresponds to a "sense" transition, $|\lambda_1 - \lambda_2| > n_0$ to a "nonsense" transition. There are corresponding relations for (λ_3, λ_4)

In the limit $t \to \infty$, and for $\alpha(s)$ near $\alpha(s_0) = n_0$, these amplitudes behave like

$$\begin{aligned} F_{ss} &\sim b_{ss} S(\alpha)\, t^{\alpha(s)}, \\ F_{sn} &\sim b_{sn} (\alpha(s) - n_0)^{\frac{1}{2}}\, S(\alpha)\, t^{\alpha(s)}, \\ F_{nn} &\sim b_{nn}((\alpha(s) - n_0)\, S(\alpha)\, t^{\alpha(s)}, \end{aligned} \tag{33}$$

where $S(\alpha)$ is the signature factor (30) and the coefficients b, which are proportional to the Regge pole residua, must satisfy the factorization condition

$$b_{ss} b_{nn} = b_{sn}^2 . \tag{34}$$

The factors $(\alpha(s) - n_0)^{\frac{1}{2}}$, etc., in Eq. (33) are a consequence of the behaviour of the rotation functions near $\alpha(s) = n_0$ (see, for example, Eq. (31)). Generally, the function $F_{sn}(s, t)$ has no branch point at $\alpha(s) = n_0$, and unless we want to introduce special fixed cuts in the angular momentum plane, we should require that $b_{sn} \sim (\alpha(s) - n_0)^{\frac{1}{2}}$ for $\alpha(s) \to n_0$. The factorization condition (34) then suggests the following possibilities, *modulo* common factors of $(\alpha(s) - n_0)$:

1. $\alpha(s)$ choosing *sense* for $\alpha(s) \sim n_0$:

$$b_{ss} \sim \text{const.} \quad b_{sn} \sim (\alpha(s) - n_0)^{\frac{1}{2}}, \quad b_{nn} \sim (\alpha(s) - n_0) .$$

2. $\alpha(s)$ choosing *nonsense* for $\alpha(s) \sim n_0$:

$$b_{ss} \sim (\alpha(s) - n_0), \quad b_{sn} \sim (\alpha(s) - n_0)^{\frac{1}{2}}, \quad b_{nn} \sim \text{const.}$$

On the other hand, if n_0 is a nonsense wrong signature point where there is a *multiplicative* fixed pole as in Eq. (32), we obtain

3. $b_{ss} \sim \text{const.} \quad b_{sn} \sim (\alpha(s) - n_0)^{-\frac{1}{2}}, \quad b_{nn} \sim (\alpha(s) - n_0)^{-1}$

as a possible behaviour of the residua.

The corresponding properties of the amplitudes can be obtained from Eq. (33). The sense mechanism 1. is, of course, not possible at a right signature point because there the signature factor $S(\alpha)$ has a pole. We do not want such a pole in $F_{sn}(s, t)$, particularly if $s_0 < 0$.

VI. Dips and Fixed Poles

We have seen that multiplicative forms of a Regge pole and a fixed pole at nonsense wrong signature points could have observable effects by filling in dips. In this section we want to see whether there are general arguments which could decide between multiplicative or additive fixed poles. For brevity we consider again the spinless case and explore the continued unitarity condition. In its general form, it does not imply any restriction on the Regge residuum for $\alpha(s)\to n_0$. The continuation $F\to F^{\mathrm{II}}$ and the limit $\lambda\to n_0$ cannot be interchanged because of the shielding cut.

Let us nevertheless explore the situation also in a perturbative fashion, making use of the fact that the residuum of the fixed pole is related to the third double spectral function, which is typical for relativistic dispersion theory and which is not present in potential models. We assume that we can expand $F(s,\lambda)$ in terms of a small third double spectral function. We write [8]

$$F(s,\lambda)=F^{(0)}(s,\lambda)+f^{(1)}(s,\lambda)+\cdots \tag{35}$$

where $f^{(1)}$ is of first order in ϱ_{ut}, and where we ignore higher-order terms. Expanding the unitarity condition (6), we have in zeroth order

$$F^{(0)\mathrm{II}}(s,\lambda)=F^{(0)}(s,\lambda)\,(1+2i\varrho(s)\,F^{(0)}(s,\lambda))^{-1}, \tag{36}$$

and the first-order equation is

$$f^{(1)}(s,\lambda)-f^{(1)\mathrm{II}}(s,\lambda)=2i\varrho(s)\,[f^{(1)}(s,\lambda)\,F^{(0)\mathrm{II}}(s,\lambda)+f^{(1)\mathrm{II}}(s,\lambda)\,F^{(0)}(s,\lambda)]\,. \tag{37}$$

In order to exhibit the relevant poles in $F^{(0)}$ and $f^{(1)}$, we write

$$F^{(0)}(s,\lambda)=\frac{\beta^{(0)}(s)}{\lambda-\alpha(s)}+\cdots \tag{38}$$

and

$$f^{(1)}(s,\lambda)=\frac{\beta^{(1)}(s)}{\lambda-\alpha(s)}+\frac{\gamma^{(1)}(s)}{\lambda-n_0}+\cdots, \tag{39}$$

where the omitted terms are regular for $\lambda\to\alpha(s)$ or $\lambda\to n_0$*.

We can solve Eq. (39) for $f^{(1)\mathrm{II}}$ and, with the help of Eq. (36), we find

$$f^{(1)\mathrm{II}}(s,\lambda)=\frac{f^{(1)}(s,\lambda)}{(1+2i\varrho(s)\,F^{(0)}(s,\lambda))^2}\,. \tag{40}$$

---

* If, at this point, one approximates $F^{(0)\mathrm{II}}$ for $\lambda\to\alpha(s)$ by a constant and $f^{(1)\mathrm{II}}$ by a fixed pole at $\lambda=n_0$, one could infer from Eq. (37) that $\beta^{(1)}(s)\sim(\alpha(s)-n_0)^{-1}$ [8]. But, actually, the residue of the fixed pole in $f^{(1)\mathrm{II}}$ vanishes at $s=s_0$.

Let us now study the limit $\lambda \to \alpha(s)$, where we find

$$F^{(0)^{\text{II}}} \to \frac{1}{2i\varrho} \tag{41}$$

and

$$f^{(1)^{\text{II}}} \to \frac{\beta^{(1)}(s)}{-4\varrho^2 \beta^{(0)2}} (\lambda - \alpha(s)) \to 0 \tag{42}$$

for $s \neq s_0$. Hence the unitarity relation is fulfilled in this limit simply in the form $f^{(1)} = f^{(1)}$.

Also of interest is the limit $\lambda \to n_0$ with $s > s_0$, where we have

$$f^{(1)^{\text{II}}}(s, \lambda) \to \frac{\gamma^{(1)}(s)}{\lambda - n_0} \, \frac{1}{(1 + 2i\varrho F^{(0)}(s, n_0))^2} \,. \tag{44}$$

Taking now in Eq. (44) the further limit $s \to s_0$ (i.e., $\alpha(s) \to \alpha(s_0) = n_0$), we see that the residue of the fixed pole in $f^{(1)^{\text{II}}}$ behaves like

$$\gamma^{(1)}\ (s) \to \frac{\gamma^{(1)}(s_0)}{-4\varrho^2(s_0)\, \beta^{(0)2}(s_0)} (s - s_0)^2 \,, \tag{45}$$

i.e., it vanishes quadratically in this limit. We see that, to first order in the third double spectral function, the fixed pole is present on both sheets of $F(s, \lambda)$. Hence there is no difficulty with the continuity theorem and no shielding cuts are required in this approximation.

We conclude that the *continued unitarity condition is compatible with either additive or multiplicative fixed poles at nonsense wrong signature points** [9]. This result can be easily generalized to amplitudes with spin where the arguments are analogous to those used above for the single-channel case. For example, if we take a situation where the trajectory chooses nonsense at $\alpha(s) = n_0$, we can write

$$F_{ij}^{(0)} = \frac{\beta_{ij}^{(0)}(s)}{\lambda - \alpha(s)} + \cdots, \quad ij = ss, sn, nn \tag{46}$$

and

$$\begin{aligned} f_{ss}^{(1)} &= \frac{\beta_{ss}^{(1)}}{\lambda - \alpha} + \cdots, \\ f_{sn}^{(1)} &= \frac{\beta_{sn}^{(1)}}{\lambda - \alpha} + \frac{\gamma_{sn}^{(1)}}{(\lambda - n_0)^{\frac{1}{2}}} + \cdots, \\ f_{nn}^{(1)} &= \frac{\beta_{nn}^{(1)}}{\lambda - \alpha} + \frac{\gamma_{nn}^{(1)}}{\lambda - n_0} + \cdots, \end{aligned} \tag{47}$$

* We may ask why multiplicative fixed poles are indicated in off-shell amplitudes related to weak or electromagnetic interactions. In these cases, the product form may be required by the equal-time commutation relations of currents, which give rise to non-Regge terms in the amplitude.

where

$$\beta_{ss}^{(0)} \sim (s - s_0), \quad \beta_{sn}^{(0)} \sim (s - s_0)^{\frac{1}{2}}, \quad \beta_{nn}^{(0)} \sim \text{const.} \tag{48}$$

for $s \to s_0$. The unitarity equations are then consistent with an additive pole where $\beta_{ij}^{(1)}(s)$ has the same behaviour as $\beta_{ij}^{(0)}(s)$ in Eq. (48), or with a multiplicative pole where

$$\beta_{ss}^{(1)} \sim \text{const.}, \quad \beta_{sn}^{(0)} \sim (s - s_0)^{-\frac{1}{2}}, \quad \beta_{nn}^{(1)} \sim (s - s_0)^{-1}. \tag{49}$$

Since the presence or absence of multiplicative fixed poles is apparently a question of dynamics, we may ask whether there are dynamical models which have special requirements in this respect. Indeed, one finds that the assumption of exchange degeneracy of Regge trajectories requires the absence of multiplicative poles [11]. The situation is explained most simply with the help of an example. We consider the contributions of ϱ- and f-trajectories to the amplitude $\pi^+\pi^+ \to \pi^+\pi^+$. They are of the form

$$\beta_\varrho \frac{1 - e^{-i\pi\alpha_\varrho}}{\sin \pi \alpha_\varrho} t^{\alpha_\varrho} + \beta_f \frac{-1 - e^{-i\pi\alpha_f}}{\sin \pi \alpha_f} t^{\alpha_f}. \tag{50}$$

The $\pi^+\pi^+$-channel is exotic, containing no ordinary resonances, and hence phenomenological duality* requires that the absorptive parts of the two Regge terms in Eq. (50) cancel for all values of the momentum transfers. This implies

$$\alpha_\varrho(s) = \alpha_f(s), \quad \beta_\varrho(s) = -\beta_f(s). \tag{51}$$

Suppose now we had a multiplicative fixed pole at a nonsense wrong signature point of the ϱ-trajectory, like the point $\alpha_\varrho(s_0) = -2$. Then the corresponding pole of the Regge residuum at $s = s_0$ implies $\beta_\varrho(s_0) = \text{const.}$, and it follows from Eq. (51) that we also have $\beta_f(s_0) = \text{const.}$ But a nonsense point for the odd-signature trajectory is a sense point for the even signature one, and hence we see that there is a pole at $s = s_0$ due to the signature factor. Since $s_0 < 0$, this pole would correspond to a ghost. We see that dual models require additive fixed poles, provided they have fixed poles to begin with.

VII. Crossing Regge Poles and Branch Cuts

In the previous sections we have discussed Regge poles and branch points as independent singularities in the complex angular momentum plane. In the following paragraphs we consider aggregates of closely correlated poles and cuts. We will see that certain such pole-cut relationships are of interest for the description of diffraction scattering but they

* For discussion and references see the preceding article in this volume.

could also play an important role in other phenomenological applications, where they may give a more concise parametrization than poles and cuts separately. But, in particular, we find that pole-cut systems can give rather natural and physically meaningful models for amplitudes which violate the Pomeranchuk theorem.

We begin by considering crossing Regge pole trajectories. We have shown in Section III that Regge poles are analytic surfaces defined by

$$F^{-1}(s, \lambda = \alpha(s)) = 0$$

or, more conveniently, by

$$F_{a^2}^{-1}(s, \lambda = \alpha(s)) = 0 \tag{52}$$

where

$$F_{a^2}(s, \lambda) = \frac{1}{\pi} \int_{a^2}^{\infty} \mathrm{d}v \frac{1}{2q^2} Q_\lambda\left(1 + \frac{v}{2q^2}\right) A(s, v) . \tag{53}$$

In Eq. (53) we can choose a^2 as large as we please, and then the left-hand cuts recede to $-\infty$. If $F_{a^2}^{-1}$ is regular in the neighbourhood of a point (s_0, α_0) belonging to the surface $\lambda = \alpha(s)$, then near $s = s_0$ the surface is of the general form

$$\alpha(s) = \alpha_0 + \sum_{j=1}^{\infty} a_j (s - s_0)^{j/n} , \tag{54}$$

where $n \geqq 1$ is some integer. This expression is a direct consequence of the preparation theorem of Weierstrass. It corresponds to n crossing Regge trajectories, and it shows that the pole surfaces $\alpha(s)$ can have a branch point at $s = s_0$ *without* $F_{a^2}(s, \lambda)$ or $F(s, \lambda)$ having a singularity at this point (O, 155).

Suppose we take the simple case $n = 2$ where we have two crossing trajectories

$$\alpha_{1,2}(s) = \alpha_0 \pm a_1 (s - s_0)^{\frac{1}{2}} + \cdots . \tag{55}$$

Near the point (s_0, α_0), the amplitude $F(s, \lambda)$ is then of the form

$$F(s, \lambda) = \frac{\beta_1(s)}{\lambda - \alpha_1(s)} + \frac{\beta_2(s)}{\lambda - \alpha_2(s)} + \cdots . \tag{56}$$

If $\beta_1(s_0) = \beta_2(s_0)$, we see that the amplitude is independent of the sign of the root in Eq. (55), and hence it does not have the branch point. It is relevant to note that the trajectories which cross at $s = s_0$ may well be regular at this point. The theorem only says that they could have a branch point without $F(s, \lambda)$ having one.

We may ask at this point whether there are corresponding possibilities in the case of a branch point trajectory $\alpha_c(s)$ crossing a pole trajectory $\alpha(s)$ so that $\alpha = \alpha_c$ at $s = s_0$. However, it is easy to see that two such singular surfaces of different character cannot collaborate in a simple manner in order to prevent the amplitude $F(s,\lambda)$ from having a branch point at $s=s_0$ if it is present in $\alpha(s)$ and $\alpha_c(s)$ [12]. For example, near the point (s_0,α_0) we may have the expansion

$$F(s,\lambda)=\frac{\beta(s)}{\lambda-\alpha(s)}+\gamma(s)\,(s-s_c(s))^{\frac{1}{2}}+\cdots. \tag{57}$$

Even if $\alpha(s)=\alpha_1(s)$, $\alpha_c(s)=\alpha_2(s)$, as given by Eq. (55), we see that we cannot arrange a cancellation of the branch point at $s=s_0$.

It is interesting to note that the situation for Regge pole residua and branch cut discontinuities can be quite different with respect to the cancellation of s-plane singularities. As an example, consider the function

$$\begin{aligned} F(s,\lambda)&=\frac{A+[\lambda-\alpha_c(s)]^{\frac{1}{2}}}{\lambda-\alpha(s)}\\ &=\frac{A+(\alpha(s)-\alpha_c(s))^{\frac{1}{2}}}{\lambda-\alpha(s)}+\frac{1}{i\pi}\int_{-\infty}^{\alpha_c(s)}\mathrm{d}\lambda'\,\frac{(\lambda'-\alpha_c(s))^{\frac{1}{2}}}{\lambda'-\alpha(s)}\,\frac{1}{\lambda'-\lambda}\\ &\quad+\text{regular terms.}\end{aligned} \tag{58}$$

If $\alpha=\alpha_c$ for $s=s_0$, there is a branch point in the Regge pole residuum which is cancelled by the cut term. We will see later further examples of this mechanism* [12].

VIII. Pole-Cut Relationships

We have seen that we can have left-hand branch lines in Regge pole surfaces only as a consequence of the crossover of two (or more) such trajectories. From the phenomenological point of view, we may like to have a left-hand cut in $\alpha(s)$, for example in order to get $\mathrm{Re}\,\alpha(s)=$ const. for $s\leqq 0$ (O, 201), but we may not want to have both branches of the trajectory in the physical sheet of the λ-plane. It is at this point that pole-cut relationships come into play, because we can use fixed or moving branch points in the complex λ-plane in order to remove one of the two crossing pole trajectories into a secondary sheet with respect to these λ-branch lines.

We do not want to discuss here the properties and the mathematical characteristics of the various types of pole-cut relationships in a general

* The invariant amplitudes for πN-scattering have a structure corresponding to Eq. (58) in the model of *Carlitz* and *Kislinger*, which introduces fixed cuts in the J-plane in order to remove the wrong parity branch of baryon trajectories into an unphysical sheet [13].

fashion*. Instead, we consider a characteristic model which involves in general a moving branch point [15]. Suppose we have a pair of pole trajectories which cross at $s=0$ and for which $\operatorname{Re}\alpha(s)=\alpha_0$ for $s \leqq 0$. For s near zero, we make the Ansatz

$$\alpha_{1,2}(s)=\alpha_0 \pm [s(a+b^2 s)]^{\frac{1}{2}} \tag{59}$$

where α_0, a and b are real and positive, and the root is defined in a cut s-plane with the branch line $-a/b^2 \leqq s \leqq 0$. For the continued partial-wave amplitude $F(s,\lambda)$ near the point $(s=0, \lambda=\alpha_0)$ we assume an expression of the form**

$$F(s,\lambda) \propto (\lambda-\alpha_0)\,\varphi(s,\lambda) \tag{60}$$

with

$$\varphi(s,\lambda)=\frac{\xi(s)+[(\lambda-\alpha_{c1}(s))\,(\lambda-\alpha_{c2}(s))]^{\frac{1}{2}}}{(\lambda-\alpha_1(s))\,(\lambda-\alpha_2(s))}\,. \tag{61}$$

If we choose $\xi(s)=A+bs$, the branch point trajectories must be such that for $\lambda=\alpha_{1,2}(s)$ the absolute value of the root equals $|\xi(s)|$. Hence we find

$$\alpha_{c1,2}(s)=\alpha_0 \pm [(a-2Ab)\,s-A^2]^{\frac{1}{2}}, \tag{62}$$

with $A<a/2b$. For $s<A^2(a-2Ab)^{-1}$, we have two complex conjugate branch points which we connect by a cut through the point $\lambda=\alpha_0$; for $s>A^2(a-2Ab)^{-1}$, the corresponding branch line is along the real axis.

For real values of s, the relative position of poles and branch points is exhibited in Fig. 2, where also the deformed contours are given, which will be useful for evaluating the asymptotic behaviour of $F(s,t)$ in the crossed channel. We note that pole and branch point trajectories coincide only for $s=-A/b$. None of the branch points of $\alpha(s)$ and $\alpha_c(s)$ is inherited by $F(s,\lambda)$. The forms of the functions $\alpha(s)$ and $\alpha_c(s)$ are, of course, approximations because we have ignored the elastic (for $\alpha(s)$) and inelastic (for $\alpha(s)$ and $\alpha_c(s)$) thresholds which must in general be present in these functions.

Taking the expression (61) for $\alpha_0=1$ as a positive signature term, we find that the high-energy limit of the amplitude $F(s,t)$ is dominated by the complex conjugate Regge poles for $-A/b<s \leqq 0$. Except for factors, we have for $t \to \infty$

$$F(s,t) \sim t\,2A\,\{i\cos(\beta(s)\log t) - \gamma(s)\sin(\beta(s)\log t)+O((\log t)^{-3/2})\} \tag{63}$$

* See Ref. [14] for such a discussion. Certain pole-cut relationships have also been considered by *F. Zachariasen* and collaborators; see Ref. [16], which contains further references.

** We ignore here several factors which are not directly relevant for our qualitative discussion.

with

$$\beta(s)=[-s(a+b^2 s)]^{\frac{1}{2}} \quad \text{and} \quad \gamma(s)=\operatorname{tgh}\left(\frac{\pi}{2}\beta(s)\right).$$

The characteristic feature here is the fixed power of t for a finite interval in the momentum transfer distribution. The oscillatory terms in the coefficient, however, make a direct application of this formula to diffraction scattering somewhat difficult. Note that for $A=a/2$, $b=1$ the branch points $\alpha_c(s)$ become fixed ones at $\alpha_c=1\pm i a/2$.

Of special interest is another degenerate case of the pole-cut relationship described above. Consider the trajectory given in Eq. (59). Since, for $s>0$, it increases sufficiently fast with increasing values of s, it is perhaps reasonable to expect that it is associated with branch point trajectories of the form

$$\alpha_c(s)=n\alpha\left(\frac{s}{n^2}\right)-n+1, \qquad n\geqq 2 \tag{64}$$

(see the discussion in Section V). If we accept this expression and insert our two-valued trajectory function (59), we find

$$\alpha_{c1,2}(s)=(n\alpha_0-n+1)\pm\left[s\left(a+\frac{b}{n^2}s\right)\right]^{\frac{1}{2}}. \tag{65}$$

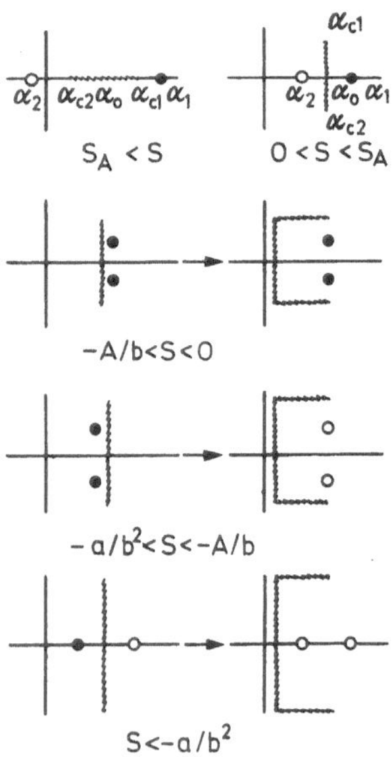

Fig. 2. Relative positions of branch points and poles of the continued partial-wave amplitude (Eqs. (59)–(62)) in the complex angular momentum plane. For $s<0$, also the deformed contours are given: $s_A=A^2(a-2Ab)^{-1}$; ● pole in the physical sheet; ○ pole in the unphysical sheet

We see that, with $\alpha_0 = 1$ and s near zero, the form

$$\alpha_{c\,1,2}(s) = 1 \pm (a\,s)^{\frac{1}{2}} \tag{66}$$

is common to all the functions (65). Let us therefore adopt this expression for $\alpha_c(s)$ in our formula $\varphi(s, \lambda)$. It corresponds to setting $A = 0$ in Eq. (62). Taking then

$$F(s, \lambda) \propto \varphi(s, \lambda)\,, \tag{67}$$

we have in particular $F(0, \lambda) \propto (\lambda - 1)^{-1}$. At first we consider the expression (67) for the positive signature amplitude. Then it follows that, for $t \to \infty$,

$$F(0, t) \sim i\,t \tag{68}$$

corresponding to a constant total cross-section. For finite, physical momentum transfers ($s \leqq s_0$, $s_0 < 0$), we see from Fig. 2 (with $A = 0$) that the branch points dominate. By deforming the branch cut as indicated in Fig. 2, we find that the poles are in the second sheet. If we retain only the tips of the cut, we obtain*

$$\begin{aligned}\varphi(s, t) \sim t(\log t)^{-\frac{3}{2}} \Big\{ &- i \cos\left((-a\,s)^{\frac{1}{2}} \log t - \frac{3\pi}{4}\right) \\ &+ \eta(s) \sin\left((-a\,s)^{\frac{1}{2}} \log t - \frac{3\pi}{4}\right)\Big\} \chi(s) + \cdots\,,\end{aligned} \tag{69}$$

where

$$\chi(s) = \left(\frac{2}{\pi}\right)^{\frac{1}{2}} \frac{(a\,s)^{\frac{1}{2}}}{b^2 s^2}\,, \quad \text{and} \quad \eta(s) = \mathrm{tgh}\left(\frac{\pi}{2}(-a\,s)^{\frac{1}{2}}\right).$$

This expression is of interest mainly because we see here the typical appearance of the asymptotic form of Bessel functions. For example, the absorption part of $\varphi(s, t)$ corresponds to the limit $\log t \gg \pi$ of

$$\iota\,\frac{a\,J_1((-a\,s)^{\frac{1}{2}} \log t)}{b^2 s(-a\,s)^{\frac{1}{2}} \log t}\,, \tag{70}$$

although the actual expression of $\varphi(s, t)$ is more involved.

In order to make the model even simpler, let us take $b = 0$ in addition to $A = 0$. Then the poles and the branch points coincide with the expression

$$\alpha(s) = \alpha_c(s) = 1 \pm (a\,s)^{\frac{1}{2}}\,, \tag{71}$$

which exactly satisfies the iteration relation (64). The function $\varphi(s, \lambda)$ degenerates to

$$\begin{aligned}\varphi_0(s, \lambda) &= [(\lambda - \alpha_1(s))\,(\lambda - \alpha_2(s))]^{-\frac{1}{2}} \\ &= [(\lambda - 1)^2 - a\,s]^{-\frac{1}{2}}.\end{aligned} \tag{72}$$

* Again we have omitted factors which are not essential for our discussion.

If now the positive signature amplitude $F_+(s, \lambda)$ is taken proportional to $\varphi_0(s, \lambda)$, we obtain for $s \leqq 0$ the asymptotic expression

$$F_+(s, t) \sim i\, t\, J_0((-a\, s)^{\frac{1}{2}} \log t)\,. \tag{73}$$

This form corresponds to an absorbing disc with logarithmically increasing radius $R \sim \sqrt{a} \log t$.

We cannot directly use the form (72) for the negative signature amplitude because it would lead to

$$F_-(s, \lambda = 1) \sim \frac{1}{(-a\, s)^{\frac{1}{2}}}\,, \tag{74}$$

and since $F_-(s, 1)$ is generally related to the physical partial-wave amplitude, we would have an unwanted pole (with superimposed branch point) at $s = 0$. However, this difficulty can very easily be avoided by an appropriate superposition of terms like $\varphi_0(s, \lambda)$. This will be discussed in a later section in connection with models for amplitudes which violate the Pomeranchuk theorem. Here we simply insert $\varphi_0(s, \lambda)$ into the Sommerfeld-Watson integral with the negative signature factor

$$S_-(\lambda) = \frac{1 - e^{-i\pi\lambda}}{\sin \pi \lambda} = i - \frac{2}{\pi}\,\frac{1}{\lambda - 1} + \text{non-singular real terms.} \tag{75}$$

Then we find the asymptotic expressions

$$\begin{aligned} \operatorname{Re} \varphi_-(s, t) &\sim -\frac{2}{\pi}\, t \log t \int_0^1 \mathrm{d}x\, J_0(x(-as)^{\frac{1}{2}} \log t) \\ &= -\frac{2}{\pi}\, t \log t \sum_{\nu=0}^{\infty} \frac{2 J_{1+2\nu}((-as)^{\frac{1}{2}} \log t)}{(-as)^{\frac{1}{2}} \log t} \end{aligned} \tag{76}$$

and

$$\operatorname{Im} \varphi_-(s, t) \sim t\, J_0((-a\, s)^{\frac{1}{2}} \log t)\,.^{\star} \tag{77}$$

In Eq. (76) we have omitted contribution which are singular at $s = 0$ and which will be eliminated later.

If taken for moderate values of the energy t, where $\log t$ is not yet very large compared with unity, an asymptotic expression like Eq. (70) or (73) gives rise to a diffraction peak due to the central maximum of

---

* A model amplitude which violates the Pomeranchuk theorem has also been constructed by *J. Finkelstein* [17]. It contains, however, logarithmic branch points (see Section IV).

After we had completed the work reported in this and in the following section, we saw a very interesting paper by *Gribov et al.* [18] where the problem of amplitudes with $\sigma \neq \bar{\sigma}$ is explored on the basis of Bessel function representations of scattering amplitudes. Starting, so to speak, from the other end of the problem, their method leads rather naturally to degenerate pole-cut relationships which are superpositions of $\varphi_0(s, \lambda)$. This paper also contains references to related work by other Russian authors. In particular, the work of *Anselm et al.* [19] is relevant in connection with our discussions.

the Bessel function. However, this peak shrinks rather rapidly because of the logarithmically increasing interaction radius. On the other hand, for $s<0$ and $\log t \gg \pi$, we have the s-independent factors $t(\log t)^{-\frac{3}{2}}$ or $t(\log t)^{-\frac{1}{2}}$ but, as in Eq. (63), there are coefficients to which may be rapidly oscillating. Unless we consider appropriate superpositions, it may therefore be difficult to use these functions directly for the construction of models for essentially non-shrinking diffraction peaks. We find the connection between our pole-cut relationships and the appearance of the typical Bessel functions in the asymptotic expressions to be of particular interest. As we see from the formal result (76), (77), these forms lead rather naturally to models for amplitudes which do not satisfy the Pomeranchuk theorem. Such models will be discussed in the following section.

IX. Amplitudes with Different Asymptotic Cross-Sections for Particle and Antiparticle Scattering

The recent experimental data obtained with the Serpukhov accelerator [20] have given an indication that for certain reactions the asymptotic total cross-sections for particle and antiparticle scattering may possibly tend to different constants. It is therefore of interest to see what kind of singular structures in the complex angular momentum plane could give rise to an asymptotic behaviour of this type.

Let us denote the amplitudes for the t-channel reactions $a+b \to a+b$ and $\bar{a}+b \to \bar{a}+b$ by $F(s,t)$ and $\bar{F}(s,t)$, respectively. If the total cross-sections have the constant asymptotic values $\sigma = \sigma_{\text{tot}}(t \to \infty)$ and $\bar{\sigma} = \bar{\sigma}_{\text{tot}}(t \to \infty)$, we find for $t \to \infty$

$$\operatorname{Im} F_{\pm}(0, t+i0) \sim \frac{\sigma \pm \bar{\sigma}}{16\pi}\, t\,, \tag{78}$$

where

$$F_{\pm}(s,t) = F(s,t) \pm \bar{F}(s,t)\,. \tag{79}$$

In view of Eq. (78), the familiar dispersion relation for the F_{-}-amplitude needs one subtraction. Substituting the asymptotic expression (78) for $\operatorname{Im} F_{-}$ into this relation, we find (O, 179) for $\operatorname{Re} F_{-}$ in the limit $t \to \infty$

$$\operatorname{Re} F_{-}(0,t) \sim -\frac{2}{\pi}\,\frac{\sigma - \bar{\sigma}}{16\pi}\, t \log t\,. \tag{80}$$

On the other hand, the dispersion relation for $F_{+}(s,t)$ shows that $\operatorname{Re} F_{+}$ increases at most like $O(t)$, and hence we have for $s=0$ and $t \to \infty$

$$\operatorname{Re} F(0,t) \sim -\operatorname{Re} \bar{F}(0,t) \sim -\frac{2}{\pi}\,\frac{\sigma - \bar{\sigma}}{16\pi}\, t \log t\,. \tag{81}$$

With the dispersive part of the amplitudes increasing faster than the absorptive part by a factor of $\log t$, we must of course have a diffraction peak which is shrinking sufficiently fast so that

$$\sigma_{\mathrm{el}} \leqq \sigma_{\mathrm{tot}} \tag{82}$$

is guaranteed.

As we have already pointed out in the last section, the highly degenerate case $\varphi_0(s, \lambda)$ of our pole cut Ansatz in Eq. (61) or (72) has many of the features which are required for amplitudes which give rise to different asymptotic cross-sections σ and $\bar{\sigma}$. Its main difficulty is the singular behaviour of $\varphi_0(s, \lambda=1)$ for $s \to 0$, which is a sense point for the negative signature amplitude. However, we can easily generalize our model by using a superposition of terms like $\varphi_0(s, \lambda)$. We write

$$F_-(s, \lambda) \propto \int_0^1 d\xi \frac{\varrho_-(\xi)}{[(\lambda - \alpha_0(s))^2 - \xi^2 a s]^{\frac{1}{2}}}, \tag{83}$$

where $\alpha_0(s) = 1 + \alpha_0'(0)\, s + \cdots$ near $s=0$. The weight function $\varrho_-(\xi)$ determines the position and character of the branch points. We assume that it is not singular so that we remain with a simple superposition of poles and square-root branch points. The essential restriction for $\varrho_-(\xi)$ is obtained for $\lambda = 1$, where we have for $s \to 0$

$$F_-(s, \lambda=1) \propto \frac{1}{(-a s)^{\frac{1}{2}}} \int_0^1 d\xi\, \varrho_-(\xi)\, \xi^{-1}. \tag{84}$$

In order to avoid this singularity we require

$$\int_0^1 d\xi\, \varrho_-(\xi)\, \xi^{-1} = 0, \tag{85}$$

which implies that $\varrho_-(\xi)$ must change sign in the interval $0 \leqq \xi \leqq 1$.

From $s=0$, we find

$$F_-(0, \lambda) \propto \frac{1}{\lambda - 1} \int_0^1 d\xi\, \varrho_-(\xi), \tag{86}$$

and hence the integral

$$\int_0^1 d\xi\, \varrho_-(\xi)$$

is proportional to $\sigma - \bar{\sigma}$.

The high-energy properties resulting from Eq. (83) are obtained from the Sommerfeld-Watson formula. Similar to Eqs. (70) and (77), we now have*

$$\operatorname{Im} F_-(s, t) \sim t^{\alpha_0(s)} \int_0^1 d\xi\, \varrho_-(\xi)\, J_0(\xi \sqrt{-a s} \log t) \tag{87}$$

---

* Again we have not written out factors like $K(s, \lambda)$ in Eq. (31).

and

$$\begin{aligned}\mathrm{Re}F_-(s,t) &\sim -\frac{2}{\pi}\,t^{\alpha_0(s)}\log t\int_0^1 \mathrm{d}\xi\,\varrho_-(\xi)\int_0^1 \mathrm{d}x\,J_0(x\xi(-as)^{\frac{1}{2}}\log t)\\ &= \frac{2}{\pi}\,t^{\alpha_0(s)}\log t\int_0^1 \mathrm{d}\xi\left(\int_0^{\xi}\mathrm{d}x\,\frac{\varrho_-(x)}{x}\right)J_0(\xi(-as)^{\frac{1}{2}}\log t)\,.\end{aligned} \tag{88}$$

Using the expression (88), we can calculate it's contribution to the elastic cross-section

$$\sigma_{\mathrm{el}}(t)\sim\frac{16\pi}{t^2}\int_{-t}^{0}\mathrm{d}s\,|F(s,t)|^2\,, \tag{89}$$

for which we find, taking $\alpha_0(s)\equiv 1$,

$$\sigma_{\mathrm{el}}(t)\sim\frac{32}{\pi a}\int_0^1 \mathrm{d}\xi\,\xi^{-1}\left(\int_0^{\xi}\mathrm{d}x\,\frac{\varrho_-(x)}{x}\right)^2 \tag{90}$$

Hence Eq. (82) can be satisfied.

There is also a restriction for the diffraction peak which can be derived from general principles [21] and which for $|F(0,t)|\sim t\log t$ and $\sigma_{\mathrm{el}}(t)\sim$ const. is of the form

$$\frac{\mathrm{d}}{\mathrm{d}s}\,\lg|F(s,t)|_{s=0}\leqq \text{const.}\,(\log t)^2\,. \tag{91}$$

The condition is satisfied by our model, which can be at the boundary, giving a $(\log t)^2$ term on the right-hand side of Eq. (91).

In the actual physical amplitude at higher energies, we must remember that possible contributions of the type (88) from negative signature singularities presumably come into play only for $\log t\gg\pi$. At lower energies, the dominant term is then given by the positive signature amplitude which therefore determines the dispersive part in the presently accessible energy region [22].

In the language of optics, the high-energy forms in Eqs. (87) and (88) may be described as superpositions of absorptive and emissive rings with logarithmically increasing radii. We cannot simply have an absorptive disc because of the constraint (85) for the weight function.

It is important to realize the limitation of our model in connection with the continued unitarity condition in the s-channel. We know that the pole trajectories $\alpha_{1,2}(s)$ and the branch point trajectories $\alpha_{c\,1,2}(s)$ cannot coincide exactly. The pole surface has to develop a branch point at the lowest threshold in the s-plane, whereas the branch point surface is associated only with the inelastic thresholds higher up along the s-axis (see Section III). Nevertheless, our expressions for the high energy behavior of amplitudes with $\sigma\neq\bar{\sigma}$ are very general as far as the neighborhood of the forward direction ($s=0$) is concerned. The questions concerning the presence or absence of oscillations in the differential cross-section will be discussed elsewhere.

In this and in the previous section, we have considered only some special examples of pole-cut relationships. There is no direct theoretical proof at present that such structures are necessary within the framework of dispersion theory, but there are some indications from potential theory (O, 164), relativistic perturbation theory (O, 167) and certain iteration schemes [16, 23] that they may be relevant.

X. Pomeranchuk Theorems

In the Scottish lecture notes we have discussed the Pomeranchuk theorems under the assumption of asymptotic forms of the type $t^{\alpha(s)}(\log t)^{\beta(s)}$ which exclude infinite oscillations* (O, 177). Here we only want to rephrase these considerations in terms of the amplitudes F and $\bar{F}$, instead of the functions $F_{\pm}$ used mainly in Ref. [1]**.

We first consider the theorem for amplitudes of elastic cross-sections. Suppose the function $F(s,t)$ has an asymptotic limit of the form

$$\frac{F(s,t+i0)}{t^{\alpha(s)}(\log t)^{\beta(s)}} \to a(s) \tag{92}$$

for $t\to\infty$ and $s\leqq 0$. Since $F(s,t)$ is analytic in the upper half of the t-plane and has a polynomial bound there, it follows from the Phragmén-Lindelöf theorem that the limit is the same for $t\to-\infty+i0$. Hence we have, assuming $\log t \gg \pi$.

$$\lim_{t\to-\infty}\frac{F(s,t+i0)}{t^{\alpha(s)}(\log t)^{\beta(s)}} = \lim_{t\to+\infty}\frac{F(s,-t+i0)}{t^{\alpha(s)}(\log t)^{\beta(s)}\,e^{i\pi\alpha(s)}} = a(s)\,. \tag{93}$$

From the reality condition

$$F^*(s^*,t^*) = F(s,t) \tag{94}$$

and the crossing relation

$$F(s,-t-i0) = \bar{F}(s,t+i0) \tag{95}$$

it follows then that

$$\lim_{t\to\infty}\frac{\bar{F}^*(s,t+i0)}{t^{\alpha(s)}(\log t)^{\beta(s)}}\,e^{-i\pi\alpha(s)} = a(s)\,, \tag{96}$$

---

* Examples of amplitudes with infinite oscillations have been considered in the last two sections.

** For a more profound proof, see Ref. [24] and references given in these papers.

and consequently

$$\frac{\lim\limits_{t\to\infty} F(s, t+i0)}{\lim\limits_{t\to\infty} \bar{F}^*(s, t+i0)} = e^{i\pi\alpha(s)}. \tag{97}$$

This relation implies that

$$\frac{d\sigma}{ds}\Big/\frac{d\bar{\sigma}}{ds} \to 1 \tag{98}$$

for $t\to\infty$. Hence both reactions have identical asymptotic diffraction peaks. But Eq. (97) does not imply the equality of total cross-section limits, except in cases where the amplitudes become pure imaginary in the high-energy limit (O, 180).

Let us now consider forward amplitudes in order to study the total cross-sections. From the Froissart bound, we have the restrictions $\alpha(0) \leqq 1$ and $\beta(0) \leqq 2$. We are mainly interested in the case $\alpha(0)=1$, $\beta(0) \geqq 0$. Writting

$$\begin{aligned} \operatorname{Im} F(s, t+i0) &\sim b\, t(\log t)^{\beta(0)}, \\ \operatorname{Im} \bar{F}(s, t+i0) &\sim \bar{b}\, t(\log t)^{\beta(0)}, \end{aligned} \tag{99}$$

the dispersive part of the amplitude $F_- = F - \bar{F}$ becomes in the asymptotic limit (O, 179)

$$\operatorname{Re} F_-(0, t) \sim -\frac{2}{\pi}(b-\bar{b})\, t(\log t)^{\beta(0)+1}. \tag{100}$$

This result follows from the dispersion relation for $F_-(0, t)$ which we may write for our purpose in the form

$$\operatorname{Re} F_-(0, t) - \frac{t}{t_0} \operatorname{Re} F_-(0, t_0) = \frac{2t(t^2-t_0^2)}{\pi} \int\limits_0^\infty dt' \frac{\operatorname{Im} F_-(0, t')}{(t'^2-t^2)(t'^2-t_0^2)}. \tag{101}$$

From the corresponding dispersion relation for $F_+(0, t)$, we find that $\operatorname{Re} F_+(0, t)$ is bounded by $(b-\bar{b})\, t(\log t)^{\beta(0)-1}$, and hence the high-energy limits of F and $\bar{F}$ are determined by $\operatorname{Re} F_-(0, t)$ as given in Eq. (100), provided, of course, that $b \neq \bar{b}$.

If $\beta(0) > 0$, the general principles of field theory and, in particular, unitarity imply the bound [22]

$$\left|\frac{\operatorname{Re} F(0, t)}{\operatorname{Im} F(0, t)}\right| \leqq \text{const.} \frac{\log t}{(\sigma_{\text{tot}}(t))^{\frac{1}{2}}}. \tag{102}$$

With $\sigma_{\text{tot}}(t) \sim (\log t)^\beta$, this gives $(\log t)^{1-\beta/2}$, whereas Eqs. (100) and (99)

$$\frac{\operatorname{Re} F(0, t)}{\operatorname{Im} F(0, t)} \sim (b-\bar{b}) \log t. \tag{103}$$

Hence we have a contradiction unless $b = \bar{b}$.

If $\beta(0)=0$, there is no general bound like Eq. (102), and hence we obtain $b=\bar{b}$ only if we *assume* that

$$\frac{\operatorname{Re}F(0,t)}{\operatorname{Im}F(0,t)\log t}\to 0\,.$$

On the other hand, if we retain $b\neq\bar{b}$ and hence accept the inequality of the asymptotic total cross-sections, the unitarity condition $\sigma_{\mathrm{el}}\leqq\sigma_{\mathrm{tot}}$ requires an effective shrinkage of the diffraction peak which goes like $(\log t)^2$. A behaviour of this type cannot be obtained with simple asymptotic forms like $t^{\alpha(s)}(\log t)^{\beta(s)}$ for $s\leqq 0$.

In fact, the examples we have discussed in the previous section have very rapid oscillations in the diffraction peak.

Acknowledgments: It is a pleasure to thank Professors *Abdus Salam* and *P. Budini*, as well as the International Atomic Energy Agency and UNESCO for their very kind hospitality at the International Centre for Theoretical Physics, Trieste.

Some parts of these lectures have also been presented at the ICTP.

This work has been supported in part by the U.S.-Atomic Energy Commission.

References

1. *Oehme, R.:* Complex Angular Momentum in Elementary Particle Scattering. In: *Strong Interactions and High Energy Physics*, pp. 129–222. Ed. R. G. Moorhouse. Edinburgh-London: Oliver and Boyd 1964. This article is frequently referred to in the form (O, page number). It contains references in the literature prior to 1964 *which we do not repeat here.*
2. — High Energy Scattering and Dispersion Theory in *Dispersion Relations and their Connection with Causality*, pp. 167–256. Ed. E. P. Wigner. New York: Academic Press 1964.
3. *Jackson, J. D.:* Rev. Mod. Phys. **42**, 12 (1970).
4. *Finkelstein, J., Jacob, M.:* Nuovo Cimento **56** A, 681 (1968); *Rivers, R. J., Saunders, L. M.:* Nuovo Cimento **58** A, 385 (1968); *Frautschi, S., Margolis, B.:* Nuovo Cimento **56** A, 1155 (1968); *Caneschi, L.:* Phys. Rev. Letters **23**, 254 (1969).
5. *Gribov, V. N., Pomeranchuk, I. Ya., Ter-Martirosyan, K. A.:* Phys. Rev. **139**, B 184 (1965); *Polkinghorne, J. C.:* preprint.
6. *Oehme, R.:* Phys. Rev. Letters **18**, 1222 (1967).
7. *Jones, C. E., Teplitz, V.:* Phys. Rev. **159**, 1271 (1967).
8. *Mandelstam, S., Wang, L. L.:* Phys. Rev. **160**, 1490 (1967).
9. *Oehme, R.:* Phys. Letters **28** B, 122 (1968).
10. *Bertocchi, L.:* Proceedings of the Heidelberg Conference on Elementary Particles, p. 187. Ed. F. Filthuth. Amsterdam: North-Holland Publ. Co 1967.
11. *Finkelstein, J.:* Phys. Rev. Letters **22**, 362 (1969).
12. *Oehme, R.:* Phys. Letters **30** B, 414 (1969).
13. *Carlitz, R., Kislinger, M.:* Phys. Rev. Letters **24**, 186 (1970).
14. *Oehme, R.:* Phys. Rev. **2** D, 801 (1970).
15. — Phys. Letters **31** B, 573 (1970).
16. *Zachariasen, F.:* Proceedings of the 1970 Coral Gables Conference (to be published).
17. *Finkelstein, J.:* Phys. Rev. Letters **24**, 172 (1970).

18. *Gribov, V. N., Kobsarev, I. Yu., Mur, V. D., Okun, L. B., Popov, V. S.*: Phys. Letters **32** B, 129 (1970).
19. *Anselm, A. A., Danilov, G. S., Dyatlov, I. T., Levin, E. M.*: Yadern. Fiz. **11**, 896 (1970).
20. *Allaby, J. V., et al.*: Phys. Letters **30** B, 500 (1969).
21. *Logunov, A. A., Nguyen van Hieu*: Proceedings of the Topical Conference on High-Energy Physics (CERN, Geneva 1968), p. 74; *Singh, V., Roy, S. M.*: Ann. Phys. (N.Y.) (to be published); *Eden, R. J.*: Cambridge University Preprint HEP-70-5 (1970).
22. See, for example, *Horn, D., Yahil, A.*: Phys. Rev. **1** D, 2610 (1970).
23. *Chew, G. F., Snider, D. R.*: Phys. Letters **31** B, 75 (1970).
24. *Martin, A.*: Nuovo Cimento **39**, 705 (1965); *Eden, R.*: Phys. Rev. Letters **16**, 39 (1966).

Professor Dr. *Reinhard Oehme*
The Enrico Fermi Institute and the Department of Physics
The University of Chicago, Chicago, Illinois, USA
and
International Centre for Theoretical Physics, Trieste, Italy.

An Introduction to Dual Resonance Models in Multiparticle Physics

H. SATZ

Contents

A. Introduction

The collision of two hadrons at laboratory energies from a few GeV on up leads about four out of five times to the production of additional hadrons, and the average number of produced particles continues to grow with increasing energy. These empirical facts [1] make the theoretical description of multiparticle reactions one of the most important tasks of present-day hadron physics.

Unitarity interrelates elastic and inelastic processes; the strong presence of production reactions thus prevents us from taking the classical road of first "solving" the dynamics in the elastic case, then going on to the more complex many-body problem. In fact, high energy elastic scattering at small momentum transfer appears to be due largely to the shadow of all inelastic reactions, and thus an understanding of multiparticle processes becomes necessary for that of an elastic two-body reaction.

Nevertheless, in many-body problems one still tries to avoid at the beginning the complicated relations provided by unitarity, and thus – hoping for a later "unitarization" – starts generally with a fixed N-body problem uncoupled from other N. Instead of the historical two- and many-body separation, it here turns out to be more fruitful to distinguish between *diffractive* and *non-diffractive* processes (Pomeron and non-

Pomeron exchange); non-diffractive two-body reactions appear more similar to non-diffractive production reactions than to elastic scattering, and inelastic diffraction dissociation is difficult to accommodate in the usual models for non-diffractive production.

Starting from finite energy sum rules, the study of non-diffractive processes has in the past two years led to the fruitful concept of duality [2] and to the development of the dual resonance model [3–8] – a crossing symmetric model which contains in a dual fashion both the low energy resonance and the high energy Regge behaviour. In this framework the inherent similarity of non-diffractive processes finds its expression in the so-called "bootstrap consistency" (factorization), which, for any N, allows the construction of the (N-1) particle amplitude from that for N particles. The multiparticle dual resonance model thus contains, in a way, "more" of the properties one expects of a correct description than any previous model – and the non-trivial nature of such a formulation is perhaps best illustrated by the fact that it is up to now not clear how to extend it from fictitious scalar bosons to physical particles in general.

The subsequent development took two main directions: on the one hand, one tries to use the fact that the narrow width approximation already yields so many of the wanted properties to embed, in a perturbative sense, the Veneziano formula as a Born term in a theory of strong interactions [9]; on the other hand, with complex trajectories as crude unitarization, the dual resonance model provides (at least in principle) a phenomenological model [10–12] incorporating many of the presently known basic empirical features. On a phenomenological level, we also find some attempts to extend duality [13, 14] and dual resonance descriptions [15] to diffractive processes.

Both roads are at present rather blocked; exponential level degeneracy leading to divergent higher order perturbative terms, ghosts, no formulation for fermions, Pomeron questions – there remains more than enough to do. If nothing else, one now knows what the next model has to achieve to be considered as progress.

These lectures are intended as an introduction to the application of dual resonance techniques in multiparticle problems. As mentioned, these applications occur both in the more formal attempts to embed the Veneziano formula in a theory of strong interactions and to obtain a basis for phenomenological considerations. I have therefore attempted to keep the general arguments (Section B) as much as possible separate from the phenomenology (Section C), so that those mostly interested in the formal approach may, at the end of Section B, go on directly to Professor *Fubini*'s lectures. Nevertheless, the twofold purpose will unavoidably have a (hopefully not entirely negative) effect upon the presentation of the subject.

B. The Multiparticle Veneziano Amplitude

1. A Review of B_4 Essentials

To illustrate the properties we would like our multiparticle amplitude to have, as well as to fix notation and terminology, we begin by briefly reviewing the essentials of B_4.

Consider a system of neutral bosons, all states of which lie on a single straight Regge trajectory $\alpha(x)$ or on daughters thereof, with lowest state $J^P = 0^+$ of mass μ

$$\alpha(x) = \alpha(0) + \alpha' x; \quad \alpha(0) = -\mu^2 \alpha'. \tag{B–1.1}$$

The Veneziano amplitude [3] for the non-diffractive reaction $0^+ + 0^+ \to 0^+ + 0^+$ then takes the form (cf. Fig. B–1.1)

$$T_4 = \beta\,[V(1234) + V(2134) + V(2314)] \tag{B–1.2}$$

with constant β and with

$$\begin{aligned} &V(1234) = B_4(-\alpha(s_{12}), -\alpha(s_{23})), \quad \text{etc.} \\ &B_4(x, y) = \int_0^1 \mathrm{d}u\, u^{x-1}(1-u)^{y-1} = \frac{\Gamma(x)\,\Gamma(y)}{\Gamma(x+y)} \end{aligned} \tag{B–1.3}$$

where $s_{ij} = (p_i + p_j)^2$.

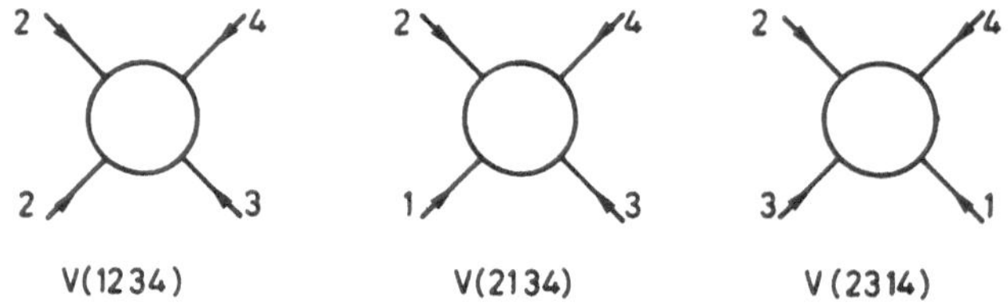

Fig. B–1.1. Graphs for Veneziano amplitude for the reaction $0^+ + 0^+ \to 0^+ + 0^+$

We note in particular:

a) $V(1234)$ is invariant under cyclic and anticyclic permutations of the external lines, since $B_4(x, y)$ remains invariant under the interchange of x and y. The $4! = 24$ permutations of $p_1, \ldots, p_4$ are thus divided into three $(24/(4+4) = 3)$ equivalence classes, each of which leaves invariant one of the terms in (B–1.2) and hence can be represented by it or by the corresponding graph of Fig. B–1.1. Because of this cyclic invariance, the fully crossing symmetric amplitude, instead of being a sum of all 24 permutations, is given by the sum (B–1.2) of the three non-equivalent terms.

b) $B_4(x, y)$ is analytic in x except for poles at $-x=0, 1, 2, \ldots$; because of a), the same is true for the dual variable y. The poles occur since for $-x=0, 1, \ldots$, the integral (B–1.3) diverges at $u=0$, and for $-y=0, 1, \ldots$ at $v=(1-u)=0$; because $u=1$ at $v=0$ and vice versa, there can, by construction, be no double (coincident) poles in x and y. We can thus expand $B_4(x, y)$ as a sum of poles

$$B_4(x, y)=\sum_{k=0}^{\infty} R_k(y)\frac{1}{k+x}, \tag{B–1.4}$$

$$R_k(y)=(-1)^k\binom{y-1}{k} \tag{B–1.5}$$

with the residue at the kth pole in x being a polynomial of degree k in y. The kth term thus corresponds to the "exchange" of a family of particles (parent and daughters) of spin $j \leqq k$. Since by symmetry a) the same holds for a pole expansion in y, we thus have a crossing symmetric resonance saturation picture.

c) In the limit $\alpha(s)=\alpha(s_{12})$ going to infinity [in both $\operatorname{Re}\alpha(s)$ and $\operatorname{Im}\alpha(s)$] with $t=s_{23}$ fixed, we find from (B–1.2) and (B–1.3)

$$T \rightarrow \frac{\pi}{\Gamma(\alpha(t)-1)}\left[\frac{1+e^{i\pi\alpha(t)}}{\sin\pi\alpha(t)}\right][\alpha(s)]^{\alpha(t)} \tag{B–1.6}$$

with analogous results for the limits in the other channels. [The specific approach to infinity with $\operatorname{Im}\alpha(x)$ also diverging is necessary to obtain (B–1.6) and is considered as a "phenomenological unitarization".] Since the amplitude T thus gives, in a crossing symmetric fashion, both low energy resonance and high energy Regge behaviour, it is in accord with what is generally considered as duality.

Summarizing: the essential points in the construction of a dual resonance amplitude for $N=4$ were invariance under cyclic and anticyclic permutations and a (trajectory determined) pole structure without coincident poles in dual variables.

In concluding this part, we make a remark on the uniqueness of the construction. We can replace $B_4(x, y)$ by

$$\begin{aligned}\tilde{B}_4(x, y)&=\int_0^1 du\, f(u)\, u^{x-1}(1-u)^{y-1}\\ &=\sum_{m=0}^{\infty}\sum_{n=0}^{\infty} \beta_{mn} B_4(x+m, y+n)\end{aligned} \tag{B–1.7}$$

and not alter the properties a) to c), provided $f(u)=f(1-u)$ and $f(u)$ is well behaved at $u=0$. The series in (B–1.7) contains $B_4(x, y)$ as first term; the absence of the higher ("satellite") terms in (B–1.3) thus is only a hopeful assumption.

2. The Construction of B_N

We now want to extend this picture to arbitrary N, still considering the system of neutral bosons introduced above. As of B_4, we require of B_N invariance under cyclic and anticyclic permutations of the external lines; the fully crossing symmetric amplitude T_N can then be written as a sum over the $(N-1)!/2$ $(=N!/(N+N))$ non-equivalent permutations P

$$T_N = \beta_N \sum_{\{P\}} B_N(P). \tag{B–2.1}$$

Before going further, we now have to generalize the concept of dual variable. For $N=4$, we had called energy and momentum transfer dual to each other; but equally well we could have defined as dual those Mandelstam variables s_{ij} which in no Feynman tree diagram can have simulataneous poles: this in fact makes s_{12}, s_{23} and s_{13} $(s, t$ and $u)$ mutually dual. With

$$s_{ij} = (p_i + p_{i+1} + \cdots + p_j)^2 \qquad i<j \tag{B–2.2}$$

the latter definition can be extended directly to general N: s_{ij} and s_{kl} are dual if no Feynman tree diagram with poles in both can be constructed. In practice, the labelling problem is easily solved using "dual diagrams"; associate with each ordered N point function $B_N(P)$ an N sided polygon $Q_N(P)$, with external lines of $B_N(P)$ corresponding to sides of $Q_N(P)$ (cf., Fig. B–2.1). All possible Mandelstam variables are then in a one-to-one correspondence with the diagonals linking non-adjacent vertices of Q_N, the diagonal (1,3) being associated with $s_{13} = (p_1 + p_2 + p_3)^2$, etc. All *intersecting* diagonals of Q_N are now dual to each other, just as the two diagonals of the square for $N=4$. For each ordering P there remain $(N-3)$ non-intersecting, and hence mutually non-dual, variables which can have coincident poles. For these, any $(N-3)$ non-intersecting variables can be chosen (cf., Figs. B–2.2 and B–2.3); the "multiperipheral" form shown in Fig. B–2.3 is, however, particularly easy to visualize and the set s_{1j}, $j=2, 3, \ldots, (N-2)$ is therefore often used. We return to this freedom of choice in subsection 4.

Introducing for each s_{ij} a trajectory

$$x_{ij} = \alpha(s_{ij}) = \alpha(0) + \alpha' s_{ij} \qquad i<j \tag{B–2.3}$$

and its conjugate variable u_{ij}, the essential problem now becomes quite clear: to write down an integral over the $(N-3)$ mutually non-dual u_{ij} such that we have: i) poles as fixed by $\alpha(x)$ for all corresponding x_{ij}; ii) symmetry under cyclic and anticyclic permutations of the external lines, and iii) no coincident poles in dual variables.

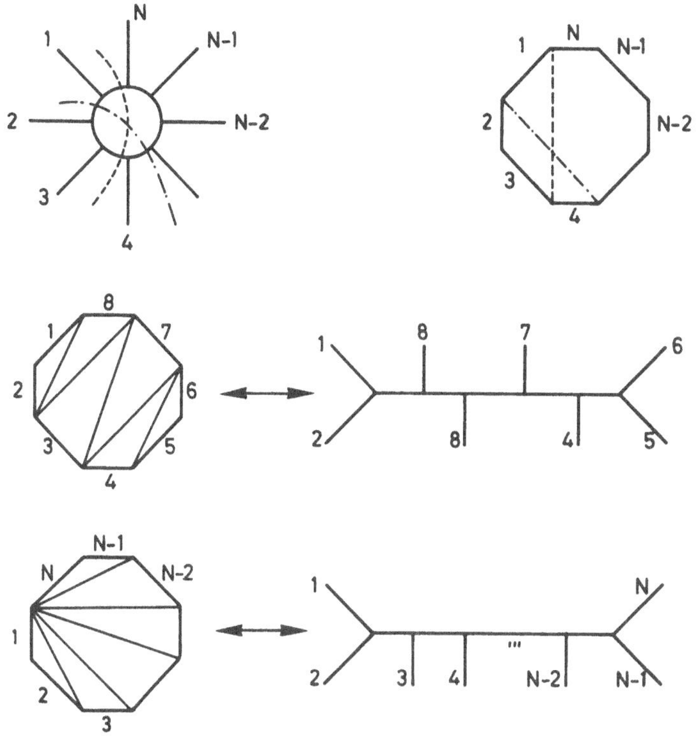

Fig. B–2.1, 2.2, 2.3. Dual diagrams

Let us try to develop a pattern of solutions. We had

$$B_4(x_{12},x_{23})=\int_0^1 \mathrm{d}u_{12}\int_0^1 \mathrm{d}u_{23}\, u_{12}^{x_{12}-1}u_{23}^{x_{23}-1}\delta(u_{12}+u_{23}-1) \qquad \text{(B–2.4)}$$

where the auxiliary variable u_{23} dual to u_{12} is fixed by forbidding coincident poles. Similarly we can write

$$B_5(x_{12},x_{23},x_{34},x_{45},x_{51})=\int_0^1 \mathrm{d}u_{12}\int_0^1 \mathrm{d}u_{23}\int_0^1 \mathrm{d}u_{34}\int_0^1 \mathrm{d}u_{45}\int_0^1 \mathrm{d}u_{51}\, *$$

$$*\, u_{12}^{x_{12}-1}u_{23}^{x_{23}-1}u_{34}^{x_{34}-1}u_{45}^{x_{45}-1}u_{51}^{x_{51}-1}G(u_{12},u_{23},u_{34},u_{45},u_{51}) \qquad \text{(B–2.5)}$$

where $G(u_{12},\dots,u_{51})$ now has to express the absence of poles in dual variables in such a way as to render B_5 invariant under cyclic and anticyclic permutations of external lines. We thus choose a particular set of the $N-3=2$ mutually non-dual u_{ij}, say u_{12} and $u_{13}=u_{45}$ (from the corresponding polygon it is easy to see that all $u_{i,i+1}$ with a common index are dual to each other). The three auxiliary variables, here u_{23}, u_{34} and u_{51}, are then in fact fixed by the requirement of no double poles in dual variables: when $u_{12}=0$ (pole in x_{12}), we must have u_{51} and $u_{23} \neq 0$, etc. The resulting conditions are fulfilled if

$$u_{12}=1-u_{51}u_{23}, \tag{B–2.6a}$$

$$u_{23}=1-u_{12}u_{34}, \tag{B–2.6b}$$

$$u_{34}=1-u_{23}u_{45}, \tag{B–2.6c}$$

$$u_{45}=1-u_{34}u_{51}, \tag{B–2.6d}$$

$$u_{51}=1-u_{45}u_{12}. \tag{B–2.6e}$$

Of these relations, only three are independent; for our choice of u_{12} and u_{45} as independent variables, we find that (B–2.6b), (B–2.6c) and (B–2.6e) alone give

$$u_{23}=(1-u_{12})/(1-u_{12}u_{45}), \tag{B–2.7a}$$

$$u_{34}=(1-u_{45})/(1-u_{12}u_{45}), \tag{B–2.7b}$$

$$u_{51}=(1-u_{45}u_{12}) \tag{B–2.7c}$$

and therewith solve, as is easily verified, all five relations (B–2.6). Hence we find as solution:

$$\begin{aligned} &G(u_{12}, u_{23}, \ldots, u_{51}) \\ &\quad =\delta(u_{23}+u_{12}u_{34}-1)\,\delta(u_{34}+u_{23}u_{45}-1)\,\delta(u_{51}+u_{45}u_{12}-1) \end{aligned} \tag{B–2.8}$$

and with it the Bardakçi-Ruegg-Virasoro form [4, 5] of B_5

$$\begin{aligned} &B_5(12345) \\ &=\int_0^1 \mathrm{d}u_{12}\,\mathrm{d}u_{45}\, u_{12}^{x_{12}-1} u_{45}^{x_{45}-1}(1-u_{12})^{x_{23}-1}(1-u_{45})^{x_{34}-1}(1-u_{12}u_{45})^{x_{51}-x_{23}-x_{34}} \end{aligned} \tag{B–2.9}$$

which was the first generalization of the Veneziano amplitude to $N>4$.

Let us now denote $x_{12}, \ldots, x_{51}$ and $u_{12}, \ldots, u_{51}$ by x_k and u_k, with $k=1, \ldots, 5$; we can then write

$$B_5(x_1, \ldots, x_5)=\int_0^1 \prod_{k=1}^{5} \mathrm{d}u_k\, u_k^{x_k-1} \prod_{\bar{k}=1}^{5}{}' \delta(u_{\bar{k}}+u_{\bar{k}-1}u_{\bar{k}+1}-1) \tag{B–2.10}$$

where we take $u_0=u_5$, $u_6=u_1$ for compact notation. The second (primed) product in the integral is defined to run over all u_k except the two mutually non-dual (and hence non-adjacent) chosen as independent variables. The argument of the delta-function is of the form "variable plus product of all dual variables minus one". This formulation, the Chan form [6] of B_5, clearly exhibits both the invariance under cyclic and anticyclic permutations as well as the absence of double poles in dual variables.

From (B–2.10) we can now obtain in a straightforward fashion the corresponding Chan form for arbitrary N [6, 7]. As for $N=4$ and 5, we write an integral over all $R_N = N(N-3)/2$ conjugate Mandelstam variables of the problem, denoted by u_k, $k=1, 2, \ldots, R_N$:

$$B_N(x_1, \ldots, x_{R_N}) = \int_0^1 \prod_{k=1}^{R_N} \mathrm{d}u_k \, u_k^{x_k - 1} \left\{ \prod_{\bar{k}=1}^{R_N}{}' \delta\left(u_{\bar{k}} + \prod_{\bar{k}=1}^{R_N}{}'' u_{\bar{k}} - 1\right)\right\} \tag{B–2.11}$$

where the primed product runs over all $\bar{k}$ except the $N-3$ mutually nondual independent variables (whichever one we have chosen), and the doubly primed over all variables dual to $\bar{k}$. As before, the $u_k^{x_k-1}$ bring in the pole structure, while the product of delta functions enforces the absence of coincident poles in dual variables, thereby determining the $(N-2)(N-3)/2$ auxiliary variables. The invariance under cyclic and anticyclic permutations is assured by the invariance of the duality relations under these operations.

To obtain a specific representation, one can now carry out the integrations over all delta-functions. Choosing as independent variables the $(N-3)$ multiperipheral u_{1j}, $j=2, 3, \ldots, N-2$ and defining

$$\omega_{qr} = u_{1,q} u_{1,q+1} \ldots u_{1,r} \qquad q<r \tag{B–2.12}$$

we arrive at the generalized Bardakçi-Ruegg form [4]

$$\begin{aligned} B_N(1\ldots N) = \int_0^1 & \mathrm{d}u_{12} \mathrm{d}u_{13} \ldots \mathrm{d}u_{1,N-2} u_{12}^{x_{12}-1} \ldots \\ & \ldots u_{1,N-2}^{x_{1,N-2}-1} * (1-u_{12})^{x_{23}-1} * (1-u_{13})^{x_{34}-1} \ldots \\ & \ldots (1-u_{1,N-2})^{x_{N-2,N-1}-1} * (1-\omega_{23})^{-2(p_2 p_4)\alpha' + \alpha_0 + \alpha'} * \ldots \\ & \ldots (1-\omega_{N-3,N-2})^{-2(p_{N-3}\, p_{N-1})\alpha' + \alpha_0 + \alpha'} * (1-\omega_{24})^{-2(p_2 p_5)\alpha'} \ldots \\ & \ldots (1-\omega_{N-4,N-2})^{-2(p_{N-4}\, p_{N-1})\alpha'} * \cdots * (1-\omega_{2,N-2})^{-2(p_2\, p_{N-1})\alpha'} \end{aligned} \tag{B–2.13}$$

which has been used extensively in the first investigations of level structure [16, 17]; we shall come to this in the next subsection, and return to the question of other equivalent representations of B_N in subscction 4.

3. Some Properties of B_N

Let us consider the simpler case $N=5$. Expanding the Bardakçi-Ruegg form (B–2.9) about $u_{12} u_{45} = 0$, we obtain after integration [18]

$$\begin{aligned} & B_5(x_{12}, \ldots, x_{51}) \\ & = \sum_{k=0}^{\infty} (-1)^k \binom{x_{51} - x_{23} - x_{34}}{k} B_4(x_{12}+k, x_{23})\, B_4(x_{45}+k, x_{34}) \end{aligned} \tag{B–3.1}$$

which, incidentally, is the form usually used for computer calculations of B_5. For $x_{12} \to 0$ it yields

$$B_5(x_{12}, \dots, x_{51})$$
$$= \frac{1}{x_{12}} \left[\sum_{k=0}^{\infty} (-1)^k \binom{x_{51}-x_{23}-x_{34}}{k} B_4(x_{45}+k, x_{34}) \frac{x_{12}}{x_{12}+k} \right.$$
$$\left. * \frac{\Gamma(x_{12}+k+1)\,\Gamma(x_{23})}{\Gamma(x_{12}+x_{23}+k)} \right]_{x_{12}=0} = \frac{1}{x_{12}} B_4(x_{45}, x_{34}) \tag{B–3.2}$$

i.e., by restricting ourselves to the first pole of any variable, we regain the Veneziano four-point formula (cf., Fig. B–3.1). Thus we first encounter here the bootstrap consistency mentioned in the Introduction: if we know the amplitude for $N=5$, it is uniquely fixed for $N=4$ as residue of the first pole.

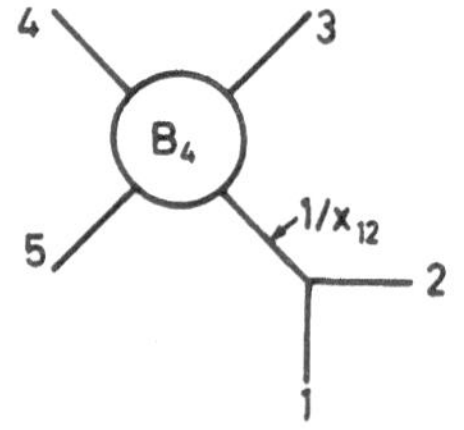

Fig. B–3.1. Relation between B_5 and B_4

In general we have with $x_{12} \to -N$

$$B_5(x_{12}, \dots, x_{51})$$
$$= \frac{1}{x_{12}+N} \sum_{k=0}^{N} \binom{x_{51}-x_{23}-x_{34}}{k} \binom{x_{23}-1}{N-k} (-1)^N B_4(x_{45}+k, x_{34}) \tag{B–3.3}$$

and can thus write analogously to (B–1.4):

$$B_5(x_{12}, \dots, x_{51}) = \sum_{k=0}^{\infty} C_k(x_{23}, \dots, x_{51}) \frac{1}{x_{12}+k} \tag{B–3.4}$$

We see from (B–3.3) that the residues $C_k(x_{23}, \dots, x_{51})$ are polynomials of degree k in the "angle" variables x_{51} and x_{23} (cf., Fig. B–3.1), and thus the kth pole corresponds, as in (B–1.5), to the exchange of a family of particles with spins from k down to zero (mother plus daughters).

By requiring bootstrap consistency in general (not only for the first poles or, equivalently, scalar external particles), we can obtain through factorization from (B–3.1) the Veneziano amplitude for three external particles of spin zero and one of spin j. For a vector and three scalar particles, we write (B–3.1)

$$[p^{\mu}_{(1)} - p^{\mu}_{(2)}]\, B^{(4)}_{\mu}(p_3, p_4, p_5) \tag{B–3.5}$$

to find for the ordering of Fig. B–3.1 through antisymmetrization in 1 and 2

$$B^{(4)}_{\mu}(p_3, p_4, p_5) = [p^{(3)}_{\mu} + p^{(5)}_{\mu}]\, B_4(x_{45}+1, x_{34}) \tag{B–3.6}$$

and, similarly for other orderings. This then yields

$$\begin{aligned} T_\mu &\equiv p^{(3)}_{\mu} A_1 + p^{(4)}_{\mu} A_2 + p^{(5)}_{\mu} A_3 \\ &= p^{(3)}_{\mu}[B_4(x_{34}, x_{45}) + B_4(x_{35}, x_{45})] + p^{(4)}_{\mu}[B_4(x_{34}, x_{35}) \\ &\quad + B_4(x_{35}, x_{45})] + p^{(5)}_{\mu}[B_4(x_{34}, x_{35}) + B_4(x_{34}, x_{45})] \end{aligned} \tag{B–3.7}$$

as the full Veneziano amplitude to be contracted with the polarization vector ε_μ of the external vector particle.

Finally, let us look at the various "high energy" limits of B_5. Using standard formulae for beta and gamma functions, one can rewrite (B–3.1) in the form [19]

$$\begin{aligned} &B_5(12345) \\ &= B_4(12, 51)\, B_4(23, 34)\, {}_3F_2(-45+12+23, 51, 34; 12+51, 23+34; 1) \end{aligned} \tag{B–3.8}$$

where ij is used to denote x_{ij} and

$$\begin{aligned} &{}_3F_2(\alpha_1, \alpha_2, \alpha_3; \beta_1 \beta_2; z) \\ &= \sum_{n=0}^{\infty} \frac{\Gamma(\alpha_1+n)}{\Gamma(\alpha_1)} \frac{\Gamma(\alpha_2+n)}{\Gamma(\alpha_2)} \frac{\Gamma(\alpha_3+n)}{\Gamma(\alpha_3)} \frac{\Gamma(\beta_1)}{\Gamma(\beta_1+n)} \frac{\Gamma(\beta_2)}{\Gamma(\beta_2+n)} \frac{z^n}{n!} \end{aligned} \tag{B–3.9}$$

is a generalized hypergeometric function. Using its limiting properties, one obtains for x_{45} and $x_{12} \to \infty$, x_{23} and $\eta = x_{12}/x_{45}$ fixed ("single Regge limit", cf., Fig. B–3.2)

$$B_5(1 \ldots 5) = B_4(45, 51)\, R(23, 34)\,. \tag{B–3.10}$$

Here $B_4(45, 51)$ is the Veneziano amplitude for the reaction $4 + 5 \to (23) + 1$ [which, as we know, gives the usual $(x_{45})^{x_{51}} * f(x_{51})$ for large x_{45}] while

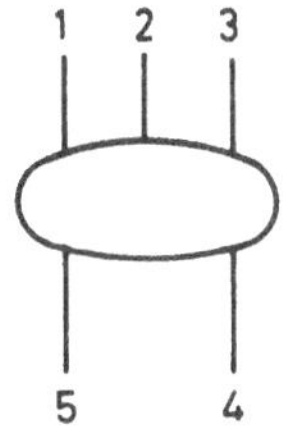

Fig. B–3.2. Graph for $2 \to 3$ reaction

the vertex function

$$R(23, 34) = \sum_{m=0}^{\infty} \frac{c_m(34, \eta)}{x_{23} + m} \qquad \text{(B–3.11)}$$

is expressed as a sum over resonances in the (23) system. We thus find in fact the correct behaviour in the single Regge limit. For x_{45}, x_{12} and $x_{23} \to \infty$ with $\varkappa = x_{12} x_{23}/x_{45}$ fixed ("double Regge limit") one finds in a similar fashion

$$\begin{aligned} B_5(1 \ldots 5) &= B_4(45, 51)\, B_4(23, 34-51)\, f(34, 51, \varkappa) \\ &\simeq x_{12}^{x_{51}}\, x_{23}^{x_{34}}\, \tilde{f}(34, 51, \varkappa) \end{aligned} \qquad \text{(B–3.12)}$$

and thus the usual double Regge form. It should be noted that both Regge limits of B_5 introduce a well-defined dependence on the Toller angle ($\sim \eta, \varkappa$); for details, see Ref. [19].

Thus our B_5 does have the essential properties we had wanted to generalize from B_4: besides the dual pole structure, we find residues polynomial in angle variables and thus correct spin structure, factorization and thus bootstrap consistency, and finally high energy Regge behaviour. This all remains true for arbitrary N; rather than repeating it here for the general case, we refer the reader to Ref. [R 1] and the original literature quoted there. Here we shall restrict ourselves to a rather qualitative introduction to some aspects of factorization.

We had seen for B_5 that at the first pole in any $x_{i,i+1}$ the residue factorizes into a part depending only on particles i and $i+1$ (here simply a constant) and a part depending on the other three particles (namely B_4). How does this look in general at higher poles? Consider the residue of B_N for arbitrary N at the pole $\alpha(s_{1i}) = -n$, $s_{1i} = (p_1 + \cdots + p_i)^2$ (cf.,

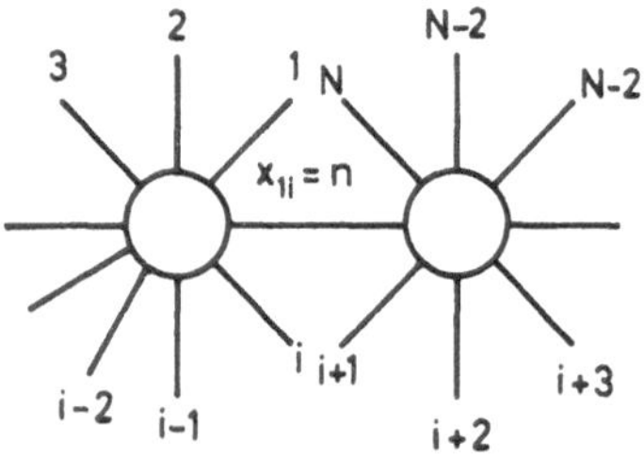

Fig. B–3.3. B_N at the pole $\alpha(s_{1i}) = -n$

Fig. B–3.3). The question is: can this residue be written as the sum of no less than $z(n)$ terms, with $z(n)$ *independent of N and* i, such that each term is the product of one factor depending only on the momenta of the external particles $1, \ldots, i$ and one depending on $i+1, \ldots, N$. If this is indeed the case, we can speak of factorization with $z(n)$-fold degenerate level structure.

It has been shown [16, 17]: a) B_N does in fact provide such a factorization; b) the number of degenerate levels grows as

$$z(n) \sim \exp a\sqrt{n} \tag{B–3.13}$$

i.e., the number of states at fixed large n grows exponentially with $\sqrt{n}$; since the number of spin states contributing to each n grows only linearly in n, the same is true generally for a spin j part of pole n; c) there are in general ghosts, that is, in the sum corresponding to the factorized residue there are negative terms (negative coupling). It should be noted that both b) and c) are not true for the parent trajectory – its contributions are ghost-free and non-degenerate.

The further discussion of these questions forms an important part of the perturbative approach [9, R 2–R 4] to a Veneziano "theory" and will presumably be considered in Professor *Fubini*'s lectures. Let me just emphasize here that both factorization and the result (B–3.13) are first instances of some type of unitarity-like relation between amplitudes of different particle numbers, e.g., between B_4 and B_N: factorization requires B_N to be of such a form as to yield B_4 as residue of the proper pole, while the level degeneracy of a four-point amplitude can be determined *only* by studying factorization of the amplitude for arbitrary N.

In closing this Section, we remark that the satellite ambiguity found for $N=4$ also persists at higher N, although factorization here imposes some further restrictions [20] on the weight corresponding to $f(u)$ in (B–1.7).

4. Koba-Nielsen Form and Duality Transformations

Let us for the moment return to the four-point function

$$B_4(x_{12}, x_{23}) = \int_0^1 \mathrm{d}u_{12}\, u_{12}^{x_{12}-1}(1-u_{12})^{x_{23}-1}\,. \tag{B–4.1}$$

We had suggested that the essential properties of this amplitude were: a) invariance under cyclic and anticyclic permutations and, b) pole structure without coincident poles in dual variables.

Consider now the function

$$\tilde{B}_4(x_{12}, x_{23}) = \int_0^1 \mathrm{d}u_{12}\, \mathrm{e}^{u_{12}(x_{12}+x_{23})}\,. \tag{B–4.2}$$

It is invariant under cyclic and anticyclic permutations of the external lines, but not dual: it allows in neither x_{12} nor x_{23} a pole expansion with residues polynomial in the dual variables, does not give Regge behaviour, etc. On the other hand

$$\tilde{B}_4(x_{12}, x_{23}) = \int_0^1 d u_{12}\, u_{12}^{x_{12}-1}(1-u_{12})^{x_{23}-1} f(u_{12}) \qquad \text{(B–4.3)}$$

with $f(u) \neq f(1-u)$ is dual in the above sense, yet not invariant under cyclic and anticyclic permutations of the external lines. Hence these properties are independent.

The invariance of (B–4.1) under the permutation corresponding to $x_{12} \rightleftarrows x_{23}$ is assured by the invariance of the integrand under the transformation $u \rightleftarrows (1-u)$. Similarly, all other permutations can be associated with transformations of the integration variable [R 4]. This connects cyclic invariance and thus finally crossing symmetry with the invariance of the integrand under transformations of a permutation group.

The second property ("duality") is satisfied since u_{23}, the conjugate variable of x_{23}, is with

$$u_{23} = 1 - u_{12} \qquad \text{("duality condition")} \qquad \text{(B–4.4)}$$

chosen such that it is unity when u_{12} vanishes. We can now ask whether it is possible to associate this property also with an invariance under some group.

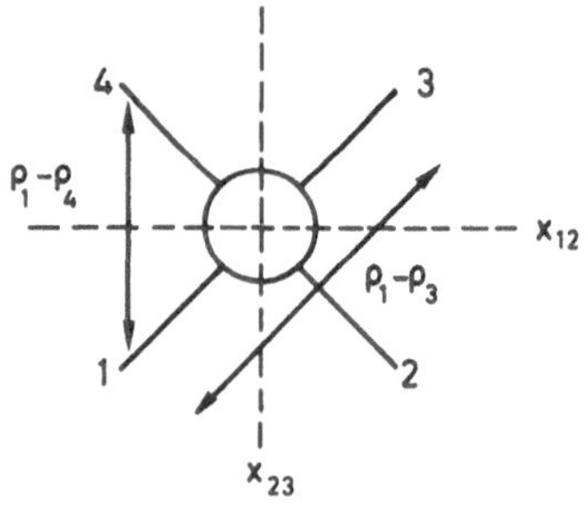

Fig. B–4.1. Koba-Nielsen form of B_4

This question finds at least a partial answer when one rewrites B_4 in terms of anharmonic ratios (Koba-Nielsen form [8]). We associate with the cut x_{23} of Fig. B–4.1 the anarmonic ratio

$$Z(23, 41) = \frac{\varrho_1 - \varrho_4}{\varrho_1 - \varrho_3} * \frac{\varrho_2 - \varrho_3}{\varrho_2 - \varrho_4} \qquad \text{(B–4.5)}$$

where each ϱ_i is an arbitrary real number associated with external particle i; in the numerator of (B–4.5) appears the difference of pairs on each side of the cut, in the denominator the "crossed" differences as seen from top or bottom.

The anharmonic ratio corresponding to the cut x_{12} then is

$$z(12,34)=\frac{\varrho_2-\varrho_1}{\varrho_2-\varrho_4}*\frac{\varrho_3-\varrho_4}{\varrho_3-\varrho_1}=1-z(23,41) \qquad \text{(B–4.6)}$$

and thus $z(12,34)$ and $z(23,41)$ fulfil the duality relation for the conjugate variables of x_{12} and x_{23}. We can therefore write

$$B_4(x_{12},x_{23})=\int_0^1 \mathrm{d}z(12,34)\,[z(12,34)]^{x_{12}-1}\,[z(23,41)]^{x_{23}-1}\,. \qquad \text{(B–4.7)}$$

Now, it can easily be shown that the anharmonic ratios are invariant under projective transformations, i.e.,

$$z(12,34)\equiv z(\varrho_1\varrho_2;\varrho_3\varrho_4)=z(\varrho_1'\varrho_2';\varrho_3'\varrho_4') \qquad \text{(B–4.8)}$$

with

$$\varrho_i'=\frac{a\varrho_i+b}{c\varrho_i+d}\,;\quad ad-bc\neq 0\,. \qquad \text{(B–4.9)}$$

Since, on the other hand, anharmonic ratios form in the Veneziano model the solution of the "duality condition", the group $L(2,R)$ of projective transformations appears to be closely related to what we consider as duality. The study of these "duality transformations" thus forms a rather central part of much present work on the formal aspects of the Veneziano model and may perhaps lead to a better understanding of what is really meant by duality.

The requirement $0\leqq z(12,34)\leqq 1$ for the region of integration implies an ordering for the ϱ_i; we must have either $\varrho_1<\varrho_2<\varrho_3<\varrho_4$ or $\varrho_4<\varrho_3<\varrho_2<\varrho_1$. Now one can always find a projective transformation which takes any given $\{\varrho_i\}$ of this order into $\varrho_1=0$, $\varrho_3=1$, $\varrho_4=\infty$; hence we can write

$$D_4(x_{12},x_{23})=\int_0^1 \mathrm{d}\varrho_2\,[z(12,34)]^{x_{12}-1}\,[z(23,41)]^{x_{23}-1}\,. \qquad \text{(B–4.10)}$$

Another, formally more general form, would be

$$B_4(x_{12},x_{23})=\frac{1}{V}\int_{-\infty}^{\infty}\prod_{k=1}^{4}\mathrm{d}\varrho_k\theta(\varrho_k-\varrho_{k-1})\,[z(12,34)]^{x_{12}-1}\,[z(23,41)]^{x_{23}-1}\,. \qquad \text{(B–4.11)}$$

Because of the invariance of anharmonic ratios under $L(2,R)$ this implies, however, an infinite normalization volume V.

In closing this subsection, we briefly sketch how the general N particle duality condition

$$\Big(u_p+\prod_{\bar p}{}'' u_{\bar p}-1\Big)=0 \qquad \text{(B–4.12)}$$

[cf., (B–2.11)] is also solved by anharmonic ratios. For this, we associate with an x_{ij} cut (cf., Fig. B–4.2) the anharmonic ratio

$$u_{ij} \equiv z(i+1, j-1; j, i) = \frac{\varrho_i - \varrho_j}{\varrho_i - \varrho_{j-1}} * \frac{\varrho_{i+1} - \varrho_{j-1}}{\varrho_{i+1} - \varrho_j} \qquad \text{(B–4.13)}$$

and obtain from it a factor $(u_{i,j})^{x_{i,j}-1}$ in the integrand of B_N. As before, a pole in $x_{i,j}$ will result from $u_{i,j}=0$. From (B–4.13) this means either $\varrho_i = \varrho_j$ or $\varrho_{i+1} = \varrho_{j-1}$. A generalization of the $\varrho_i > \varrho_{i-1}$ ordering found for $N=4$ then requires $\varrho_i = \varrho_j = \varrho_r$, $i \leqq r \leqq j$. This means for a dual variable, e.g., $x_{i+1,j+1}$ (cf., Fig. B–4.2) that

$$u_{i+1,j+1} = \frac{\varrho_{i+1} - \varrho_{j+1}}{\varrho_{i+1} - \varrho_j} * \frac{\varrho_{i+2} - \varrho_j}{\varrho_{i+2} - \varrho_{j+1}} = 1 \qquad \text{(B–4.14)}$$

and hence prevents any simultaneous poles in variables dual to $x_{i,j}$. The $u_{i,j}$ defined by (B–4.13) thus fulfil by construction the N particle duality condition.

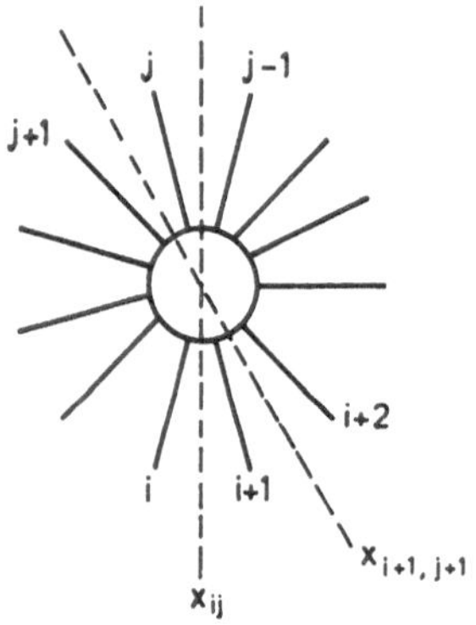

Fig. B–4.2. N-particle Koba-Nielsen form

For the further development of the N particle Koba-Nielsen form (construction of the integration measure), we refer the reader to Refs. [8, 21].

5. Unitarity and Other Questions

The most striking shortcoming of the Veneziano amplitude is that as narrow resonance approximation it violates unitarity already in giving rise to unbounded partial waves. Essentially three different approaches towards an improvement of the unitarity situation for N point functions have so far been suggested.

a) "Complex trajectories": this is what is used in Veneziano phenomenology – one chooses the trajectory $\alpha(x)$ complex when x is an energy, with $\operatorname{Im}\alpha(x)$ increasing with x and determined such as to reproduce correctly all observed resonance widths. This trick of course takes the poles off the real axis in the physical region and at the same time assures correct high energy Regge behaviour.

b) "Smoothing Veneziano": one can consider B_4 as a distribution to be smeared out by convolution with a suitable weight function ϕ to obtain, e.g., [22]

$$W_4(x_{12}, x_{23}) = \int_{\varkappa_0}^{1} \mathrm{d}\varkappa\, \phi(\varkappa)\, B_4(\varkappa y_{12}, \varkappa y_{23})$$

$$y_{12} = \alpha(s_{12} - 4/3) = x_{12} - 4\alpha'/3, \quad \text{etc.} \qquad \text{(B–5.1)}$$

as new four-point amplitude. While this procedure also moves the poles out of the physical region, it modifies the high energy Regge behaviour.

Both a) and b) are essentially phenomenological approaches in the sense that although they make the partial waves finite, they do not indicate how the actual unitarity condition expressing the interrelation of amplitudes for all particle numbers is to be solved.

c) "Perturbative unitarization": this much more ambitious approach considers, as pointed out in the Introduction, the Veneziano amplitude as a skeleton theory in the Born term sense [9], for which one should now calculate self-energy-like corrections – which should then in principle provide both finite width resonances and the right unitarity coupling of all channels. Whether or not this can be done, appears at present completely open.

In summarizing, it can be noted that unitarity remains perhaps the most fundamental open problem in the multiparticle Veneziano model – just as it was in previous attempts (multiperipheral model, multi-Regge model).

The other problems we want to consider here are concerned with the extension of the dual resonance picture to physical particles: the fermion question and the problem arising from spin one lowest particle states.

So far, no (fully satisfactory) formulation of a Veneziano model for a reaction involving fermions has been given (e.g., for pion-nucleon scattering). One can, of course, take a Veneziano form for an invariant amplitude free of kinematical singularities, similar to what is done in a Regge picture. While this, in fact, leads to the desired pole structure, the straight line trajectories in s, t and u give rise to an amplitude invariant under change of sign of the energy $W = \sqrt{s}$ and thus by *MacDowell* symmetry [23] to parity doubling. On the other hand, trajectories of the form $\alpha(x) = \alpha_0 + \alpha' x^{\nu}$ with $\nu < 1$ no longer give rise to residues polynomial in the dual variable and hence destroy the spin structure ("ancestors"); they also modify the high energy Regge behaviour.

Finally we want to illustrate, by considering a specific example, the type of difficulties that arise when one tries to extend the B_N constructed above even to physical bosons. Let us look at the amplitude for the $\sigma 4\pi$ system [24–26], where σ denotes a $J^P = 0^+$ boson; the relevant trajectories are shown in Fig. B–5.1. If we take $B_5(\alpha_\pi^{12}, \alpha_\varrho^{23}, \alpha_\varrho^{34}, \alpha_\varrho^{45}, \alpha_\pi^{51})$ as Veneziano form for this amplitude, with $\alpha\varrho(x) = \alpha_0 + \alpha x$, $\alpha_0 = 1 - \alpha' m_\varrho^2$ and with physically reasonable values of α' and m_ϱ, then we introduce a spurious state of negative mass at $\alpha_\varrho(x) = 0$. If we try to circumvent this by starting the rho trajectory at one, $B_5(\alpha_\pi^{12}, \alpha_\varrho^{23} - 1, \alpha_\varrho^{34} - 1, \alpha_\varrho^{45} - 1, \alpha_\pi^{51})$, then we have to multiply, e.g., by α_π^{12} to obtain indeed spin one at the $\alpha_\varrho^{23} = 1$ pole; but then the $\alpha_\pi^{51} = 0$ pion pole is not of spin zero, as it should be.

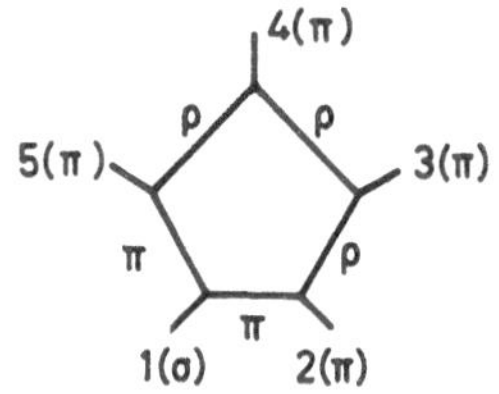

Fig. B–5.1. Amplitude for the $\sigma\, 4\pi$ system

A possible solution is given by [24–26]: $\alpha_\varrho^{34} B_5(\alpha_\pi^{12}, \alpha_\varrho^{23} - 1, \alpha_\varrho^{34}, \alpha_\varrho^{45} - 1, \alpha_\pi^{51})$, since the multiplicative factor kills the $\alpha_\varrho^{34} = 0$ ghost and provides in fact the correct spin for all the rho poles without altering that of the pion poles.

We can see from this, and similar work on N pion amplitudes [27], that even the construction of the Veneziano N point function for general physical bosons is quite far from being a trivial extension of (B–2.11).

C. Multiparticle Veneziano Phenomenology

1. General Aspects

One of the main reasons for the great interest in the Veneziano model is, as already mentioned, that by introducing a crude "unitarization" in form of complex trajectories, one arrives at a model in principle directly applicable to the description of experiments – a model which contains low energy resonance and high energy Regge behaviour in a dual (and crossing symmetric) picture and thus exhibits already many features that one would want to find in a theoretical description of production processes.

The question then becomes: applicable to what reactions, and with what further approximations? That some further ones are necessary is evident: all measurable hadron processes involve fermions, for which, as we saw above, no Veneziano model free from parity doubling has so far been constructed. So we know there will be *spin approximations.*

Furthermore, we have the problem of how to deal with diffractive processes in such a picture, i.e., how to combine *dual resonance models* and *Pomeron exchange.* The reason why this question is of such considerable importance in the actual analysis of reactions is that there are very few processes in which Pomeron exchange contributions can be excluded; and these are not the most common or the best measured ones.

That the Pomeron plays a different role in any dual picture is clear from the fact that it is also dominantly present in reactions which, to our present knowledge, do not contain direct channel resonances (e.g., elastic PP or K^+P scattering) – so it cannot be built up from resonances as are the normal Regge trajectories. There are attempts [13, 14] to incorporate the Pomeron in a dual scheme by relating it to non-resonant background, e.g., in FESR form

$$\begin{aligned}
&\int_{\nu_0}^{N} \mathrm{d}\nu \, \mathrm{Im}\, A(\nu,t) = \int_{\nu_0}^{N} \mathrm{d}\nu \, [\mathrm{Im}\, A_{\mathrm{Res}}(\nu,t) + \mathrm{Im}\, A_{\mathrm{Bg}}(\nu,t)] \,, \\
&\int_{\nu_0}^{N} \mathrm{d}\nu \, \mathrm{Im}\, A_{\mathrm{Res}}(\nu,t) \sim \beta(t)\, N^{\alpha_x(t)+1} \,, \qquad \text{(C–1.1)} \\
&\int_{\nu_0}^{N} \mathrm{d}\nu \, \mathrm{Im}\, A_{\mathrm{Bg}}(\nu,t) \sim \gamma(t)\, N^{\alpha_{\mathbb{P}}(t)+1}
\end{aligned}$$

with α_x and $\alpha_{\mathbb{P}}$ denoting normal and Pomeron trajectory, respectively. This conjecture is in accord with the approximate constancy of total PP and K^+P cross-sections, with reality properties of some amplitudes at high energies, etc.; nevertheless, it is more a reformulation of where one should expect resonance saturation to work, more a restriction on the "usual" duality picture, than an actual tool in the analysis of Pomeron dominated reactions, the non-resonant background being a rather elusive thing. Moreover, it has recently been shown that in a crossing symmetric picture, the duality between Pomeron and non-resonant background is not compatible with resonance saturation [28]; if non-resonant background is present in Pomeron channels, crossing puts it also in channels where no Pomeron is possible, i.e., channels which should be resonance saturated.

So we find that up to now the role of the Pomeron in the duality picture remains rather mysterious. In the remainder of this more general part we want to consider the consequences of the absence of exotic states and then the extension of B_N to particles with isospin.

By exotic states we mean mesons which cannot be accommodated in the usual $SU(3)$ classification scheme [mesons as $SU(3)$ singlets or octets and baryons as singlets, octets, or decuplets] – or, in other words, mesons or baryons that cannot be represented as quark-antiquark ($q\bar{q}$) or three quark (qqq) system, respectively. The fact that such states have so far not been observed leads us to postulate their absence in all channels and derive from this conditions on non-diffractive scattering amplitudes [29, 30]. Since in a resonance saturation picture the imaginary part of the amplitude is built up from direct channel resonances, the amplitude must be purely real if the direct channel is exotic. Since the real part of the amplitude is built up from the resonances in crossed channels, the amplitude must vanish if these are also exotic.

For actual checks, the so-called duality diagrams of *Harari* [29] and *Rosner* [30] prove to be quite helpful. One represents each particle in the process by its quark picture (oppositely directed arrows for q and $\bar{q}$). If the reaction can be drawn such that: i) no quark lines cross, ii) no lines start and end at the same particle, and iii) any "cut" results in a permissible quark state ($q\bar{q}, qqq, \bar{q}\bar{q}\bar{q}$), then the corresponding permutation contributes to the amplitude, otherwise not (cf., Fig. C–1.1). For the above reaction $K^+N \to K^0P$, we find thus as only possible diagram the one for $K^-P \to \bar{K}^0N$ (cf., Fig. C–1.2), i.e., the one obtained through $s-u$ crossing. Similarly one can in higher particle number amplitudes exclude such permutations as shown in Fig. C–1.3; as is evident from the duality diagram, all orderings are excluded in which K and $\bar{K}$ are not adjacent. We shall see further down how this is used to greatly reduce the number of permutations in the analysis of three-body final states.

Let us consider now the question of how to bring isospin into the Veneziano picture, that is, we want to determine an isospin projection coefficient $C_N(P)$ multiplying the $B_N(P)$ corresponding to permutation P of external particles

$$T_N = \sum_{\{P\}} C_N(P)\, B_N(P) \qquad \text{(C–1.2)}$$

such that: i) $C_N(P)\, B_N(P)$ remains invariant under cyclic and anticyclic permutations; ii) factorization is retained, and iii) no exotics occur in any channel.

We shall treat here only the case of N external isovector particles; condition iii) then requires the absence of states with $I > 1$. A solution fulfilling all requirements is given by [31, 32]

$$C_N(12 \ldots N) = \tfrac{1}{2}\,\mathrm{Tr}(\tau_{a_1}\tau_{a_2} \ldots \tau_{a_N}) \qquad \text{(C–1.3)}$$

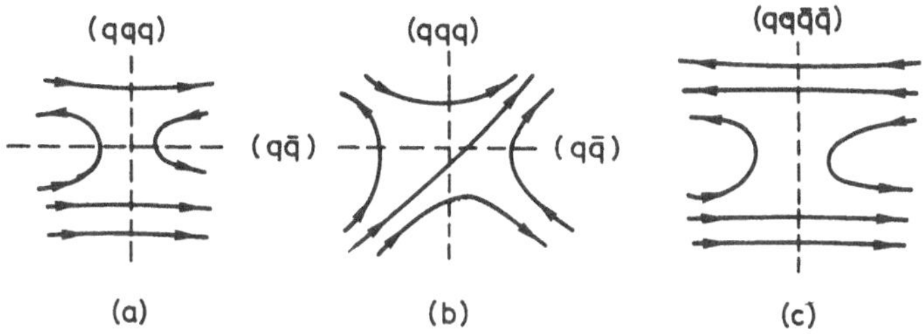

Fig. C–1.1. Duality diagrams – (*a*) and (*b*) allowed, (*c*) forbidden

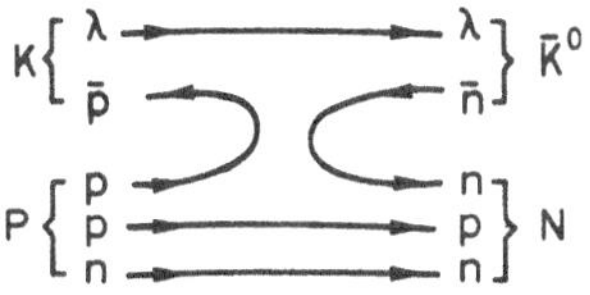

Fig. C–1.2. Duality diagram for $K^- P \to \bar{K}^0 N$

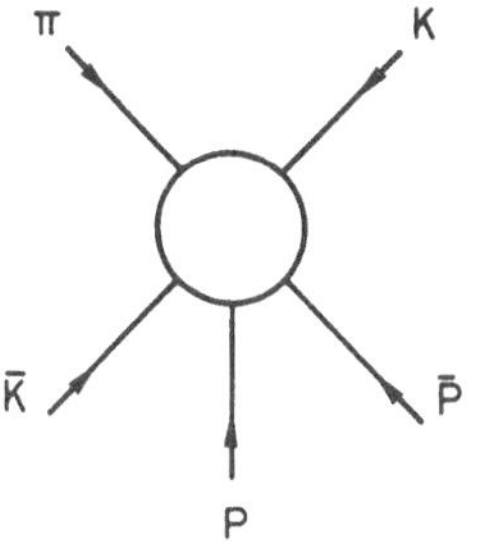

Fig. C–1.3. $K\bar{K}P\bar{P}\pi$ diagram

where τ_{a_i} denotes the 2×2 Pauli matrix representing the isospin state a_i of external particle i. From

$$\mathrm{Tr}\, ABC = \mathrm{Tr}\, CAB \tag{C–1.4}$$

we immediately have i), while iii) follows from the closure under multiplication of the 2×2 Pauli matrices. Finally one can prove

$$\begin{aligned} \tfrac{1}{2}\,\mathrm{Tr}(\tau_{a_1} \dots \tau_{a_N}) = {} & \tfrac{1}{4}\,[\mathrm{Tr}(\tau_{a_1} \dots \tau_{a_M})\,\mathrm{Tr}(\tau_{a_{M+1}} \dots \tau_{a_N})] \\ & + \tfrac{1}{4} \sum_{x=1}^{3} [\mathrm{Tr}(\tau_{a_1} \dots \tau_{a_M} \tau_x)\,\mathrm{Tr}(\tau_x \tau_{a_{M+1}} \dots \tau_{a_N})] \end{aligned} \tag{C–1.5}$$

and thus factorization. The first term here corresponds to a singlet (isospin zero) and the sum of the three terms to a triplet (isospin one) intermediate state. We thus find in general each pole contributing both to $I=0$ and $I=1$, i.e., we have an isospin degeneracy. Upon summing over all permutations, it turns out, however, that the two states have different signature, so that we have in effect found the $\varrho - f_0$ degeneracy from the isospin factor.

Such considerations can be extended to isospin zero and one-half external particles, as well. Let us perhaps note again finally, however, that all this is only a solution of the pure isospin problem, *not* an answer to the question of how to construct Veneziano amplitudes for physical particles, e.g., pseudoscalar mesons!

2. Classification of Reactions

We want here to briefly look at production processes under the aspect of where Pomeron exchange can be excluded, and how.

a) No Pomeron exchange

Hyperon production by the interaction of nucleons and antikaons

$$\bar{K}N \to Ys\pi, \quad s=1,2,3,\ldots \tag{C–2.1}$$

and baryon-antibaryon annihilation into mesons

$$N\bar{N} \to s\pi + r(K+\bar{K}) \tag{C–2.2}$$

are the only observable cases with arbitrary particle number where no Pomeron exchange is possible, because of strangeness and baryon number exchange, respectively. In the three-particle final state process

$$\pi N \to K\bar{K}N \quad \Delta Q_N \neq 0 \tag{C–2.3}$$

the Pomeron is forbidden if the nucleon changes its charge; charge exchange plays a similar role in

$$\pi N \to YK\pi \quad \Delta Q_\pi \neq 0, \tag{C–2.4}$$

$$NN \to YKN \quad \Delta Q_N \neq 0. \tag{C–2.5}$$

This essentially exhausts the list of measured reactions where Pomeron exchange is strictly forbidden.

b) Probably no Pomeron exchange

Current concepts of diffraction dissociation generally include the so-called *Gribov* [33] – *Morrison* [34] rule, which permits only natural parity changes for the diffractively produced system

$$\Delta P = (-1)^{\Delta J} \tag{C–2.6}$$

with ΔP and ΔJ denoting the parity and spin change between incident particle and diffractively excited final system. This rule excludes Pomeron exchange in several other reactions; in particular

$$KN \to KN\pi \quad \Delta Q_K \neq 0, \tag{C–2.7}$$

$$\bar{K}N \to \bar{K}N\pi \quad \Delta Q_{\bar{K}} \neq 0, \tag{C–2.8}$$

$$\pi N \to K\bar{K}N \tag{C–2.9}$$

are then non-diffractive because of the transition from one to two pseudoscalar mesons.

c) Pomeron exchange

The bulk of the measured processes, in particular such reactions as

$$\pi N \to N s \pi \qquad s = 2, 3, \ldots \tag{C–2.10}$$

$$N N \to N N (s-1)\pi \tag{C–2.11}$$

$$K N \to N K (s-1)\pi \tag{C–2.12}$$

$$\bar{K} N \to N \bar{K} (s-1)\pi \tag{C–2.13}$$

can thus contain Pomeron exchange, and various investigations [35–37] have indicated that even at intermediate energies they do in fact contain it to a considerable extent. On the other hand, one can generally not exclude ordinary Regge exchanges in channels where Pomeron exchange is possible, except at very high energies (e.g., $P_{\text{lab}} \sim 20$–30 GeV/c for three-body final states, higher for more). Thus one really needs both the N point Veneziano amplitude and a dual model for diffraction dissociation in order to obtain a dual resonance description for production processes in general. In the following subsection we shall survey the present phenomenological attempts to solve this. We shall not consider here baryon-antibaryon annihilation investigations [10, 38]; for reviews of work on these processes, we refer, e.g., to the contribution of Dr. *Lusignoli* at this meeting.

3. Analyses of Production Experiments

A detailed discussion of all applications of Veneziano amplitudes to production reactions is certainly outside the scope of these lectures; we shall therefore instead concentrate on some interesting aspects in one or two applications of a pure B_5 model, of a dual resonance model for diffraction dissociation, and of a combination of these two. For further details and other applications, we shall attempt to list the original references.

a) A B_5 model (no Pomeron exchange)

There are, in our opinion, essentially two features in which a Veneziano analysis is expected to be an improvement over previous multi-Regge descriptions: the Veneziano amplitude is applicable to the entire reaction, not just resonance production or background or high energy behaviour; because of crossing symmetry, the same amplitude (with fixed parameters) should describe the whole set of reactions obtainable from each other by crossing.

The B_5 amplitude was first applied in the analysis of a production experiment by *Petersson* and *Törnqvist* [11], who studied the reaction

$$K^- P \rightarrow \Lambda \pi^+ \pi^- \tag{C–3.1}$$

which appears to be quite suitable for such an analysis because of the absence of both Pomeron and pion exchange, the position of the latter in a duality scheme also being somewhat unclear. The resulting fits are remarkably good and encouraged further applications [39]. Since reaction (C–3.1) could at least in principle have Pomeron contributions in a crossed channel and since the actual analysis encounters some difficulties with the hypothesis of no exotics, we prefer to consider in more detail a subsequent analysis of a different set of reactions. *Chan, Raitio, Thomas* and *Törnqvist* [12] have investigated the $(K^\pm \bar{K}^0 \pi^\pm P \bar{P})$ system, which via Gribov-Morrison rule appears to be the only case with no Pomeron in any possible channel. We thus want to look at reactions of the kind

$$K^+ P \rightarrow K^0 \pi^+ P, \tag{C–3.2}$$

$$K^- P \rightarrow \bar{K}^0 \pi^- P, \tag{C–3.3}$$

$$\pi^- P \rightarrow K^0 K^- P, \tag{C–3.4}$$

$$\pi^+ P \rightarrow \bar{K}^0 K^+ P, \tag{C–3.5}$$

$$P \bar{P} \rightarrow K^- K^0 \pi^+ \tag{C–3.6}$$

which are different channels of the same five-point amplitude.

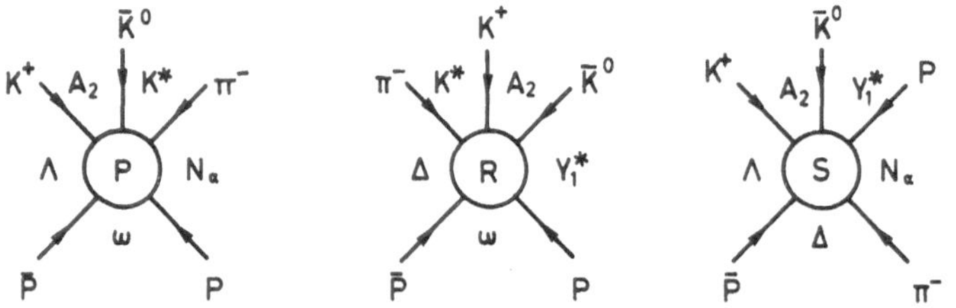

Fig. C–3.1. $K\bar{K}'P\bar{P}\pi$ diagrams

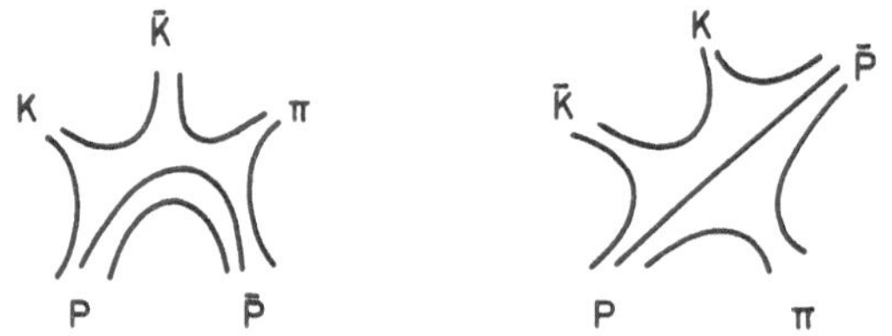

Fig. C–3.2. $K\bar{K}'P\,\bar{P}\pi$ diagrams

Let us see first how many diagrams in fact contribute. We know that there are $(N-1)!/2=12$ inequivalent permutations of external lines; of these, all leading to exotic channels should, however, vanish. This leaves only the three diagrams shown in Fig. C–3.1, as is easily seen by looking at the corresponding *Harari-Rosner* duality diagrams; only the two topologies of Fig. C–3.2 are permitted. It becomes clear here how restrictive the absence of exotics turns out to be.

The choice of trajectories to be taken into account in each diagram is, analogously to that question in Reggeology, dictated to a large extent by experiment. We shall not enter here into this discussion, but simply indicate in Fig. C–3.1 the choice used in Ref. [12]. With this choice, one then takes, e.g., for the first diagram of Fig. C–3.1 the form

$$\begin{aligned}(P) = {} & \varepsilon_{\mu\nu\varrho\sigma}\, p_\pi^\mu p_K^\nu p_{\bar{K}}^\varrho p_{\bar{P}}^\sigma \,* \\ & B_5(1-\alpha_{K^*}, 1-\alpha_{A_2}, \tfrac{1}{2}-\alpha_\Lambda, 1-\alpha_\omega, \tfrac{1}{2}-\alpha_N)\,.\end{aligned} \tag{C–3.7}$$

This form gives us poles at the right places, and because of the ε factor with the right angular behaviour (meson trajectories start with p wave resonances) without destroying the cyclic symmetry. The spin of the external baryons, however, is totally neglected.

Finally, one fixes the relative weights of the three terms by requiring the wrong signature contributions of N and Δ to vanish, obtaining

$$T = \beta\,[(P)+(R)+(S)]\,. \tag{C–3.8}$$

Since the trajectory parameters are known, we are left with only one open parameter, the over-all normalization β.

In Figs. C–3.3—C–3.5 we show now representative results obtained in Ref. [12]. While there are discrepancies in details (in particular somewhat too flat momentum transfer distributions in channels where the possible pion exchange has been neglected), the over-all agreement with the data is rather promising, keeping in mind the roughness of the approximation. The reason for the over-all discrepancy of roughly a factor two in the reaction C–3.4 is not clear; since the over-all values for reaction (C–3.6) appear, with very sparse data, also rather much larger in theory than in experiment, it may be that the present forms of the model encounter difficulties with very "central" interactions ($K\bar{K}$ formation, $P\bar{P}$ annihilation).

By taking the amplitude (C–3.8) at the nucleon or Λ pole, one can now get predictions for two-body reactions of the kind [40]

$$K^- P \to \bar{K}^0 N\,, \tag{C–3.9}$$

$$\pi^- P \to K^0 \Lambda\,. \tag{C–3.10}$$

The resulting agreement is fairly good.

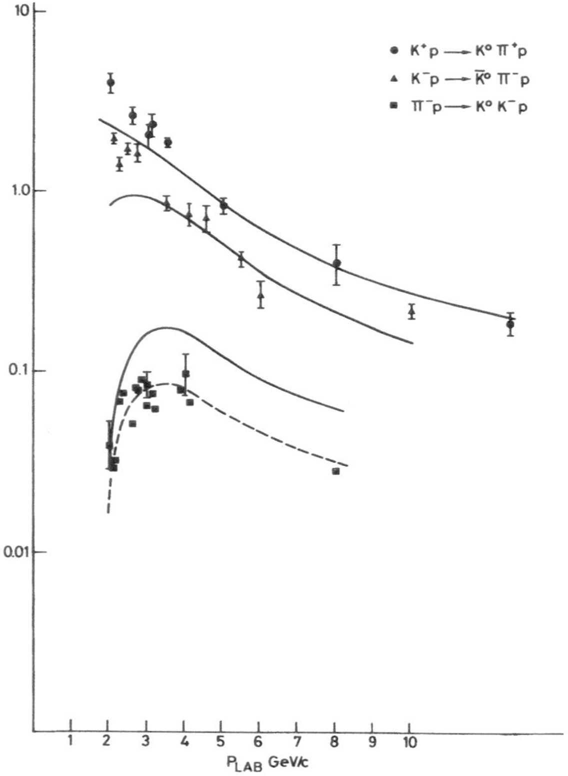

Fig. C–3.3. Cross-sections for $K^{\pm} P \to K^0(\bar{K}^0)\,\pi^{\pm} P$ and $\pi^- P \to K^0 K^- P$, from Ref. [12]

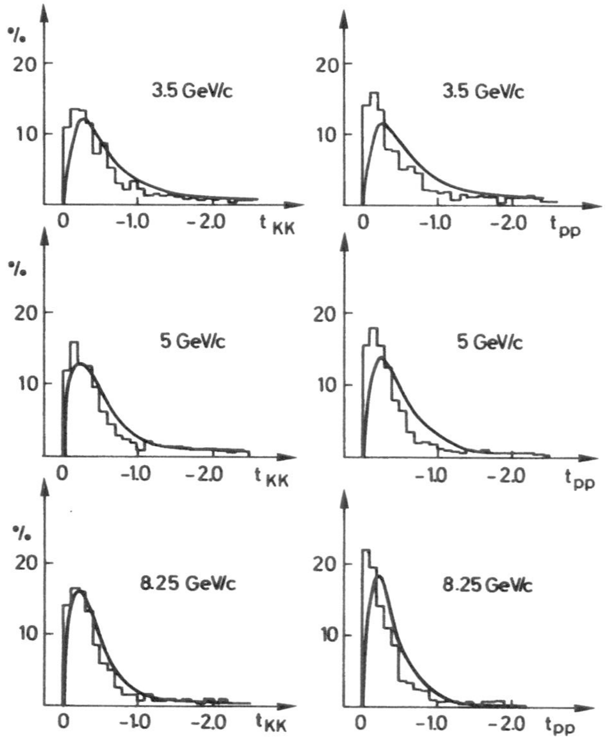

Fig. C–3.4. Proton-proton and kaon-kaon momentum transfer distributions in $K^+ P \to K^0 \pi^+ P$, from Ref. [12]

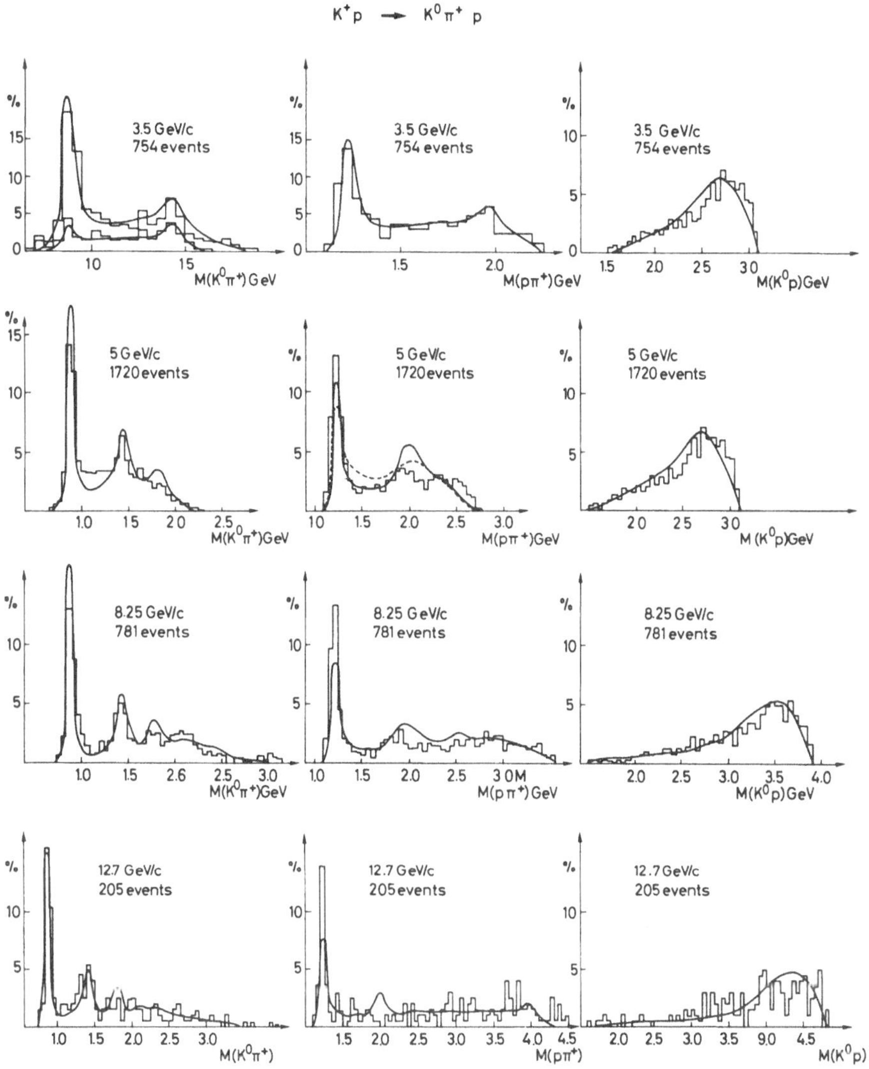

Fig. C–3.5. Invariant mass distributions in $K^+P \to K^0\pi^+P$, from Ref. [12]

In conclusion, let us note other analyses and extensions. In Ref. [41] a similar analysis to the one in Ref. [12] was carried out. The problem of baryon spins was taken up in considerable detail in Refs. [42 and 43]; while not solving the problem of parity doubling of baryon resonances, one may hope that this will eventually lead to a fuller understanding of spin effects and of the relations between B_5 and B_4 subreactions [44].

b) A dual resonance approach to diffraction dissociation

The next classes of processes we want to consider are those where the Pomeron can be taken as dominant whenever it can occur, i.e., high energy diffraction dissociation (cf., Fig. C–3.6) such as

$$NN \to N(\pi N)\,, \qquad \text{(C–3.11)}$$

$$\pi N \to \pi(\pi N)\,, \qquad \text{(C–3.12)}$$

$$KN \to K(\pi N)\,, \qquad \text{(C–3.12)}$$

$$\pi N \to (3\pi)\,N\,; \quad KN \to (K2\pi)\,N \qquad \text{(C–3.14)}$$

and also, assuming the photon to behave essentially as a hadron

$$\gamma N \to (2\pi)\,N\,. \qquad \text{(C–3.15)}$$

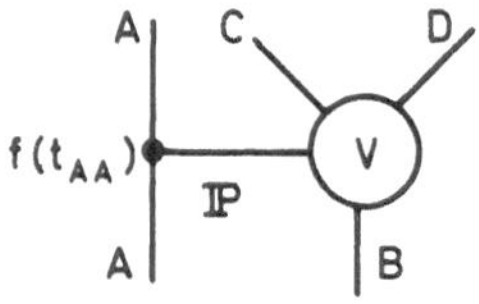

Fig. C–3.6. Diagram for diffraction dissociation

A conclusive test for strong Pomeron dominance is, of course, the constancy of the cross-section; given a measurement at one energy only, one can, however, also compare different charge configurations [45] – thus Pomeron dominance implies for reaction (C–3.11) the ratio

$$\sigma(PP \to PN\pi^+)/\sigma(PP \to PP\pi^0) = 2/1\,.$$

In the diffraction dissociation vertex V (cf., Fig. C–3.6), one observes in general both direct and crossed channel exchange effects: while in, e.g., (C–3.15) the direct channel ϱ is strongly dominant, there are clearly observable shifts in the two-pion mass distribution indicating the presence of pion exchange contributions; on the other hand, while, e.g., (C–3.11) can be rather well described by reggeized pion exchange in V, there are N^* resonance effects which in this fashion are neglected ($PP \to PN^*_{1470}$, PN^*_{1518}).

Pokorski and the present author [15] have therefore attempted to describe diffraction dissociation reactions by an amplitude of the form (cf., Fig. C–3.6)

$$T \sim f(t_{AA})\,\bar{s}\,V_{\mathbb{P}B \to CD} \qquad \text{(C–3.16)}$$

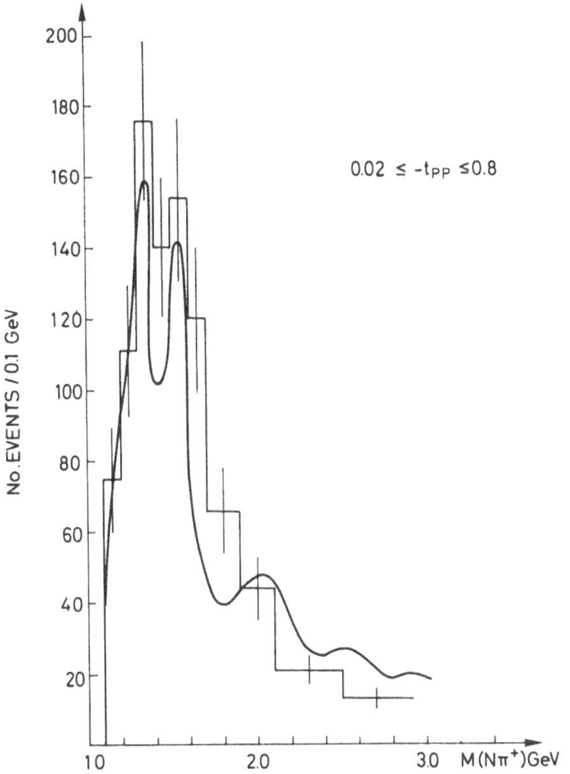

Fig. C–3.7. $N\pi^+$ mass distribution in $PP \to PN\pi^+$, from Ref. [15]

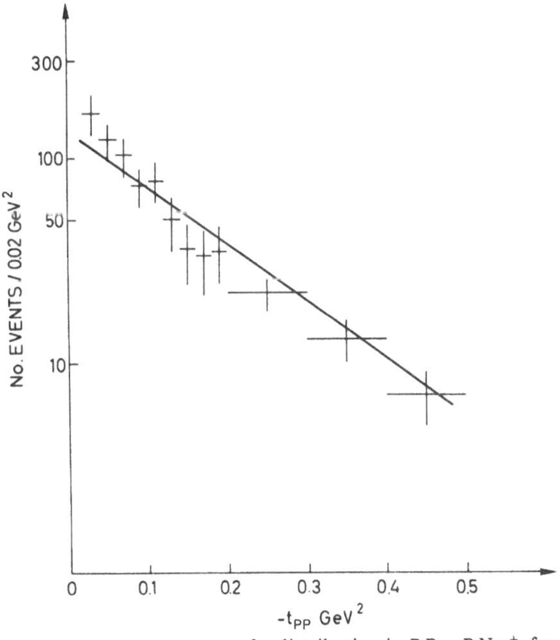

Fig. C–3.8. Proton-proton momentum transfer distribution in $PP \to PN\pi^+$, from Ref. [15]

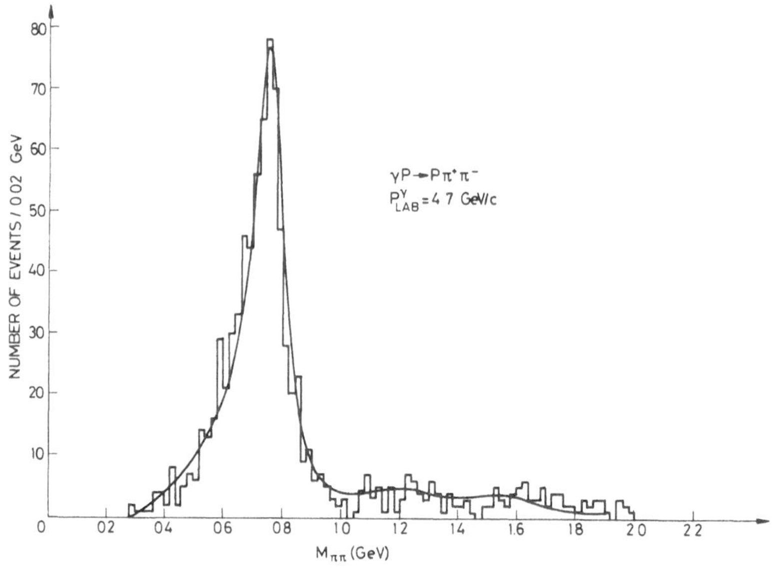

Fig. C–3.9. Two-pion mass distribution in $\gamma P \rightarrow P\pi^+\pi^-$, from Ref. [46]

where $f(t_{AA})$ denotes a form factor for the hadron-hadron-Pomeron vertex which can, by factorization, be determined from elastic scattering; the factor $\bar{s} = (p_A + p_B)^2$ is to account for the Pomeron propagator ($\bar{s}^{\alpha_{\mathbb{P}}} \approx \bar{s}$); finally, $V_{\mathbb{P}B \rightarrow CD}$ denotes the amplitude for the "reaction" $\mathbb{P} + B \rightarrow C + D$. For the latter, we want to take a dual resonance form and thus obtain a Veneziano description of diffraction dissociation (DRDD).

Assuming the Pomeron to behave roughly as a $J^P = 0^+$ particle, we thus take for (C–3.15) for example the form [46]

$$V_{\mathbb{P}\gamma \rightarrow \pi^+\pi^-} \sim [B_4(1 - \alpha_p^s, -\alpha_\pi^t) + B_4(1 - \alpha_p^s, -\alpha_\pi^u)] \qquad \text{(C–3.17)}$$

with ϱ and π trajectories dominant in direct and crossed channels of $\gamma\mathbb{P} \rightarrow \pi^+\pi^-$, respectively.

In Figs. C–3.7—C–3.10 we again show some representative results obtained in such an approach; again the results are on the whole quite satisfactory.

Besides the application to $NN \rightarrow N(\pi N)$ [15], $\pi N \rightarrow \pi(\pi N)$ [47], and two-pion [46] as well as two-kaon [48] photoproduction, work is in progress also on the diffractive dissociation into three particles (taking $V \sim B_5$) and appears to yield rather good agreement for $KN \rightarrow (K2\pi)N$ [49].

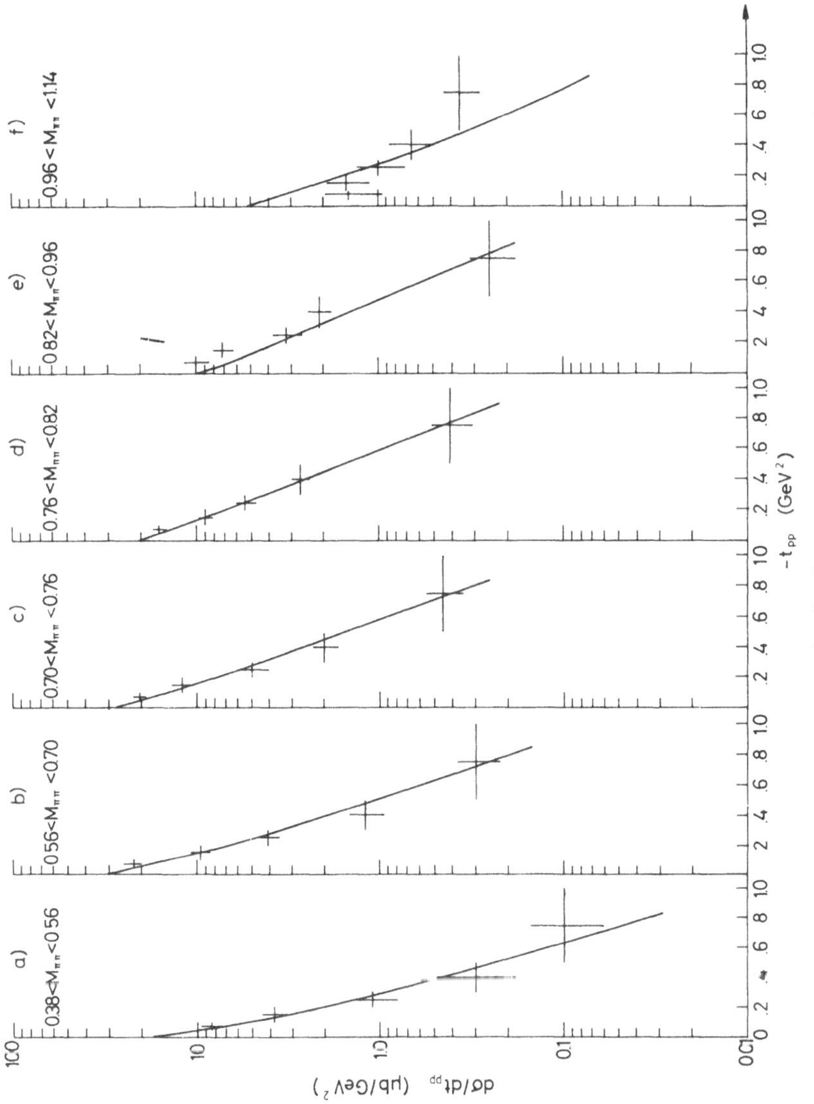

Fig. C–3.10. Proton-proton momentum transfer distribution for various two-pion masses, in $\gamma P \rightarrow P\pi^+\pi^-$ from Ref. [46]

c) Models B_5 plus DRDD

As already mentioned, one has to combine models such as those described under C–3.a) and 3.b) to obtain a description applicable to production reactions in the intermediate energy range. A number of such analyses is presently in progress; work on $K^\pm P \rightarrow K^\pm \pi^0 P$ [50] and on $K^- P \rightarrow K^- \pi^+ N$ [51] has just appeared and seems to give fairly good agreement in most instances. In the second of these analyses pion exchange has been included and is found to give in fact strong contributions.

While there are still many open ends, it thus appears that one now has on a crude phenomenological level an over-all outline for the dual resonance analysis of production processes; only further investigations can tell us how critical the approximations are that we have made to obtain this scheme, and to what extent particularly spin effects can eventually be taken into account.

4. Present Problems and Outlook

In closing this introduction to dual resonance models in multiparticle phenomenology, we want to emphasize once more what the main unsolved problems are.

As in previous descriptions, the unitarity question remains completely open, and primitive unitarizations such as the complex trajectories used above may soon encounter at least two difficulties: it appears that more detailed analyses will require different widths for parents and daughters [52, 53], and it is not clear what this will do in general to the properties of B_N; unitarizations neglecting the interrelation between various particle number channels probably will not give rise to the smooth energy behaviour of the over-all inelastic cross-section observed experimentally. Moreover, it is not clear to what extent the separation into diffractive and non-diffractive processes is possible in the long run; here one encounters such questions as whether the Pomeron is built up, through unitarity, mainly from non-diffractive many-body reactions, or whether diffraction dissociation will provide significant parts in this build-up; (looking at low energy K^+P scattering, which is already strongly Pomeron dominated in spite of essentially no inelastic Pomeron reactions, one might hope for a construction via bootstrap from non-diffractive reactions only).

Secondly, the fermion question will certainly become more and more important as the analyses become more detailed. It remains to be seen if the present attempts to remove parity doublets in Regge theory can, in a workable fashion, be extended to dual models – i.e., if parity doublets can be removed in a crossing symmetric fashion.

Thirdly, as we saw in reactions involving $K\bar{K}$ and $P\bar{P}$ systems, it is not clear if the present form of the Veneziano amplitude is applicable both to very peripheral and to very central reactions.

Finally, we should emphasize the fact that the complexity of the expression and consequently the amount of technical effort increase greatly when one just increases the particle number by one or two. It is difficult to conceive calculations of six or seven-body final states with the present formulae. Thus we would consider an equally important question that of how to simplify the Veneziano form for large N without losing its essential aspects.

Acknowledgements: It is a great pleasure to thank many colleagues at CERN, in particular *Chang Hong-Mo, K. Kajantie, B. Petersson, H. Ruegg* and *K. Schilling*, for stimulating conversations on various aspects of "Venezianology".

Literature

Reviews:

R 1. *Chan Hong-Mo:* CERN preprint TH. 1057, July 1969 (Royal Society Meeting Lecture).
R 2. *Lovelace, C.:* CERN preprint TH. 1123, January 1970 (Irvine Regge Pole Meeting Lecture).
R 3. *Alessandrini, V., Amati, D., Le Bellac, M., Olive, D.:* CERN preprint TH. 1160, May 1970.
R 4. *Veneziano, G.:* July 1970 (Erice Lectures).

References

1. Cf., e.g., *Czyzewski, O.:* Rapporteur's talk, Proceedings of Vienna conference (1968).
2. *Dolen, R., Horn, D., Schmid, C.:* Phys. Rev. **166**, 1768 (1968).
3. *Veneziano, G.:* Nuovo Cimento **57** A, 190 (1968).
4. *Bardakçi, K., Ruegg, H.:* Phys. Letters **28** B, 342 (1968); — — Phys. Rev. **182**, 1884 (1969).
5. *Virasoro, M.:* Phys. Rev. Letters **22**, 37 (1969).
6. *Chan Hong-Mo:* Phys. Letters **28** B, 425 (1969); — *Tsou Sheung Tsun:* Phys. Letters **28** B, 485 (1969).
7. *Goebel, C. L., Sakita, B.:* Phys. Rev. Letters **22**, 259 (1969).
8. *Koba, Z., Nielsen, H. B.:* Nucl. Phys. B **10**, 633 (1969).
9. *Kikkawa, K., Sakita, B., Virosoro, M. A.:* Phys. Rev. **184**, 1701 (1969).
10. *Lovelace, C.:* Phys. Letters **28** B, 264 (1968).
11. *Petersson, B., Törnqvist, N. A.:* Nucl. Phys. B **13**, 629 (1969).
12. *Chan Hong-Mo, Raitio, R. O., Thomas, G. H., Törnqvist, N. A.:* Nucl. Phys. B **19**, 173 (1970).
13. *Harari, H.:* Phys. Rev. Letters **20**, 1395 (1968).
14. *Freund, P. G. O.:* Phys. Rev. Letters **20**, 235 (1968).
15. *Pokorski, S., Satz, H.:* Nucl. Phys. B **19**, 113 (1970).
16. *Fubini, S., Veneziano, G.:* Nuovo Cimento **64** A, 811 (1969).
17. *Bardakçi, K., Mandelstam, S.:* Phys. Rev. **184**, 1640 (1969).
18. *Hopkinson, J. L., Plahte, E.:* Phys. Letters **28** B, 489 (1969).
19. *Bialas, A., Pokorski, S.:* Nucl. Phys. B **11**, 479 (1969).
20. *Gross, D.:* CERN preprint TH. 1048 (1969).
21. *Koba, Z., Nielsen, H. B.:* Z. Physik **229**, 243 (1969).
22. *Martin, A.:* Phys. Letters **29** B, 431 (1969).
23. *MacDowell, S. W.:* Phys. Rev. **116**, 774 (1959).
24. *Savoy, C.:* Nuovo Cimento Letters **2**, 870 (1969).
25. *Waltz, R. E.:* Nucl. Phys. B **18**, 61 (1970).
26. *Dethlefsen, J.:* A_1 parameters from a modified five-point function. Copenhagen preprint (1969).

27. *Dorren, D. J., Rittenberg, V., Rubinstein, H. R.*: CERN preprint TH. 1192 (1970), and further references therein.
28. *Del Giudice, E., Veneziano, G.*: Nuovo Cimento Letters **3**, 363 (1970).
29. *Harari, H.*: Phys. Rev. Letters **22**, 563 (1969).
30. *Rosner, J. L.*: Phys. Rev. Letters **22**, 689 (1969).
31. *Paton, J. E., Hong-Mo, Chan*: Nucl. Phys. B **10**, 516 (1969).
32. *Neville, D. E.*: Phys. Rev. Letters **22**, 494 (1969).
33. *Gribov, V. N.*: Yad. Fiz. USSR **5**, 197 (1967).
34. *Morrison, D. R. O.*: Phys. Rev. **165**, 1699 (1968).
35. *Satz, H.*: Nucl. Phys. B **14**, 366 (1969).
36. Scandinavian Collaboration, Phys. Letters **30** B, 369 (1969).
37. *Van Hove, L.*: CERN preprint TH. 1178 (1970); cf., also: *Bialas, A., et al.*: Nucl. Phys. B **11**, 479 (1969).
38. *Altarelli, G., Rubinstein, H. R.*: Phys. Rev. **183**, 1469 (1969); *Rubinstein, H. R., Squires, E. J., Chaichian, M.*: Phys. Letters **30** B, 191 (1969).
39. *Hoyer, P., Petersson, B., Törnqvist, N. H.*: CERN preprint TH. 1159 (1970).
40. *Petersson, B., Thomas, G. H.*: Nucl. Phys. B **20**, 451 (1970).
41. *Bartsch, J., et al.*: Nucl. Phys. B **20**, 63 (1970).
42. *Bender, I., Dosch, H. G., Müller, V. F., Rothe, H. J.*: Z. Physik **237**, 192 (1970); and: Production of $\pi\pi$ resonances in a dual model.... Heidelberg preprint (July, 1970).
43. *Hirshfeld, A. C., Schmidt, M. G.*: The dual $\bar{K}N\bar{\Lambda}\pi\pi$ amplitude with spin. Heidelberg preprint (July, 1970).
44. — Relations between scattering processes derivable from a dual five-point amplitude with spin. Heidelberg preprint (July, 1970).
45. *Satz, H.*: Phys. Letters **29** B, 38 (1969).
46. — *Schilling, K.*: Nuovo Cimento **67** A, 511 (1970).
47. *Schilling, P. K.*: DESY Int. Rep. R 1/70/3 (February, 1970).
48. *Satz, H., Schilling, K.*: Nuovo Cimento Letters **3**, 723 (1970).
49. *Otter, G., Goldsack, S. J.*: Veneziano model calculations for the decay $Q \to K\pi\pi$. Imperial College London preprint (1970); and —, Private communication.
50. *Kajantie, K., Papageorgiou, S.*: CERN preprint TH. 1170 (1970).
51. Aachen-Berlin-CERN-London-Vienna collaboration, CERN/D. PH. II/Phys. 70–13, (June, 1970).
52. *Hoyer, P., Lee, A., Paton, J. E., Petersson, B., Thomas, G. H.*: to appear soon.
53. *Pokorski, S., Szeptycka, M., Znieminski, A.*: to appear soon.

Dr. *Helmut Satz*
CERN Theory Division
CH-1211 Genève 23, Switzerland

Physical N-Pion Functions

H. R. RUBINSTEIN

Contents

Our purpose here is to discuss the problem of constructing physical N point functions for pions [1]. We use the word physical as a pictorial way to condense a large number of attributes the amplitude must possess. The list includes, for the generalized Born term or tree approximation with any number of legs: 1. the proper spectrum of particles; 2. absence of tachyons (particles of $J=0$ lying on trajectories with positive intercept); 3. factorization properties as expected from poles; 4. positivity conditions of partial waves and more, positivity of every residue for elastic channels (no ghosts); 5. the bootstrap condition that demands that at a π pole for example, the residue function is the $(N-2)$ π function; 6. Regge behaviour for the full amplitude in all relevant domains; 7. simplicity and maximum duality. This last condition will be defined in Section 1 and below.

In a paper written in collaboration with *Rittenberg* [2], we discussed the most general way one can use to attack this problem provided the standard form of the N point function is used. Let us look at these forms first. The integral forms that generalize the Euler function are indeed very appealing but have a very fundamental drawback. If the trajectories have positive intercept, they lack full Regge behaviour in some channels. This is due to the duality correlation between resonances and asymptotic behaviour in the dual channels. To illustrate better this point we consider the simplest physical example: $\pi^+\pi^+$ scattering. A beta function gives:

$$B(1-\alpha(s), 1-\alpha(t)) = \frac{\Gamma(1-\alpha(s))\,\Gamma(1-\alpha(t))}{\Gamma(2-\alpha(s)-\alpha(t))}$$

and one obtains for large s, $\alpha(s)^{\alpha(t)-1}$. One must then introduce a function $B(1-\alpha(s); \alpha(t))$ and multiply by $\alpha(t)$. It is clear that the obvious procedure to try is to look for N point functions of the form

$$\sum P_i(\alpha_{nm} \cdots)\, B_N^i(\alpha_{nm} - m_{nm}, \cdots)$$

where P_i is a polynomial and m_{ij} are numbers that ought to be fixed in accordance to the physical requirements stated above.

That such a programme is feasible and that it works for whatever type of trajectory is not obvious. It is our purpose to discuss it here and show its advantages and limitations.

One positive feature is that the solution can be presented in terms of propagator rules [3].

Another interesting aspect is that it does carry enough complexity as to allow the construction of *any* amplitude with mesons of any spin. These are achieved by forming bound states and constructing the relevant tensors by means of suitable combinations of momenta. However, though all possible mesonic amplitudes can be found this way, one should keep in mind that if some substructure of the quark type is important, there might be other solutions to the integer spin problem that we are completely overlooking. This possibility should not be neglected since our solutions are far from perfect and since nucleons must be fed into the theory [4]. So, we emphasize that one should not neglect the essential repercussions that half integral spin might have upon our amplitudes.

It is reassuring, however, that qualitative duality effects involving baryons, that seem to depend on the structure of these functions, are in agreement with experiment [5–7].

The functions presented here have interest for phenomenology. Most important, they test to what extent the simplest models have the same properties the more complicated have.

As it turns out, two main features appear:

a) leading trajectories are now populated by more than one state. This property is appealing since it may be related to the by now famous A_2 splitting, however, the ϱ is also doubled;

b) as a consequence of the absence, term by term, of cyclical symmetry, the factorization properties of the amplitude are complicated and radically different from the case studied by *Fubini* and *Veneziano*, and *Bardakçi* and *Mandelstam* [8]. Though the leading trajectory has a degeneracy that is independnet of the number of legs, the daughter structure at a mass $\alpha(s)=J$ demands the analysis of graphs with n_J legs, where n_J is an increasing function of J.

The seriousness of this problem is difficult to evaluate. Though degeneracies of this type are not impossible, they amount in fact to a breakdown of factorization [9], an unpleasant feature indeed. Loop calculations [10] to enforce unitarity are ruled out. Whether the over-simplified problem, that is trouble free, is a better approximation to a hitherto unknown amplitude, is a question that we cannot answer.

It is possible that there is a more complicated way to solve the problem, other than Eq. (1). This we have tried without success.

Dorren, Rittenberg and myself [11] have analyzed the N pion function with the $\pi - A_1$ unnatural parity trajectory in the $G = -1$ channels and the $\varrho - f_0$ trajectory in the $G = +1$ channels. Positivity along the leading trajectory is well verified but particles with imaginary coupling constants (ghosts) persist at the daughter level. These functions seem useful for practical applications and a vigorous program is under way [12].

Natural parity $G = -1$ trajectories have also been studied using epsilon tensor couplings at the level of the six-pion function [13–14] and it seems that its generalization for N pion do not present special problems. We will discuss this case in some detail here for the six-pion case to emphasize the new points that characterize this case.

1. Unnatural Parity $G = -1$ Trajectories

Our method to construct N point functions is best described by means of graphical techniques. Using dual graphs, we will be able to establish a one-to-one correspondence between the factors composing the N point function and the geometrical structure of the graphs. These graphs allow also a simple understanding of the properties of the leading trajectories in all channels.

These rules were established by painful calculations but the results embodied in these rules are very simple and allow for the construction of the desired amplitude in a very simple fashion.

A typical term of the amplitude with N pions is assumed to have the structure

$$A(p_1, \dots p_N) = \int \cdots \int \prod_{i,j} u_{ij}^{-\alpha_{ij}-1} p(\alpha_n, \dots \alpha_{ij} \dots \alpha_{N-2,N-1}; u_{12}, \dots u_{N-2,N-1}) \, \mathrm{d} V_N \tag{1}$$

where $\alpha_{ij} = a^{(K)} - b(p_i + p_{i+1} + \cdots + p_j)^2$, $a^{(K)}$ $(K = 1, 2)$ being the intercept of the two possible types of trajectory, n_{ij} are the usual $N(N-3)/2$ Chan variables, $\mathrm{d} V_N$ is the volume element which includes δ-functions to ensure duality. $P(\alpha_{ij}, u_{ij})$ is a polynomial in α_{ij} and u_{ij}. The main function of this polynomial is to eliminate the unwanted poles at $\alpha_{ij} = 0$ if the intercept is positive. Other functions could be tried but the strict requirements of meromorphy and absence of ancestors make any other choice difficult, if not impossible. A natural choice like some simple denominator, for example, produces inevitably ancestors in some dual channel.

Let us now see the effect each "building block" has on the amplitude and associate to it a geometrical pattern that can be used in the dual graph:

1. $\alpha_{ij}(1-u_{ij}) \rightarrow$ —— (continuous line in dual graph)

This factor clearly eliminates the $\alpha_{ij}=0$ pole in the amplitude and on all the trajectories dual to (i,j) because of the $(1-u_{ij})$ factor. Spin structure and Regge behaviour is not affected since both factors have compensating effects.

2. $(1-u_{ij}) \rightarrow$ - - - - (dotted line dual graphs)

This term eliminates the $\alpha_{ij}=0$ on the trajectories dual to the channel (i,j) and shifts these trajectories to the first daughter level. The term has non-leading Regge behaviour for the variables dual to (i,j).

3. $u_{ij} \rightarrow + + + + +$ (lines with crosses in the dual diagrams)

These factors eliminate the $\alpha_{ij}=0$ and shift the trajectory to the first daughter level.

We then associate a graph to a given polynomial. To clarify how the procedure works, we describe a simple example. In a six-pion function a possible polynomial is

$$P=(1-u_{24})\,\alpha_{34}(1-u_{34})\,\alpha_{14}(1-u_{14})\,. \tag{2}$$

Using the afore-mentioned rules, one obtains the graph depicted in Fig. 1. This graph then is associated to an amplitude that can be explicitly

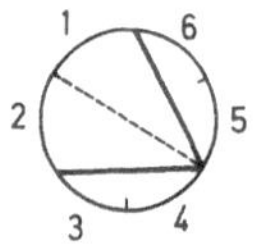

Fig. 1. Graph illustrating the polynomial given by Eq. (2) for the six-pion function. Continuous line represents a $(1-u)\,\alpha$ factor; dotted line represents a $(1-u)$ factor

constructed and whose properties are, using the afore-mentioned rules: non-leading behaviour in the channels (1,2), (4,5), (1,3), (3,5) since they cross the dotted line and hence are dual to it; non-leading behaviour in (3,4), (2,4), etc., (they do not cross the dotted line). The $\alpha=0$ pole is missing in all $G=+$ (two-body) channels.

Because of the factors we have introduced, cyclical symmetry is *not* preserved term-by-term. The total amplitude is then written

$$A=\sum_{\text{permutations}} \operatorname{Tr}(\tau_1, \ldots \tau_N)\, A(p_1, \ldots p_N) \tag{3}$$

where $\mathrm{Tr}(\tau_1, \ldots \tau_n)$ is the Chan-Paton isospin factor and whose presence ensures isospin invariance and absence of exotics. The sum is understood to be over all cyclical and non-cyclical permutations.

It is both amusing and rewarding that the properties of the amplitude, at the level of the leading trajectory, can be simply expressed in terms of geometrical properties of the graphs. These rules are true as one can verify by direct computation, but we omit here all these tedious calculations.

To illustrate the method, we go back to the example given by Eq. (2). We fix our attention on the (1,3) pole and consider the effect of all cyclical permutations when we study this particular singularity. As seen in Fig. 2, we get two graphs that contribute. The other four possible graphs

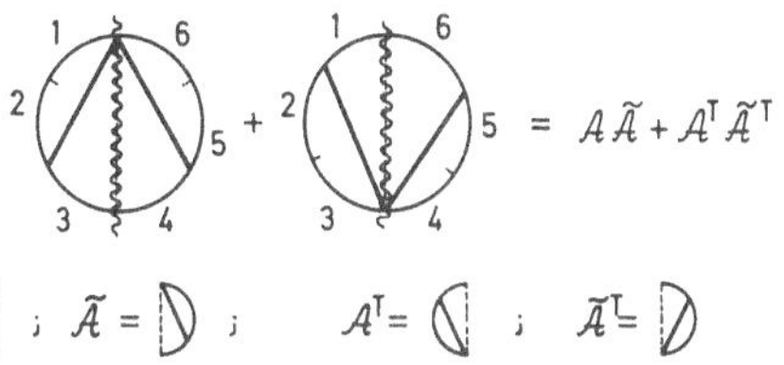

Fig. 2. The contribution of the cyclically permuted graph shown in Fig. 1 to the (1,3) leading trajectory (wiggled line)

have no leading trajectory in that channel since the (1,3) pole goes across a line of type 2. The contributions to the amplitude from these graphs will be further specified as follows:

$$\mathscr{A}\tilde{\mathscr{A}} + \mathscr{A}^T\tilde{\mathscr{A}}^T$$

where $\mathscr{A}$ is the half left figure, $\tilde{\mathscr{A}}$ is the reflected figure $\mathscr{A}$ with respect to the pole axis, $\mathscr{A}^T$ is the inverted figure with respect to the orthogonal direction. These definitions can be best understood by looking at Fig. 2. $\mathscr{A}$ can represent a sum of diagrams as well in the statements made below. The power of this technique stems from the following simple theorem:

a) if the sum of cyclically permuted graphs has the structure

$$\mathscr{A}\tilde{\mathscr{A}} + \mathscr{A}^T\tilde{\mathscr{A}}^T$$

then the leading trajectory is ghost free and has for each J both $I=0$ and $I=1$, except for $J=0$ where $I=1$ only obtains;

b) the sum reads

$$(\mathscr{A} + \mathscr{A}^T)(\tilde{\mathscr{A}} + \tilde{\mathscr{A}}^T)$$

In this case the leading trajectory is ghost free, for even J one gets isospin 1 (if $G=-$) or 0 (if $G=+$). The situation is reversed for odd J;

c)

$$(\mathscr{A} - \mathscr{A}^T)(\tilde{\mathscr{A}} - \tilde{\mathscr{A}}^T)$$

Again, the leading trajectory is ghost free, for even J we get isospin 0 for $G=+$ and 1 for $G=-$ and the opposite for $J=\text{odd}$.

If the amplitude contribution *cannot* be expressed in any of these three forms, then ghosts on the leading trajectory are inevitable. We must emphasize once more that the study of ghosts is a lengthy unpleasant calculation, but these rules follow naturally once these expressions are written down. Once the amplitude has been brought to these forms, the spin-isospin part of the theorem is trivial. To have the full amplitude we must add the non-cyclical permutations given by the twisting operation to obtain

$$(\mathscr{A} \pm \mathscr{A}^T)(\mathscr{A} \pm \mathscr{A}^T) \qquad \text{(case a)}$$

$$[(\mathscr{A} + \mathscr{A}^T) \pm (\mathscr{A} + \mathscr{A}^T)]\,(\tilde{\mathscr{A}} + \tilde{\mathscr{A}}^T) \qquad \text{(case b)}$$

$$[(\mathscr{A} - \mathscr{A}^T) \mp (\mathscr{A} - \mathscr{A}^T)]\,(\tilde{\mathscr{A}} - \tilde{\mathscr{A}}^T) \qquad \text{(case c);}$$

since under twisting $\mathscr{A} \to \mathscr{A}^T$ or $\tilde{\mathscr{A}} \to \tilde{\mathscr{A}}^T$ up to a sign. In fact, the signs depend on the spin-isospin structure. This comes as no surprise since we know that twisting is intimately connected to signature.

Each graph has a different analytic structure and factorizes by itself. However, if we accept a general solution with differnet diagrams contributing, we will increase the number of levels accordingly.

By means of these tools we analyze the possible solutions.

The Four-pion Function

This is the Lovelace formula which is obtained choosing

$$P(\alpha_{ij}, \dots u_{ij} \dots) = \alpha_{12}(1-u_{12}) \tag{4}$$

in the expression (3):

$$A(p_1 \dots p_N) = \int_0^1 \int_0^1 u_{12}^{-\alpha_{12}-1} u_{23}^{-\alpha_{23}-1} P(\alpha_{12}, u_{12})\,\delta(1-u_{12}-u_{23})\,\mathrm{d}u_{12}\,\mathrm{d}u_{23} \tag{5}$$

The graph associated with the polynomial (4) is shown in Fig. 3, as well as the properties of the leading trajectory. We obtain two "*b*"-type graphs, so that the amplitude can be written as

$$\mathscr{A}\tilde{\mathscr{A}} + \mathscr{B}\tilde{\mathscr{B}}$$

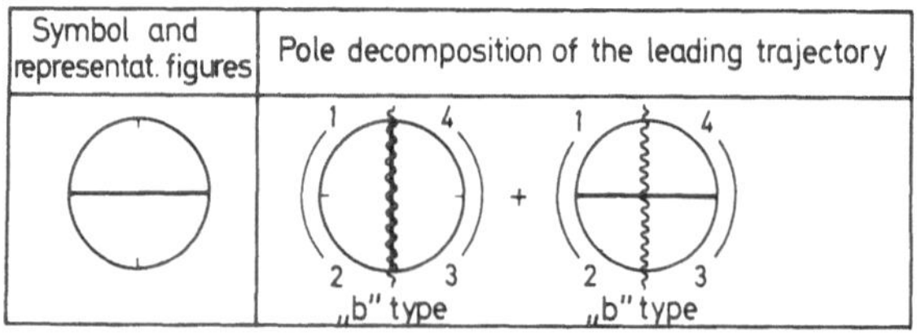
Symbol and representat. figures
Pole decomposition of the leading trajectory
1
4
2
3
„b" type
+
1
4
2
3
„b" type

Fig. 3. Graphical interpretation of the four-pion function

where $\mathscr{A} = \mathscr{A}^T$; $\mathscr{B} = \mathscr{B}^T$. In this case $\mathscr{A}$ and $\mathscr{B}$ are analytically the same, even if the graphical structure is different (this is true only for the four-pion function) and thus we have a simple $G = +1$ trajectory of "b"-type, i. e., the $\varrho - f^0$ trajectory.

The Six-pion Functions

A systematic study considering the different possibilities to build the six-pion function is shown in Fig. 4.

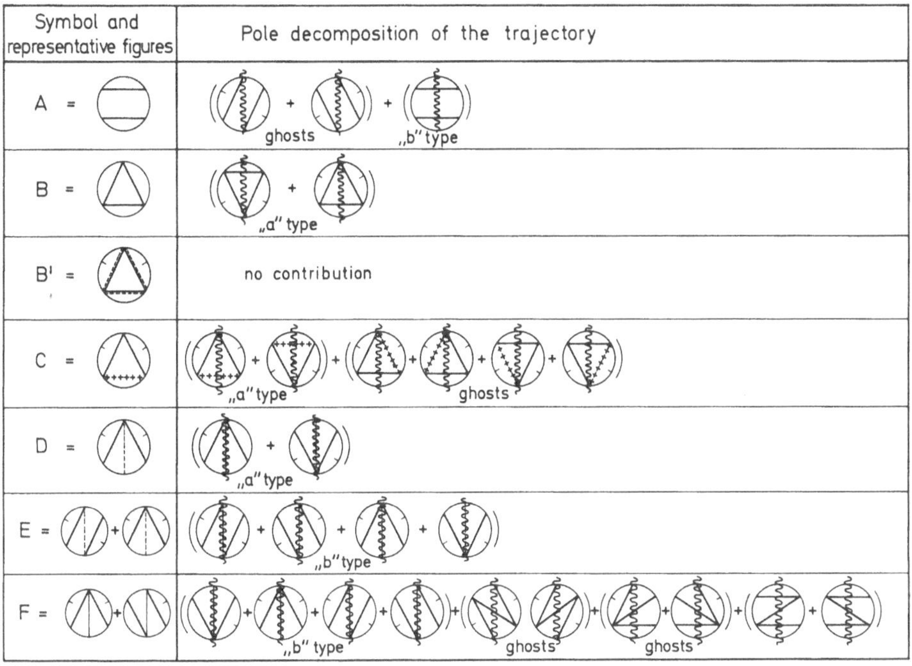

Fig. 4. Analysis of the $G = -1$ trajectory for different solutions of the six-pion functions. Continuous line represents $\alpha(1-u)$ factors, dotted lines represent $(1-u)$ factors, lines with crosses represent u factors, the wiggled lines represent thc polc linc

Let us first consider those solutions which contain only continuous lines. They have maximum Regge behaviour term-by-term.

"A" has ghosts (this solution was in fact considered in Ref. [2], it cannot be cast in the forms a, b, or c);

"B" is of "a" type, has not the π pole, has no ghosts and has, as can be seen by direct calculation, pure d wave coupling for the $A_1 \to \varrho\pi$ decay;

"F" has ghosts.

Thus the single solution which is leading term-by-term and has no ghosts would be "B". But "B" has no π pole and we have thus to consider amplitudes which have no leading Regge behaviour in all channels term-by-term.

"B'" has no leading behaviour in any $G=-1$ channel but has the $\varrho-\varrho-\varrho$ coupling;

"C" is a "maximum duality" type solution (maximum duality means maximum asymptotic behaviour compatible with the pole structure for each term) considered in Ref. [2], it has also ghosts;

"D" has no ghosts, is of "a" type and has s wave $A_1 \to \varrho\pi$ coupling;

"E" if of "b" type and thus has no $J=1$, A_1 pole. This is the solution considered in Ref. [3];

2"D"-"E" is a "c" type solution, has no π pole but has the $I=1$, A_1 pole.

Thus the general solution for the six-pion amplitude should be

$$c_1 B + c_2 B' + c_3 D + C_{21}(2D - E) + c_5 E\,. \tag{6}$$

As we shall see, most of these solutions are ruled out by the bootstrap principle at the eight-pion level.

Eight-pion Function

The "B" solution may be extended at the eight-pion level, as in Fig. 5. Unfortunately, it has ghosts and is thus ruled out. The same is valid for the $(2D-E)$ solution. The "D" solution has no ghosts and gives

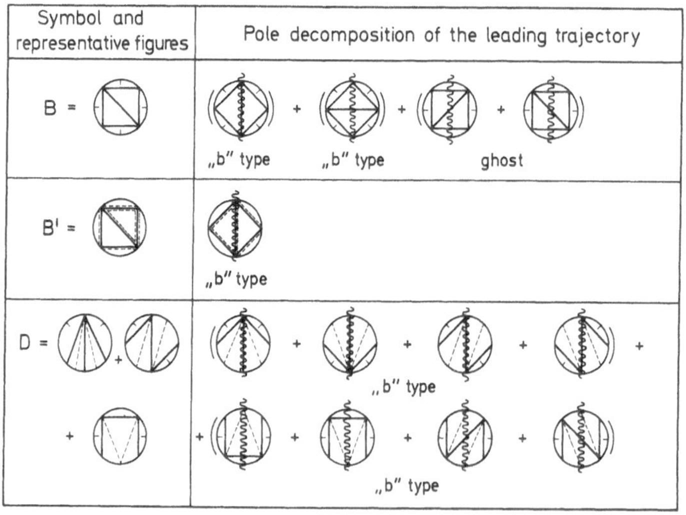

Fig. 5. Analysis of $G=+1$ trajectory for different solutions of the eight-pion functions. The different lines have the same meaning as in Fig. 4

two "b" type trajectories, as well as the "E" solution. Since, however, the "E" solution has not the A_1 trajectory and since the "$2D-E$" solution which gives the A_1 trajectory does not work for $G=+1$ poles, it follows that the eight-pion function should be written as

$$c_2 B' + c_3 D \tag{7}$$

the "B'" solution may be maintained if we want to have the $\varrho-\varrho-\varrho$ coupling which does not exist either in the "D" solution or in "E".

N-pion Function

This will be (7) where "B'" and "D" have to be defined for N pions. The "D" solution is defined in the general case through those graphs where the $G=+1$ lines are continuous and open (there are no $G=+1$ "loops") and where the $G=-1$ lines are dotted and have at most one extremity touching a $G=+1$ line. The total number of lines per graph being $N-3$. The "B'" solution is defined in the general case by the condition that it does not contain $G=-1$ lines. The total number of lines per graph being $N-3$.

The solution (7) has the $\pi-A_1$ trajectory singled and exchange degenerated and the $\varrho-f^0$ trajectory doubled ($c_2=0$) or tripled ($c_2 \neq 0$).

Leg Dependence of the Level Structure

The degeneracy of the parent trajectory is easily seen to be leg independent. By looking at the diagrams, it is seen that only one configuration always contributes for the π trajectory and two for the $\varrho-f^0$. Other cuts invariably cross a line that make the contribution non-leading.

For the daughters the situation is essentially different. At every level of the graph structure *new* possible cuts appear and these affect the degeneracy of some states. However, for a given spin, the number of lines that can be crossed is finite and so knowledge of these contributions completely determines the degeneracy of that level. However, the complete amplitude lacks factorization in the sense that some daughter states have an infinite degeneracy. This precludes their use for simple-minded loop calculations.

2. Natural Parity $G=-1$ Trajectories: Application to $\pi\varrho$ Scattering

We shall confine ourselves here to the six-point functions. Propagator rules of very similar nature of the ones described in the previous section can be written and a complete theory of $N\pi$ or K functions is thus completed.

The main feature of this case is the need for antisymmetric couplings. The B functions being inherently symmetric cannot provide them by themselves. The method calls for external polynomials that, at the relevant poles, factorize, into products of epsilon tensors times the momenta of each side of the graph. The interesting new aspect is that the spin content of the object in all channels is tricky. It can be analyzed by tensorial methods [14] or by use of the Gramm determinant [13].

The same method as before yields an amplitude

$$\begin{aligned} &\mathrm{Tr}(\tau_1, \dots \tau_N)\, \varepsilon_{\mu\nu\theta\varphi}\, P_1^{\nu} p_2^{\theta} p_3^{\varphi} \varepsilon_{\mu\alpha\beta\gamma} p_4^{\alpha} p_5^{\beta} p_6^{\gamma} \times \\ &B(1-\alpha_{12}, 1-\alpha_{23}, 1-\alpha_{45}, 1-\alpha_{56}, 2-\alpha_{34}, 2-\alpha_{25}, 1-\alpha_{13}, \times \\ &3-\alpha_{24}, 3-\alpha_{35})\,. \end{aligned} \tag{8}$$

The choice $2-\alpha_{24}$ and $3-\alpha_{35}$ is not the one that gives superficially the maximum duality. However, using the propagator rules or direct computation, one sees that any other leads to ghosts. Eq. (8) can be analyzed and one can see that the $G = -$ trajectory is simple, has alternate isospins and can be identified with the $\omega - A_2$ one. The total amplitude is seen to be the sum of (8) and the six-point function generated in Section 1, since it does not contain the $\pi - A_1$ trajectory. The degeneracy of the states may be altered by the higher point functions. This problem can be solved. Also, the leading trajectory can be shown to have the natural parity component only.

It is interesting to deduce the $\pi\varrho$ scattering [13] amplitude. The invariant functions that one obtains do not have leading Regge behaviour one-by-one, though the helicity amplitudes still do. In this respect the construction of $N\pi$ functions offers new possibilities and shows the limitation of previous efforts along this line. By demanding full Regge behaviour to each invariant amplitude and factorization with single states along the leading trajectories, one is led to solutions that are not acceptable from the bootstrap condition point of view and positivity.

References and Footnotes

1. The simplest N point functions with identical particles lying on isospinless trajectories and with negative intercepts are discussed by Satz in his lectures. We refer to his paper for references to the original work.
2. *Rittenberg, V., Rubinstein, H. R.:* Phys. Rev. Letters **25**, 191 (1970).
3. For an early attempt along these lines, see: *Olive, D., Zakrewski, W. J.:* Nucl. Phys. (to be published). Their rules however are not general enough and their solutions are not acceptable.
4. The attempts to introduce fermions have not been successful.
5. *Altarelli, G., Rubinstein, H. R.:* Phys. Rev. **183**, 1469 (1967); — *Rubinstein, H. R., Squires, E., Chaichian, M.:* Phys. Letters **30** B, 189 (1969).

6. For production amplitudes see the work reported by M. Jacob at Lund and the work of the CERN group.
7. *Rubinstein, H. R.*: Phys. Letters **32** B, 370 (1970).
8. *Fubini, S., Veneziano, G.*: Nuovo Cimento **64** A, 811 (1969); *Bardakçii, K., Mandelstam, S.*: Phys. Rev. **184**, 1640 (1969).
9. One may have an infinity of states at some mass.
10. See, e. g.: *Alessandrini, V., Amati, D., Le Bellac, M., Olive, D.*: CERN preprint TH. 1160 (1970).
11. *Dorren, D., Rittenberg, V., Rubinstein, H. R.*: Nuclear Physics B **20**, 663 (1970).
12. For example $p\bar{p} \to 4\pi$ at rest can be studied in full detail.
13. *Dorren, D., Rittenberg, V., Rubinstein, H. R., Chaichian, M., Squires, E.*: Nuovo Cimento (to be published).
14. *Gabarro, J., Gonzales Mestres, L.*: Nuovo Cimento Lettere, 1970.

Dr. *H. R. Rubinstein*
CERN
CH-1211 Genf 23
and
Weizmann Institute
Rehovot, Israel

Application of Harmonic Analysis to Inelastic Electron-Proton Scattering

W. RÜHL

Contents

1. Introduction

The aim of harmonic analysis is to map a class of functions on another class by means of Fourier transformations and to characterize the classes related in this fashion. A typical example of such relation is expressed in the classical Paley-Wiener theorem. Let us consider the class of infinitely differentiable functions on R_1 with support in the interval $[-\alpha, \alpha], \alpha > 0$. We denote this class by $C^\infty[-\alpha, \alpha]$. Fourier transforms of functions of this class are entire analytic. In fact, the Paley-Wiener theorem asserts [1, 2]: The restriction of the entire analytic function $g(s)$ to the real axis is the Fourier transform of a function $f(x) \in C^\infty[-\alpha, \alpha]$, if and only if for every $n = 0, 1, 2 \ldots$ a constant C_n exists such that for all complex s

$$|s^n g(s)| \leqq C_n e^{\alpha |\mathrm{Im}\, s|} . \tag{1}$$

It follows that $g(s)$ is of exponential order one and type α if α or $-\alpha$ is really a boundary point of the support of $f(x)$. Another version of the Paley-Wiener theorem can be formulated for distributions with support in the interval $[-\alpha, \alpha]$ (α may also be zero now). We denote this class of distributions by $D'[-\alpha, \alpha]$. We have: The restriction of the entire analytic function $g(s)$ to the real axis is the Fourier transform of a distribution $f(x) \in D'[-\alpha, \alpha]$, if and only if a constant C and an integer N exists such that for all complex s

$$|g(s)| \leqq C(1+|s|)^N e^{\alpha|\mathrm{Im}\, s|}. \tag{2}$$

Theorems which relate a class of functions that are characterized by their singularities in a compact domain with the Fourier transformed class specified by the asymptotic behaviour at infinity, can be called "theorems of Paley-Wiener type". In most applications of harmonic analysis to elementary particle physics it is the correspondence between these two properties which is of interest.

As an explicit and very important example we study the distribution

$$f_\varrho(x) = \frac{x_+^\varrho}{\Gamma(\varrho+1)} \tag{3}$$

where the subscript plus or minus denotes for arbitrary $f(x)$

$$f(x)_\pm = |f(x)|\, \theta(\pm f(x)), \quad \theta(x) = \begin{cases} 1, x>0, \\ 0, x \leqq 0. \end{cases}$$

The distribution (3) is defined to be entire analytic, that is to say, we define the regularization procedure for $\mathrm{Re}\,\varrho \leqq -1$ such that $f_\varrho(x)$ is entire analytic in ϱ. We denote such regularization "analytic regularization". In particular we have for $\varrho = -1, -2, -3 \ldots$

$$f_{-n-1}(x) = \delta^{(n)}(x), \; n = 0, 1, 2, \ldots. \tag{4}$$

Its Fourier transform is [3]

$$\begin{aligned} g_\varrho(s) &= \int_{-\infty}^{+\infty} \mathrm{d}x\, e^{ixs} f_\varrho(x) \\ &= i[e^{i\frac{\pi}{2}\varrho} s_+^{-\varrho-1} - e^{-i\frac{\pi}{2}\varrho} s_-^{-\varrho-1}]. \end{aligned} \tag{5}$$

We see that $g_\varrho(s)$ goes smoother to zero at infinity, the smoother $f_\varrho(x)$ behaves at $x=0$.

Instead of Fourier transformations on R_n we can study harmonic analysis on simple Lie groups. Again there exist theorems of Paley-Wiener type. As an example we consider the homogeneous Lorentz group $SL(2, C)$. The Fourier transform of a function $f(a)$ on $SL(2, C)$ is an operator function $T^\lambda(f)$ where λ varies over the imaginary axis (the principal series) and the operator acts in a direct sum Hilbert space

$$\sum_M{}^{\oplus} H_M.$$

In this sum M runs over all integers and half integers. For an element $a \in SL(2, C)$ we define

$$a = \begin{pmatrix} a_{11} & a_{12} \\ a_{21} & a_{22} \end{pmatrix}, \qquad |a|^2 = \sum_{i,j=1,2} |a_{ij}|^2. \tag{6}$$

The Fourier transform of the function

$$f_\sigma(a) = |a|^{2\sigma}, \quad \mathrm{Re}\,\sigma < -1 \tag{7}$$

has a nonvanishing component only in the subspace H_0 (for $M=0$) and this operator in H_0 maps the spin-zero vector on itself and all vectors orthogonal to it on zero. The matrix element between the spin-zero vectors is [4]

$$K_\sigma(\lambda) = \frac{1}{32\pi(-\sigma-1)} B\left(\tfrac{1}{2}(\lambda-\sigma-1), -\tfrac{1}{2}(\lambda+\sigma+1)\right). \tag{8}$$

This Fourier transform (8) is meromorphic in λ and can be continued to an entire analytic function of σ. The asymptotic property of $f_\sigma(a)$ for $|a| \to \infty$ reflects itself in two sequences of poles in the complex λ plane. If $\mathrm{Re}\,\sigma < -1$, $f_\sigma(a)$ is square integrable on $SL(2, C)$ with respect to the Haar measure. In this case there exists a strip of holomorphy around the imaginary λ axis of width $\mathrm{Re}\,\sigma + 1$. If we continue in σ beyond the boundary of the half plane $\mathrm{Re}\,\sigma < -1$, $K_\sigma(\lambda)$ generates a linear functional which operates on entire functions of λ by means of contour integrals. These contour integrals are obtained by analytic continuation of a formula of Parseval's kind in σ.

2. The Kinematics of Electron-Proton Scattering

We want to apply the ideas presented in the Introduction to electron-proton scattering. This forces us to deal with first the kinematics of this process. In the one-photon-exchange approximation the elastic plus inelastic differential cross section can be deduced from the diagonal matrix element

$$W_{\mu\nu}(x) = \tfrac{1}{2} \sum_{s=\pm\frac{1}{2}} \langle p, s | [j_\mu(x), j_\nu(0)] | p, s \rangle \tag{9}$$

of the commutator of the electromagnetic current between proton states. We normalize these states as

$$\langle p_1, s_1 | p_2, s_2 \rangle = (2\pi)^3 \frac{p_0}{M} \delta^{(3)}(\boldsymbol{p}_1 - \boldsymbol{p}_2)\, \delta_{s_1 s_2}.$$

From the definition (9) and time reversal invariance we find the symmetries

$$W_{\mu\nu}(x) = -W_{\nu\mu}(-x) = W_{\nu\mu}(x) \tag{10}$$

and the causality restriction

$$W_{\mu\nu}(x)=0 \quad \text{for} \quad x^2<0\,. \tag{11}$$

By a Fourier transformation we obtain

$$\begin{aligned}\tilde{W}_{\mu\nu}(k)&=\frac{1}{2\pi}\int d^4x\, e^{ikx}\, W_{\mu\nu}(x)\\ &=\left(\frac{k_\mu k_\nu}{k^2}-g_{\mu\nu}\right)W_1(k^2,\nu)+\left(p_\mu-\frac{\nu}{k^2}k_\mu\right)\left(p_\nu-\frac{\nu}{k^2}k_\nu\right)W_2(k^2,\nu)\end{aligned} \tag{12}$$

with

$$\nu=kp,\ p^2=M^2=1\,.$$

The whole dynamics is contained in the structure functions W_1 and W_2. Only spacelike momenta k are of interest in electron-proton scattering. For the expression of the differential cross section for electron-proton scattering and the total cross sections σ_t and σ_l of virtual photoproduction in terms of W_1 and W_2 see Ref. [5].

Another decomposition of $\tilde{W}_{\mu\nu}(k)$ into invariant functions is [6]

$$\begin{aligned}\tilde{W}_{\mu\nu}(k)&=(k_\mu k_\nu-g_{\mu\nu}k^2)\,\tilde{V}_1(k^2,\nu)\\ &\quad+[(p_\mu k_\nu+p_\nu k_\mu)\,\nu-p_\mu p_\nu k^2-g_{\mu\nu}\nu^2]\,\tilde{V}_2(k^2,\nu)\end{aligned} \tag{13}$$

where $\tilde{V}_{1,2}(k^2,\nu)$ are Fourier transforms (with the same normalization as in (12)) of causal functions $V_{1,2}(x)$

$$V_{1,2}(x)=0 \quad \text{for} \quad x^2<0 \quad \text{and} \quad V_{1,2}(-x)=-V_{1,2}(x)\,. \tag{14}$$

The functions $\tilde{V}_{1,2}$ can be expressed by $W_{1,2}$ and vice versa as

$$\begin{aligned}W_1&=k^2\tilde{V}_1+\nu^2\tilde{V}_2\,,\\ W_2&=-k^2\tilde{V}_2\,.\end{aligned} \tag{15}$$

In the domain of spacelike vectors k the linear combinations

$$\varepsilon(\nu)\,[\tilde{V}_1+\tilde{V}_2]$$

$$\varepsilon(\nu)\,[k^2\tilde{V}_1+\nu^2\tilde{V}_2]$$

are nonnegative. The support of the functions $\tilde{V}_{1,2}$ or $W_{1,2}$ for spacelike k is the union of an elastic part

$$2|\nu|+k^2=0$$

and an inelastic part

$$2|\nu|+k^2\geqq(1+m_\pi)^2-1\,.$$

3. Harmonic Analysis in R_4

3.1. The Scaling Law

If the quantities k^2 and ν tend to infinity such that the ratio

$$\xi = \frac{-k^2}{2\nu}, \; k^2 < 0$$

stays fixed we speak of the "Bjorken limit" [7]. We define this limit only for $\xi \neq 0$. The existence of the Bjorken limits

$$\begin{aligned} B\text{-lim}\, W_1(k^2, \nu) &= F_1(\xi), \\ B\text{-lim}\, \nu W_2(k^2, \nu) &= F_2(\xi) \end{aligned} \tag{16}$$

is called the scaling law. Due to the support restriction on $W_{1,2}$, the support of these limit functions $F_{1,2}(\xi)$ is restricted to the interval

$$-1 \leqq \xi \leqq 1 .$$

The functions $F_{1,2}(\xi)$ have the symmetry

$$\begin{aligned} F_1(\xi) &= -F_1(-\xi), \\ F_2(\xi) &= +F_2(-\xi). \end{aligned} \tag{17}$$

From (15) and (16) we have also

$$\begin{aligned} B\text{-lim}\, \nu \tilde{V}_1 &= \frac{1}{2\xi}\left[-F_1(\xi) + \frac{1}{2\xi} F_2(\xi)\right], \\ B\text{-lim}\, \nu^2 \tilde{V}_2 &= \frac{1}{2\xi} F_2(\xi). \end{aligned} \tag{18}$$

The elastic contributions to $F_{1,2}(\xi)$ are proportional to the delta-function $\delta(1-\xi)$. If the proton form factors of Sachs type, $G_E(k^2)$ and $G_M(k^2)$, behave as

$$\lim_{-k^2\to\infty} (-k^2)^{-\frac{1}{2}} G_E(k^2) = \lim_{-k^2\to\infty} G_M(k^2) = 0$$

the elastic contributions are zero. In the sequel we shall therefore assume that the limit functions $F_{1,2}(\xi)$ are continuous for $\xi \neq 0$. The positivity constraints yield finally

$$F_2(\xi) \geqq 2\xi F_1(\xi) \geqq 0 .$$

We want to show now that the scaling law results if the causal functions $V_{1,2}(x)$ exhibit a certain kind of singularities on the light cone. For this purpose we Fourier transform the distributions

$$F_\varrho(x) = \frac{1}{2\pi i}\, \varepsilon(x_0)\, g(x_0)\, \frac{(x_+^2)^\varrho}{\Gamma(\varrho+1)} \tag{19}$$

where ϱ is arbitrary complex and $g(x_0)$ is even and infinitely differentiable and tentatively assumed to be rapidly decreasing. For physical reasons we shall later have to modify the last premise. With

$$g(x_0) = \int_{-\infty}^{+\infty} d\sigma\, e^{i\sigma x_0} G(\sigma) \tag{20}$$

we get

$$\begin{aligned} \tilde{F}_\varrho(k) &= \frac{1}{2\pi} \int d^4x\, e^{ikx} F_\varrho(x) \\ &= \int_{-\infty}^{+\infty} d\sigma\, G(\sigma)\, \mathscr{F}_\varrho(k_0+\sigma, \boldsymbol{k}). \end{aligned} \tag{21}$$

Here we used the notation

$$\mathscr{F}_\varrho(k) = 2^{2\varrho+2} \varepsilon(k_0) \frac{(k_+^2)^{-\varrho-2}}{\Gamma(-\varrho-1)}. \tag{22}$$

If we denote

$$(k_0+\sigma)^2 - \boldsymbol{k}^2 = (\sigma-\sigma_+)(\sigma-\sigma_-) \tag{23}$$

so that

$$\begin{aligned} \sigma_+ &= -k_0 + |\boldsymbol{k}|, \\ \sigma_- &= -k_0 - |\boldsymbol{k}| \end{aligned} \tag{24}$$

we can write finally

$$\tilde{F}_\varrho(k) = \frac{2^{2\varrho+2}}{\Gamma(-\varrho-1)} \left(\int_{\sigma_+}^{\infty} - \int_{-\infty}^{\sigma_-} \right) d\sigma\, G(\sigma)\, [(\sigma-\sigma_+)(\sigma-\sigma_-)]^{-\varrho-2}. \tag{25}$$

As usual we understand the integral to be regularized analytically if necessary. For the particular value $\varrho = -1$ we find

$$\begin{aligned} F_{-1}(x) &= \frac{1}{2\pi i} \varepsilon(x_0)\, g(x_0)\, \delta(x^2), \\ \tilde{F}_{-1}(k) &= \frac{1}{2|\boldsymbol{k}|} [G(\sigma_+) - G(\sigma_-)], \end{aligned} \tag{26}$$

and for $\varrho = 0$

$$\begin{aligned} F_0(x) &= \frac{1}{2\pi i} \varepsilon(x_0)\, g(x_0)\, \theta(x^2), \\ \tilde{F}_0(k) &= -\frac{1}{|\boldsymbol{k}|^2} [G'(\sigma_+) + G'(\sigma_-)] \\ &\quad + \frac{1}{|\boldsymbol{k}|^3} [G(\sigma_+) - G(\sigma_-)]. \end{aligned} \tag{27}$$

We shall call ϱ the order of the light cone singularity.

In the rest system of the proton we have

$$k_0 = \nu, \; |\boldsymbol{k}| = (\nu^2 - k^2)^{\frac{1}{2}} \tag{28}$$

so that the Bjorken limits yield (signs are correlated!)

$$\underset{\nu \to \pm\infty}{B\text{-lim}} \, \sigma_\pm = \xi, \quad \underset{\nu \to \pm\infty}{B\text{-lim}} \, \sigma_+ = \mp\infty \, . \tag{29}$$

In the Regge limit $\nu \to \pm\infty$, k^2 fixed, we have correspondingly

$$\underset{\nu \to \pm\infty}{R\text{-lim}} \, \sigma_\pm = 0, \quad \underset{\nu \to \pm\infty}{R\text{-lim}} \, \sigma_\mp = \mp\infty \, . \tag{30}$$

If we identify the most singular part on the light cone of

$$V_1(x) \quad \text{with} \quad F_{-1}(x), g(x_0) = g_1(x_0)$$

and of

$$V_2(x) \quad \text{with} \quad F_0(x), g(x_0) = g_2(x_0)$$

(always in the rest system of the proton) we obtain the scaling laws with the identifications

$$\begin{aligned} G_1(\xi) &= \frac{1}{\xi}\left[-F_1(\xi) + \frac{1}{2\xi} F_2(\xi)\right], \\ G_2'(\xi) &= -\frac{1}{2\xi} F_2(\xi) \, . \end{aligned} \tag{31}$$

First we recognize that the supports of $G_1(\xi)$ and $G_2'(\xi)$ are restricted to the interval

$$-1 \leqq \xi \leqq +1 \tag{32}$$

so that both $g_1(x_0)$ and $x_0 g_2(x_0)$ are entire functions. The second function being odd, we may assume that $g_2(x_0)$ is itself entire. We neglect this way a possible deltafunction $\delta(x_0)$ in $g_2(x_0)$ which would lead to the undefined expression $\delta(x_0)\,\varepsilon(x_0)$ in (19). This is equivalent to setting $G_2(\xi)$ equal zero outside the interval (32). Finally we want to use physical information to determine the behaviour of $g_{1,2}(x_0)$ at infinity which by *Paley-Wiener* arguments is connected with the smoothness of the functions $G_1(\xi)$ and $G_2'(\xi)$. Since we want our arguments that led to the formulae (13) to hold true, we have to assume at least that both functions $G_1(\xi)$ and $G_2'(\xi)$ are continuous for $\xi \neq 0$. This means that $F_{1,2}(\xi)$ should be continuous for $\xi \neq 0$, which every physicist will certainly accept. At $\xi = 0$, however, we must allow for distribution singularities, which can be determined by means of the information supplied by the Regge limits.

The Regge limit of a distribution $\tilde{F}_\varrho(k)$ (25) is easily evaluated and solely determined by the singularity of $G(\sigma)$ at $\sigma = 0$ only if the support of the distribution $\mathscr{F}_\varrho(k)$ shrinks to the light cone, i.e. if the order of the light cone singularity of $F_\varrho(x)$ is one of the integers $\varrho=-1,0,1,\ldots$. The Regge limit in the nonintegral case can best be treated with the method of harmonic analysis on $SL(2,C)$ due to Toller and we postpone it therefore till Section 4.3. We cannot expect that the most singular part of $V_{1,2}(x)$ on the light cone that is responsible for scaling in the Bjorken limit, contains simultaneously the most singular function $g_{1,2}(x_0)$ at infinity which is responsible for the Regge limit. Instead we must be aware of the possibility that $V_{1,2}(x)$ contain terms with lower order singularities on the light cone but with the same asymptotic property of their weight functions $g_{1,2}(x_0)$ at $x_0\to\infty$. Having this in mind we make the simplest possible model ansatz

$$\begin{aligned} V_1(x) &= \sum_{n=0}^{N} F_{-1+n}(x)+R_{1,N}(x)\,, \\ V_2(x) &= \sum_{n=0}^{N} F_n(x)+R_{2,N}(x) \end{aligned} \tag{33}$$

as an expansion in a sequence with decreasing orders of light cone singularities. Each $F_{-1+n}(x)\,(F_n(x))$ contains a weight function $g_{1,n}(x_0)$ $(g_{2,n}(x))$. If $G_{1,n}^{(n)}(\xi)$ and $G_{2,n}^{(n+1)}(\xi)$ are continuous for $\xi \neq 0$, we obtain from (25)

$$\begin{aligned} \tilde{V}_1(k) &= \sum_{n=0} \frac{2^{n-1}(-1)^n}{|\boldsymbol{k}|^{n+1}} G_{1,n}^{(n)}(\sigma_+) - \begin{array}{l}\text{terms obtained by the}\\ \text{replacement } k\to -k,\end{array} \\ &\quad +\tilde{R}_{1.N}(k) \\ \tilde{V}_2(k) &= \sum_{n=0}^{N} \frac{2^{n}(-1)^{n+1}}{|\boldsymbol{k}|^{n+2}} G_{2,n}^{(n+1)}(\sigma_+) - \begin{array}{l}\text{terms obtained by the}\\ \text{replacement } k\to -k\end{array} \\ &\quad +\tilde{R}_{2,N}(k) \end{aligned} \tag{34}$$

where we kept only the lowest powers of $|\boldsymbol{k}|^{-1}$ for each n. This expansion (34) has to be reconciled with the Regge limits [8] (see also our Section 4.2)

$$\begin{aligned} W_1(k^2,\nu) &\cong \beta_1(k^2)\,\nu^{\alpha(0)}, \\ W_2(k^2,\nu) &\cong k^2\,\beta_2(k^2)\,\nu^{\alpha(0)-2} \end{aligned} \tag{35}$$

where $\alpha(0)=1$ for Pomeranchuk exchange. For $k^2 \neq 0$ we have from (15) and (35)

$$\begin{aligned} \underset{\nu\to+\infty}{R\text{-lim}}\,\nu^{-1}\tilde{V}_1 &= \frac{1}{k^2}\,[\beta_1(k^2)+\beta_2(k^2)]\,, \\ \underset{\nu\to+\infty}{R\text{-lim}}\,\nu\,\tilde{V}_2 &= -\beta_2(k^2)\,. \end{aligned} \tag{36}$$

Both (34) and (36) can be made to agree if we assume that

$$
\begin{aligned}
G_{1,n}(\xi) &= \frac{\Delta_{1,n}}{(n+1)!\,\xi^2} + \text{less singular distribution at } \xi=0\,, \\
G'_{2,n}(\xi) &= \frac{\Delta_{2,n}}{n!\,\xi} + \text{less singular distribution at } \xi=0
\end{aligned}
\tag{37}
$$

where the main terms are to be understood as a principal value and its distribution theoretic derivative [9]. From (37) we deduce

$$
\begin{aligned}
&\lim_{\xi\to 0} \xi^{n+2} G^{(n)}_{1,n}(\xi) = (-1)^n \Delta_{1,n}\,, \\
&\lim_{\xi\to 0} \xi^{n+1} G^{(n+1)}_{2,n}(\xi) = (-1)^n \Delta_{2,n}
\end{aligned}
\tag{38}
$$

and finally

$$
\begin{aligned}
\beta_1(k^2)+\beta_2(k^2) &= -\tfrac{1}{2}\sum_{n=0}^{N} \Delta_{1,n}\left(\frac{4}{-k^2}\right)^{n+1} + \varrho_{1,N}(k^2)\,, \\
\beta_2(k^2) &= +\tfrac{1}{2}\sum_{n=0}^{N} \Delta_{2,n}\left(\frac{4}{-k^2}\right)^{n+1} + \varrho_{2,N}(k^2)\,.
\end{aligned}
\tag{39}
$$

This result is still a rigorous consequence of our ansatz (33). It is, however, a new hypothesis to interpret these expansions as asymptotic expansions for $-k^2\to\infty$, as has been done first by *Abarbanel et al.* [8]. From (37) we find for the behaviour of $g_{1,n}(x_0)$ and $g_{2,n}(x_0)$ at infinity

$$
\begin{aligned}
g_{1,n}(x_0) &= -\pi\Delta_{1,n}|x_0| + \text{less singular terms}, \\
g_{2,n}(x_0) &= -\pi\Delta_{2,n}|x_0|^{-1} + \text{less singular terms}.
\end{aligned}
\tag{40}
$$

The light cone singularity of highest order which is responsible for scaling (the term $n=0$ in (33)) and the additional hypothesis on the asymptotic character of the expansion (39) give the prediction

$$
\lim_{-k^2\to\infty} k^2\beta_{1,2}(k^2) < \infty \tag{41}
$$

which is therefore independent of the specific structure of the ansatz (33).

Our presentation of the material of this section is the result of an attempt to formulate the ideas of Refs. [10, 11] in mathematically more concise form. The material presented in the sequel is another product of these entertainments on the field of mathematics. For a review of the experimental situation concerning the scaling law see Refs. [5, 12].

3.2. Time Dependent Light Cone Singularities

The distribution $F_\varrho(x)$ (19) considered in Section 3.1 possesses a time-independent order ϱ of the singularity on the light cone. We may ask ourselves whether any physical restrictions can exclude the possibility that a causal function exhibits a singularity of the type $(x_+^2)^{\varrho(x_0)}$. Physical restrictions are the spectrum conditions which are normally used to derive spectral representations for the causal functions. The constraints on the light cone singularities are then implicitly contained in these spectral representations. To make them explicit turns out to be another and a very beautiful application of harmonic analysis indeed.

The best suited spectral representation for the investigation of the spin-averaged diagonal matrix element of the current commutator is the Deser-Gilbert-Sudarshan representation [13]. For a causal function $V(x)$ of the type $V_{1,2}(x)$ and in the rest system of the proton the most popular form of this spectral representation is

$$V(x) = -i\int \mathrm{d}s\,\Delta(x,s)\,\varphi(x_0,s) \tag{42}$$

where $\Delta(x,s)$ is the Pauli-Jordan function

$$\Delta(x,s) = i(2\pi)^{-3}\int \mathrm{d}^4k\,\varepsilon(k_0)\,\delta(k^2-s)\,\mathrm{e}^{-ikx}\,. \tag{43}$$

$\varphi(x_0,s)$ has support on $s \geqq 0$ and obeys itself the integral representation

$$\begin{aligned}\varphi(x_0,s) &= \int_{-1}^{+1} \mathrm{d}\sigma\,\mathrm{e}^{i\sigma x_0}\tilde{\varphi}(\sigma,s)\,,\\ \tilde{\varphi}(\sigma,s) &= \tilde{\varphi}(-\sigma,s)\end{aligned} \tag{44}$$

which allows us to apply Paley-Wiener type arguments to it. Since (42) is an integral over the product of possibly two distributions it has to be handled with care. The alternative form

$$V(x) = (2\pi)^{-3}\int \mathrm{d}^4k\,\varepsilon(k_0)\,\mathrm{e}^{-ikx}\varphi(x_0,k^2) \tag{45}$$

is well defined whenever $\varepsilon(k_0)\,\varphi(x_0,k^2)$ is a distribution in k for fixed x_0. Of course, $V(x)$ ought to be a distribution of a specific kind, too, say temperate or more general. For the distributions $F_\varrho(x)$ (19) the DGS spectral representation is valid with

$$\varphi(x_0,s) = g(x_0)\,2^{2\varrho+2}\,\frac{s_+^{-\varrho-2}}{\Gamma(-\varrho-1)}\,. \tag{46}$$

In this case $\tilde{\varphi}(\sigma,s)$ is proportional to $G(\sigma)$ (20) which explains the support restriction in (44). On the other hand we learn from this example that the DGS representation does not take full account of the support conditions in momentum space. In fact the support of $\tilde{F}_\varrho(k)$ (25) is

$$2|\nu|+k^2 \geqq -1 \quad \text{instead of} \quad 2|\nu|+k^2 \geqq 0 .$$

Our aim is to construct a class of distributions $V(x)$ which have a similar form as $F_\varrho(x)$ (19) but with a time-dependent order $\varrho(x_0)$. In other words we want to replace (46) by

$$\varphi(x_0, s) \underset{s\to\infty}{\cong} \Phi(x_0)\, s_+^{-\varrho(x_0)-2} \tag{47}$$

where $\Phi(x_0)$ is an appropriately chosen function which makes $\varphi(x_0, k^2)$ satisfy (44) and which has zeros whenever

$$\varrho(x_0)=n-1, \qquad n=0,1,2,\ldots . \tag{48}$$

Even, entire functions in x_0 which depend parametrically on s and in the limit $s\to\infty$ behave as required in (47) can be constructed by means of the Mittag-Leffler functions [14]

$$E_\gamma(z)=\sum_{n=0}^{\infty} \frac{z^n}{\Gamma(1+\gamma n)} . \tag{49}$$

These functions are entire, have exponential order γ^{-1} and are of type one. For $|z|\to\infty$ in the direction $\theta=\arg z$ they behave as

$$\begin{aligned} &E_\gamma(z)=\gamma^{-1}\sum_k \exp[|z|^{1/\gamma} e^{i(2\pi k+\theta)\gamma^{-1}}]+O(|z|^{-1}) \\ &k \text{ integral} \qquad \left|k+\frac{\theta}{2\pi}\right| \leqq \tfrac{1}{4}\gamma . \end{aligned} \tag{50}$$

We introduce therefore the function

$$\begin{aligned} G(z,s)&=\theta(s)\, s^\beta E_{2/\gamma}([\alpha\,|\lg s|]^{2/\gamma} z^2), \\ \alpha&\geqq 0, \beta \quad \text{arbitrary}, \quad \gamma>0 \end{aligned} \tag{51}$$

which for real $z=x_0$ is asymptotically equal to

$$G(x_0, s) \underset{s\to\infty}{\cong} \tfrac{1}{2}\gamma\, s^{\alpha|x_0|^\gamma+\beta} . \tag{52}$$

It is even, entire, has exponential order γ and the type $\alpha|\lg s|$. In order that it fulfills the Paley-Wiener theorem connected with the integral representation (44) we must restrict γ to $\gamma<1$. Since we have then still an exponential increase along the real axis in z, $\tilde{\varphi}(\sigma, s)$ is not a distribution. This disease can, however, be cured by a multiplier $\Phi(x_0)$ such that

$$\varphi(x_0, s)=\Phi(x_0)\, G(x_0, s) \tag{53}$$

if $\Phi(x_0)$ is chosen as an even, entire function of type and order one (i.e. it satisfies (44) itself) with a sufficient exponential fall off along the real axis. Such a multiplier exists as a theorem by *Mandelbrojt* [15] asserts (because of lack of space we give the theorem rather than an explicit example).

Given a nondecreasing function $C(u) \geqq 0$ for $u \geqq 0$ such that

$$\int_1^\infty \frac{C(u)}{u^2} \mathrm{d}u < \infty . \tag{54}$$

Then there exists a nontrivial even, entire analytic function $\Phi(z)$ such that

$$|\Phi(z)| \leqq \mathrm{e}^{a|\mathrm{Im}\, z| - C(|z|)} \tag{55}$$

for any $a > 0$. Wee choose δ in the interval

$$\gamma < \delta < 1$$

and set

$$C(u) = u^\delta + c,\ c \geqq 0,\ a = 1 . \tag{56}$$

The resulting function $\Phi(x_0)$ has all properties required by (47) and (53) after an eventual multiplication with a polynomial, that gives us the finite number of zeros required in (48). The weight function $\tilde{\varphi}(\sigma, s)$ constructed in this fashion is even infinitely differentiable in σ. The resulting quantity $V(x)$ has as highest singularity on the light cone

$$V(x)_{\mathrm{h.s.}} = \frac{\gamma}{16\pi i} \varepsilon(x_0)\, \Phi(x_0) \frac{\Gamma(-\varrho(x_0) - 1)}{\Gamma(\varrho(x_0) + 1)} (\tfrac{1}{4} x_+^2)^{\varrho(x_0)} \tag{57}$$

with

$$-\varrho(x_0) = \alpha |x_0|^\gamma + \beta + 2 . \tag{58}$$

It is a distribution but not temperate.

Replacing $\alpha |\lg s|$ in (51) by any other positive function $f(s)$ which satisfies

$$\lim_{s \to \infty} f(s) = +\infty \tag{59}$$

yields other classes of distributions with time dependent order of singularity on the light cone. Among them there exist also temperate distributions. It would be quite interesting to investigate the Bjorken and Regge limits of several such classes.

4. Harmonic Analysis on $SL(2, C)$

4.1. Elementary Kinematics

The commutator (9) is the absorptive part of the virtual Compton forward scattering amplitude with the ingoing and outgoing photon lying on the same mass shell. As any elastic forward scattering amplitude it can be submitted to a Toller decomposition, i.e. a harmonic analysis on the little group $SL(2, C)$ of the inhomogeneous group $SL(2, C) \times T_4$ for vanishing four-momentum.

On the mass shell

$$k^2 = -\mu^2, \ \mu > 0$$

we take as reference momentum

$$k^R = (0, 0, 0, \mu) \tag{60}$$

whose little group is $SU(1, 1)$. Similarly the little group for the reference momentum

$$p^R = (1, 0, 0, 0) \tag{61}$$

on the proton mass shell is $SU(2)$. We denote elements of $SL(2, C)$ by a, elements of $SU(1, 1)$ by v, and elements of $SU(2)$ by u. By definition we have

$$\begin{aligned} u^+ u &= e\,, \\ v^+ \sigma_3 v &= \sigma_3\,. \end{aligned} \tag{62}$$

By

$$\begin{aligned} \mathbf{p}^R &= a(p)\, \mathbf{p}\, a(p)^+, \qquad a(p) \in SL(2, C) \\ \mathbf{p} &= p_0 e + \boldsymbol{p\sigma} \end{aligned} \tag{63}$$

and in the same fashion for k, we define cosets $u\,a(p)$ of $SU(2)$ respectively $v\,a(k)$ of $SU(1, 1)$ in $SL(2, C)$. Instead of p and k we can therefore use these cosets in $SL(2, C)$ as variables on which our scattering amplitude is defined, or equivalently replace each coset by one element of itself, the boost $a(p)$ or $a(k)$. The physical meaning of the rotation u or v which distinguishes the elements within one coset, is that of a rotation applied to the particle in its reference state p^R (61) or k^R (60).

Next we consider the amplitude not as a function on the pair of cosets or boosts, but we want to extend its definition on all elements of the cosets, thereby from three-dimensional manifolds of momenta, boosts or cosets onto the sixdimensional group manifolds of $SL(2, C)$. The simplest prescription would be to require constancy on each coset. It is, however, more convenient (though all prescriptions are basically equivalent) to require that the value of the amplitude at two different points of the same coset, say $u_1 a(p)$ and $u_2 a(p)$, differs by the rotation $u_1 u_2^{-1}$ applied to the rest state, such that in going from one element of the coset to another element, the spin indices are linearly transformed. We may summarize the substitution of arguments by

$$\begin{aligned} p &\xrightarrow[\text{one-to-one}]{} a(p) \xrightarrow[\text{one-to-}SU(2)]{} u\,a(p) = a_1\,, \\ k &\xrightarrow[\text{one-to-one}]{} a(k) \xrightarrow[\text{one-to-}SU(1,1)]{} v\,a(k) = a_2\,. \end{aligned}$$

Because of Lorentz invariance we have as effective arguments only

$$a_2 a_1^{-1} = v\, a(k)\, a(p)^{-1} u^{-1} \tag{64}$$

in other words: the amplitude depends effectively only on the double cosets

$$a = v n u \tag{65}$$

of $SU(1, 1)$ and $SU(2)$ in $SL(2, C)$. The representatives n of these double cosets form a one-dimensional manifold, which by definition is parametrized by the "Toller angle". In our case it can be shown that the parametrization

$$n = e^{\frac{1}{2}\eta\sigma_3}, \quad -\infty < \eta < \infty \tag{66}$$

is possible.

Let us now keep the proton in its rest state $p = p^R$. Comparison of (64) with (65) shows that the k-boost must necessarily be decomposable as

$$a(k) = v n u\,. \tag{67}$$

The rotation u is determined such that it rotates the three momentum $\boldsymbol{k}$ into the positive third axis

$$|\boldsymbol{k}|\, \sigma_3 = u(\boldsymbol{k}\boldsymbol{\sigma})\, u^{-1} \tag{68}$$

Inserting (66) to (68) into (63) yields

$$\mu\sigma_3 = e^{\frac{1}{2}\eta\sigma_3}(k_0 e + |\boldsymbol{k}|\, \sigma_3)\, e^{\frac{1}{2}\eta\sigma_3} \tag{69}$$

or finally (see (28))

$$k_0 = v = -\mu\, \mathrm{sh}\,\eta, \ |\boldsymbol{k}| = \mu\, \mathrm{ch}\,\eta\,. \tag{70}$$

The fact that the Toller angle varies over the whole real axis instead of over the positive half axis is obviously due to the one-shell form of the hyperboloid $k^2 = -\mu^2$. Finally we mention that the projection of the Haar measure onto the manifold of double cosets (67) gives the measure

$$\mathrm{const} \times \mathrm{ch}^2\eta\, d\eta\,.$$

4.2. Toller Analysis

The idea of harmonic analysis is to expand the amplitude into coordinate functions (matrix elements) of unitary irreducible representations of $SL(2, C)$. In order to accomodate the double coset splitting (67) it is convenient to use coordinate functions with respect to two different bases. One basis reduces the $SL(2, C)$ representation restricted to $SU(2)$ and is discrete, the other reduces the $SL(2, C)$ representation restricted to $SU(1, 1)$ and in general has both a discrete and a continuous part [16, 17]. However, the spin of the photon does not transform as a unitary representation of the little group $SU(1,1)$, but as a finite dimensional nonunitary representation. This necessitates an analytic continuation in the photon spin, which gives rise to a "discrete series" of $SL(2,C)$ (see below). The whole formalism is too complicated to be presented here. Instead we quote therefore only the resulting formulae. We shall concentrate our interest on the function $\tilde{V}_2$ (or W_2), for which the formulae are most compact. The general investigation is contained in the literature [18–20].

We define a Fourier transform of $\tilde{V}_2(\eta,\mu^2)$

$$\tilde{V}_2(\eta,\mu^2)=\tilde{V}_2(k), \qquad \tilde{V}_2(-\eta,\mu^2)=-\tilde{V}_2(\eta,\mu^2) \tag{71}$$

on $SL(2,C)$ by

$$H_2(\lambda)=\frac{2}{(\lambda^2-1)(\lambda^2-4)}\int_{-\infty}^{+\infty} d\eta\, \mathrm{ch}\,\eta\, e^{-\eta\lambda}\,\tilde{V}_2(\eta,\mu^2)\,K(\lambda,\eta) \tag{72}$$

with the kernel

$$K(\lambda,\eta)=(\lambda+1)(\lambda+2)\,e^{2\eta}+2(\lambda^2-4)+(\lambda-1)(\lambda-2)\,e^{-2\eta}. \tag{73}$$

This Fourier transformation (72) can be inverted by

$$\tilde{V}_2(\eta,\mu^2)=(2\,\mathrm{ch}\,\eta)^{-5}\,\frac{i}{2\pi}\int_{C_\pm} d\lambda\, e^{-\eta\lambda}\,H_2(\lambda)\,K(\lambda,\eta)\,. \tag{74}$$

The general proof of (74) was given in Ref. [18], however, it is easy to establish a direct proof which we postpone to the Appendix 1. The group theoretical meaning of the decomposition (72), (74) cannot be inspected from such direct proof of formula (74) but only in the general context.

A function $\tilde{V}_2(\eta,\mu^2)$ which falls off faster than any power of $\mathrm{ch}\,\eta$ at $|\eta|\to\infty$, has a Fourier transform (72) for which $(\lambda^2-1)\,H_2(\lambda)$ is an entire analytic function of λ. Holomorphy at $\lambda=\pm 2$ follows from the odd parity of $\tilde{V}_2(\eta,\mu^2)$ (71). The contour $C_\pm$ consists of the upward oriented imaginary axis and a positively (negatively) oriented circle around $+1(-1)$. The contributions of the circles (the "discrete series") appear in the course of the analytic continuation of the representation of the little group $SU(1,1)$. It is easy to show that both contours are in fact equivalent (see Appendix 1).

If the function $\tilde{V}_2(\eta,\mu^2)$ falls off like a power of $\mathrm{ch}\,\eta$, say like

$$\tilde{V}_2(\eta,\mu^2)\cong c(\mathrm{ch}\,\eta)^{-3-\varepsilon}, \qquad \varepsilon>0 \tag{75}$$

then $(\lambda^2-1)\,H_2(\lambda)$ is holomorphic in a strip of half width ε around the imaginary axis. A physical amplitude can always be regularized by multiplication with an appropriate power of $\mathrm{ch}\,\eta$, say $(2\,\mathrm{ch}\,\eta)^\sigma$, such that a strip of holomorphy results. The Fourier transform can be expected to be analytic in σ. Inserting this Fourier transform of the regularized amplitude into the inversion formula (74) and continuing both sides in σ along an appropriately chosen path, forces us in general to deform the contours $C_\pm$ into $C'_\pm$.

The well known relation

$$\lambda = \alpha(0) + 1 \tag{76}$$

between a Toller singularity and its parent Regge singularity $\alpha(t)$ at $t=0$ gives us $\lambda = +2$ for the upper edge of the Pomeranchuk singularity (the contour is deformed such that this singularity still is on the left of it). There is also a mirror singularity at $\lambda = -2$ due to the symmetry

$$H_2(\lambda) = -H_2(-\lambda)\,. \tag{77}$$

In order to obtain the contribution of the Pomeranchuk singularity in the limit $\nu \to +\infty$ or $\eta \to -\infty$ (see (70)) we move the contour $C'_\pm$ to the left. The highest order term $e^{-2\eta}$ in $K(\lambda, \eta)$ is, however, multiplied with the factor $\lambda - 2$. A single pole in $H_2(\lambda)$ at precisely $\lambda = 2$ can therefore not cause the expected Pomeranchuk behaviour (see (36))

$$\tilde{V}_2(k) \cong -\beta_2(k^2)\, \nu^{-1}\,. \tag{78}$$

This must be due to a more complicated singularity, a cut or a double pole. In the first case additional factors depending logarithmically on ν will occur. In the second case (78) is reproduced by

$$H_2(\lambda) = \frac{2\beta_2(k^2)}{\mu(\lambda-2)^2} + O((\lambda-2)^{-1})\,. \tag{79}$$

4.3. The Connection between Harmonic Analysis in R_4 and on $SL(2, C)$

Instead of the complicated integral transformation (72), we study now the singularity structure of the integral transform

$$I(\lambda) = \int_{-\infty}^{+\infty} d\eta\, \mathrm{sh}\,\eta\lambda\, V(\eta) \tag{80}$$

where $V(\eta)$ is an odd function of η, for example $V(\eta) = \tilde{V}_2(\eta, \mu^2)$. In the latter case we have

$$\begin{aligned} &H_2(\lambda) \\ &= -[(\lambda^2-1)(\lambda^2-4)]^{-1}\{(\lambda+1)(\lambda+2)\, I(\lambda-3) + 3(\lambda-1)(\lambda+2)\, I(\lambda-1) \\ &\quad + 3(\lambda+1)(\lambda-2)\, I(\lambda+1) + (\lambda-1)(\lambda-2)\, I(\lambda+3)\} \end{aligned} \tag{81}$$

which proves that knowledge of $I(\lambda)$ is sufficient to determine $H_2(\lambda)$. In particular we want to identify $V(\eta)$ with $\tilde{F}_\varrho(k)$ (21). This gives

$$I(\lambda)=\tfrac{1}{8}\left(\frac{2}{\mu}\right)^{2\varrho+4}\int_{-\infty}^{+\infty} d\sigma\, G(\sigma)$$
$$\times\left\{\frac{\Gamma(\lambda+\varrho+2)}{\Gamma(\lambda+1)}\left|\frac{\sigma}{\mu}\right|^{\lambda} {}_2F_1\left(\varrho+2,\lambda+\varrho+2;\lambda+1;-\left(\frac{\sigma}{\mu}\right)^2\right)\right.$$
$$\left.-\text{ the same expression with } \lambda \text{ replaced by } -\lambda\right\} \tag{82}$$

As in Section 3.1 we assume that $G(\sigma)$ is continuous for $\sigma \neq 0$ and has the support $-1 \leqq \sigma \leqq +1$. The singularities of $I(\lambda)$ can then have two origins. If $G(\sigma)$ is infinitely differentiable at $\sigma=0$, then

$$I(\lambda)=\Gamma(\lambda+\varrho+2)\,f(\lambda)-\Gamma(-\lambda+\varrho+2)\,f(-\lambda) \tag{83}$$

where $f(\lambda)$ is entire in λ. Global properties of $G(\sigma)$ go into the residues of these poles. If $G(\sigma)$ is an arbitrary distribution at $\sigma=0$ but ϱ assumes one of the values $-1+n$, $n=0,1,2,\ldots$, (in which case the support of $\mathscr{F}_\varrho(k)$ (22) shrinks to the light cone) it is alone the singularity of $G(\sigma)$ at $\sigma=0$ which determines the singularity structure of $I(\lambda)$. This was the case of our model ansatz (33). In general both kinds of singularities are present.

Explicit expressions of $I(\lambda)$ (82) in terms of $g(x_0)$ are very complicated. They have the general form

$$I(\lambda)=-\frac{\sin\pi\lambda/2}{4\pi}\left(\frac{2}{\mu}\right)^{2\varrho+3}\int_{-\infty}^{+\infty} dx_0\, g(x_0)\, M(\lambda,\varrho,\mu x_0). \tag{84}$$

At least an asymptotic formula can be given for the kernel

$$M(\lambda,\varrho,x_0)$$
$$\underset{x_0\to\infty}{\cong}\sum_{m=0}^{\infty}\frac{\Gamma(\varrho+m+2)\,\Gamma(\lambda+\varrho+m+2)\,\Gamma(\lambda+2m+1)}{m!\,\Gamma(\varrho+2)\,\Gamma(\lambda+m+1)}|x_0|^{-\lambda-2m-1}$$
$$+\text{ terms obtained by the replacement } \lambda\to-\lambda. \tag{85}$$

The singularities of $I(\lambda)$ induced by the asymptotic property of $g(x_0)$ are clearly determined by the series (85). However, the poles at $\pm\lambda=-\varrho-2-n$ (see (83)) can also be obtained from this asymptotic series (85), since only a finite number of terms of this series add to the residue, and these terms are exact. In that case $|x_0|^{\pm\lambda-2m-1}$ has to be regarded as a distribution with analytic regularization.

Finally we consider $V(\eta)$ as a superposition of distributions $\tilde{F}_\varrho(k)$ with $\varrho=0,1,2\ldots$ as in Section 3.1 Eqs. (34) to (41). If the functions $g_{2,n}(x_0)$ have the asymptotic behaviour (40), Eqs. (82) or (84), (85) yield a pole of first order in $I(\lambda)$ at the position $\lambda = -1$. Its residue is

$$-\frac{1}{2}\sum_{n=0}^{N} \Delta_{2,n}\left(\frac{2}{\mu}\right)^{2n+3} + \text{remainder}\,.$$

Due to (81) this induces a pole of second order in $H_2(\lambda)$ at $\lambda = +2$ and comparison with (79) gives the asymptotic expansion for $\mu \to \infty$

$$\beta_2(-\mu^2) = \frac{1}{2}\sum_{n=0}^{N}\left(\frac{2}{\mu}\right)^{2n+2} \Delta_{2,n} + \varrho_{2,N}(-\mu^2) \tag{86}$$

which we know already from (39). With the help of Eq. (82) we could now easily modify our ansatz (33) such that it allows for nonintegral orders of light cone singularities.

4.4. Conclusion

A harmonic analysis on $SL(2, C)$ which permits us to deal with the Bjorken limit can also be invented [20, 21]. Since it involves problems falling outside the scope of these lectures I will not discuss it here. The connection between the Fourier transforms of Refs. [19–21] and the notation $H_2(\lambda)$ used in these lectures is explained in Appendix 2.

Acknowledgement: We thank Dr. *B. Nagel* for the suggestion to use *Mandelbrojt*'s theorem for the construction of the multiplier function $\Phi(x_0)$ in Section 3.2.

Appendix

A.1. Proof of the Inversion Formula

We have actually to prove the "completeness relation"

$$\frac{i}{2\pi}\int_{C_\pm} d\lambda\, e^{-\eta\lambda} K(\lambda,\eta)\frac{e^{-\eta'\lambda}K(\lambda,\eta') - e^{\eta'\lambda}K(\lambda,-\eta')}{(\lambda^2-1)(\lambda^2-4)}$$
$$= (2\,\mathrm{ch}\,\eta)^2(2\,\mathrm{ch}\,\eta')^2[\delta(\eta-\eta') - \delta(\eta+\eta')]\,. \tag{A–1}$$

We split the product of the two kernels as

$$K(\lambda,\eta)\,K(\lambda,\eta')$$
$$= (\lambda^2-1)(\lambda^2-4)(2\,\mathrm{ch}\,\eta)^2(2\,\mathrm{ch}\,\eta')^2 + R(\lambda,\eta,\eta') \tag{A–2}$$

such that $R(\lambda,\eta,\eta')$ is a polynomial in λ of degree three. Inserting the first part of (A–2) into (A–1) gives already the desired result. We have therefore to prove that the remainder of (A–2) inserted into (A–1) gives zero. Whatever the result of inserting the functions $R(\lambda,\eta,\eta')$ into (A–1) is, it must be a measurable function of η'. It suffices therefore to prove that this function vanishes almost everywhere.

From (73) we find the identities

$$\begin{aligned} e^{-\eta-\eta'} R(1,\eta,\eta') &= e^{\eta+\eta'} R(-1,\eta,\eta'), \\ e^{-2(\eta+\eta')} R(2,\eta,\eta') &= e^{2(\eta+\eta')} R(-2,\eta,\eta') = 144. \end{aligned} \tag{A–3}$$

The first identity implies the equivalence of the contours $C_{\pm}$. Let us consider then the first part of the integral (A–1) with $K(\lambda,\eta)\,K(\lambda,\eta')$ replaced by $R(\lambda,\eta,\eta')$ and antisymmetrize it in η' later. If $\eta+\eta'<0$ we start from C_- and pull the contour to $-\infty$. There is at most a contribution from the pole at $\lambda=-2$. If $\eta+\eta'>0$, we start from C_+ and shift the contour to $+\infty$. Again we get a contribution at most from $\lambda=+2$. Due to the second identity (A–3) both contributions of $\lambda=\pm 2$ are identical. Antisymmetrization in η' yields zero. This completes the proof.

A.2. A Notation

Our Fourier transform $H_2(\lambda)$ (72) can be expressed in terms of the Fourier transforms $F_\alpha(\chi,\zeta)$ of Ref. [19] as

$$H_2(\lambda) = \frac{1}{2^{\frac{1}{2}}\lambda(\lambda^2-4)\,\mu^2}\left[3(\lambda^2-1)\,F_{(0.0)}(\lambda,+)+2F(\lambda,+)\right]. \tag{A–4}$$

References

1. *Gel'fand, I. M., Shilov, G. E.:* Generalized Functions vol. 1, Chapter II, Section 1.1. New York: Academic Press 1964.
2. *Donoghue, W. F.:* Distributions and Fourier Transforms, Section 43. New York: Academic Press 1969.
3. Ref. [1], Table 1, Eq. 21.
4. *Rühl, W.:* The Lorentz Group and Harmonic Analysis, Section 4.6. New York: Benjamin 1969.
5. *Nauenberg, M.:* Inelastic electromagnetic nucleon form factors, Talk presented at the Rencontre de Moriond, Meribel, France, March 1970, CERN preprint TH 1157.
6. *Meyer, J. W., Suura, H.:* Phys. Rev. **160**, 1366 (1967).
7. *Bjorken, J. D.:* Phys. Rev. **179**, 1547 (1969).
8. *Abarbanel, H. D. I., Goldberger, M. L., Treiman, S. B.:* Phys. Rev. Letters **22**, 500 (1969).
9. Ref. [1], Chapter I, Section 3.3.
10. *Leutwyler, H., Stern, J.:* Singularities of current commutators on the light cone, CERN preprint TH 1138 (1970).
11. *Brandt, R. A.:* Electroproduction structure functions, integral representations, and light cone commutators, Rockefeller University preprint (1970).
12. *Gilman, F. J.:* Proceedings of the 4th International Symposium on Electron and Photon Interactions at High Energies, Daresbury Nuclear Physics Laboratory 1969, p. 177.
13. *Deser, S., Gilbert, W., Sudarshan, E. C. G.:* Phys. Rev. **115**, 731 (1959).
14. *Sansone, G., Gerretsen, J.:* Lectures on the Theory of Functions of a Complex Variable, vol. 1, Section 6.13. New York: S.-H. Service Agency, 1960.

15. *Mandelbrojt, S.*: Fonctions Entières et Transformées de Fourier, Applications, 1967, to appear in the Collection of the Mathematical Society of Japan.
16. *Sciarrino, A., Toller, M.*: J. Math. Phys. **8**, 1252 (1967).
17. *Rühl, W.*: Commun. Math. Phys. **6**, 312 (1967).
18. *Rühl, W.*: Nuovo Cimento **68** A, 213 (1970).
19. *Rühl, W.*: Nuovo Cimento **68** A, 235 (1970).
20. *Rühl, W.*: The 0(3,1) analysis of currents and current commutators, Talk presented at the IX. Internationale Universitätswochen für Kernphysik, Schladming, Austria, Acta Physica Austriaca, Suppl. VII, 392 (1970).
21. *Rühl, W.*: Nuovo Cimento **69** A, 231 (1970).

Professor Dr. *W. Rühl*
Universität Trier-Kaiserslautern
Naturw.-Techn. Fakultät
D-6750 Kaiserslautern

Small-Distance Behaviour in Field Theory

K. Symanzik

Contents

Introduction

This is a seminar on recent material [1] related to, but also differing from, the renormalization group. The title has been chosen to allude respectfully to the fundamental paper [2] of *Gell-Mann* and *Low*, in which they showed that the intriguing structure of quantum electrodynamics as a renormalizable quantum field theory also offers access to the large momenta behaviour of its propagators. This approach was extended by *Bogoliubov* and *Shirkov* [3], and others, to vertex and higher Green's functions, and to other renormalizable theories. Ultimately, the results of all this work, and of related investigations by *Landau* and Coworkers [4], were, due to the lack of a technique that leads really beyond perturbation theory, inconclusive as far as arbitrarily large momenta are concerned. This is so also with the results to be described here, which I believe, however, to be at least a technical improvement over the ones mentioned. What this improvement consists of I will summarize at the end.

For calculational details I refer to the original paper [1]. In this presentation, I will make extensive use of the concepts and results concerning renormalization that *Sidney Coleman* [5] explained so brilliantly.

I. Large-Momenta Behaviour of Feynman Integrals

We consider the theory of a hermitean scalar field in quartic self interaction already discussed by *Coleman* [5], with Lagrangian density, in customary notation,

$$\begin{aligned} L(\phi, \partial\phi) &= \tfrac{1}{2}\partial_\mu\phi\,\partial^\mu\phi - \tfrac{1}{2}m^2\phi^2 - \tfrac{1}{24}g\phi^4 \\ &+ (Z_3-1)\left[\tfrac{1}{2}\partial_\mu\phi\,\partial^\mu\phi - \tfrac{1}{2}m^2\phi^2\right] + Z_3\tfrac{1}{2}\delta m^2\phi^2 - (Z_1-1)\tfrac{1}{24}g\,\phi^4, \end{aligned} \tag{I.1}$$

where m is the particle mass and g the renormalized coupling constant. The counter terms have divergent coefficients and would have to be defined via regularization; more pragmatically, they symbolize the recipe [5] by which from L finite perturbation theoretical results are extracted.

We consider the connected Green's functions*

$$G(x_1 \dots x_{2n}) = \langle(\phi(x_1)\dots\phi(x_{2n}))_+\rangle_{\text{conn}}$$

and their amputated one-particle irreducible parts, the vertex functions $\Gamma(x_1 \dots x_{2n})$, where $\Gamma(x_1 x_2) = -G^{-1}(x_1 x_2)$, and the Fourier transforms $(p_1 + \dots + p_{2n} = 0)$

$$\Gamma(p_1 \dots p_{2n}) = \int dx_1 \dots dx_{2n-1}\,\Gamma(x_1 \dots x_{2n})\exp(ip_1x_1 + \dots + ip_{2n}x_{2n}).$$

For these, we have in perturbation theory

$$\begin{aligned} \Gamma(p_1 \dots p_{2n}) = \sum_{\substack{\text{Graphs}\\ N \geq n}} \Bigg\{ &\text{const}\, g^N \int dk_1 \dots dk_l \\ &\times \prod_{r=1}^{L}\left[\left(\sum_{i=1}^{l} a_{ri}k_i + \sum_{j=1}^{2n-1} b_{rj}p_j\right)^2 - m^2\right]^{-1} \\ &+ \text{renormalizing terms}\Bigg\} + i(p_1^2 - m^2)\,\delta_{n1} - ig\,\delta_{n2}\,. \end{aligned} \tag{I.2}$$

The matrices $a\dots$ and $b\dots$ describe the flow of the $l = N - n + 1$ loop momenta and of the $2n-1$ independent external momenta through the $L = 2N - n$ (internal) lines of the graph.

For the superficial (or "naive") divergence degree of Feynman integrals, we have [5] (omitting Fermion arguments)

$$D = 4 - B - \sum_{i=1}^{N}(4 - d_i - b_i) \tag{I.3}$$

where B is the number of external (Boson) lines and the sum goes over the N vertices in the graph, d_i being the number of derivatives and b_i the number of (Boson) arguments of the ith vertex. Thus, in (I.2), $D_\Gamma = 4 - 2n$ since only ϕ^4 vertices are used, the counter terms figuring only in the subtraction integrals.

* The concepts and notation used here are described in Ref. [6].

We wish to examine $\Gamma(p_1 \dots p_{2n})$ for large momenta. It is convenient to introduce the scaling factor λ and to consider $\Gamma(\lambda p_1 \dots \lambda p_{2n})$ for $\lambda \to \infty$. For dimensional reasons, we have

$$\Gamma(\lambda p_1 \dots \lambda p_{2n})(m^2, g) = \lambda^{4-2n} \Gamma(p_1 \dots p_{2n})(m^2 \lambda^{-2}, g) \tag{I.4}$$

such that it is suggestive to investigate existence and properties of the zero-mass theory. As $m^2 \to 0$, the integrals in (I.2) may develop infrared (UR) divergences, such that in the formal expansion

$$\begin{aligned}\Gamma(p_1 \dots p_{2n})(m^2 \lambda^{-2}, g) = \sum_{\substack{\text{Graphs} \\ N \geqq n}} \{&\text{const}\, g^N (\textstyle\int \mathrm{d}k \dots \Pi[(\dots p + \dots k)^2]^{-1} \\ &+ m^2 \lambda^{-2} \textstyle\int \mathrm{d}k \dots [(\dots p + \dots k)^2]^{-1} \Pi[(\dots p + \dots k)^2]^{-1} + \cdots) \\ &+ \text{renormalizing terms}\} + i(p^2 - m^2 \lambda^{-2})\, \delta_{n1} - ig\, \delta_{n2}\end{aligned} \tag{I.5}$$

the first integral, which we shall later refer to as the leading integral, and afortiori all further integrals, do not exist, their sum going like $(\ln \lambda^2)^\beta$, $\beta > 0$, in the case of logarithmic, like $\lambda^2 (\ln \lambda^2)^{\beta'}$ in the case of quadratic UR divergence etc. In addition, there may be ultraviolet (UV) divergences such that the effects of the renormalizing terms must be considered.

Weinberg [7] studied the behaviour of convergent Feynman integrals for large Euclidean momenta, and his analysis was complemented by *Fink* [8]. *Kinoshita* [9] examined the singularities of convergent Feynman integrals with Minkowski momenta at vanishing mass*. Rephrased for the present model, their results are: For momenta λp_i, $i = 1 \dots 2n$, the Feynman integral behaves, if $\lambda \to \infty$, as $\lambda^\alpha (\ln \lambda)^\beta$, where α and β depend on the structure of the graph as well as on the momenta. For general momenta, in our case, $\alpha = 4 - 2n$, $\beta = 0$, but for exceptional momenta, β and even α may be larger due to logarithmic and stronger UR divergence, respectively, of the leading integral in (I.5). An example of logarithmic UR divergence is given in Fig. 1a, of quadratic UR divergence in Fig. 1b.

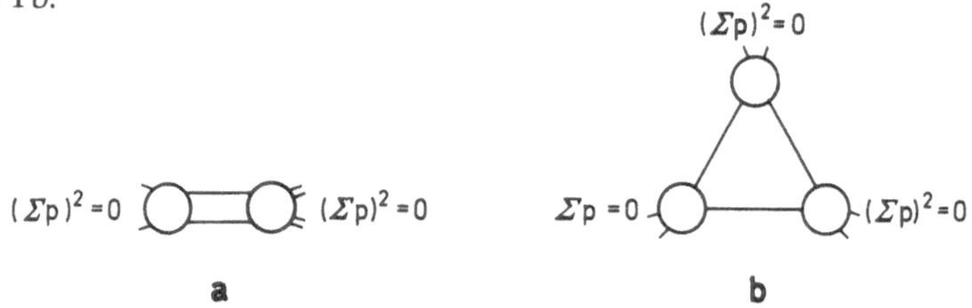

Fig. 1. Examples of a) logarithmic; b) quadratic UR divergence of leading integral

* *Kinoshita* [9] discussed also the cancellation of mass singularities in transition amplitudes, as did *Lee* and *Nauenberg* [10]. While the last authors include renormalization, they do not consider Feynman amplitudes.

Exceptional momenta are characterized by the fact that a partial sum of an (in the present model) even number of momenta, or several such partial sums, lie on the light cone, whereby in the second case for stronger than logarithmic UR divergence some partial sums must in addition be parallel. In *Weinberg*'s [7] Euclidian treatment the partial sums must vanish, and in this case the origin of the divergence can easily be understood intuitively: If partial sums of momenta vanish, there are in general sets of lines in the graph where no external momenta need flow through, such that the corresponding denominators in the Feynman integral need not become large as the external momenta increase, thus failing to contribute to the decrease of the Feynman integral, cp. Fig. 1b.

We turn to UV divergences. They require renormalizing terms in (I.2). As a consequence, UR divergences in the integrals in (I.5) as $\lambda \to \infty$ arise even for nonexceptional momenta, since in the renormalizing terms in (I.2) momenta occur that are not scaled up as $\lambda \to \infty$, thus becoming equivalent to zero momenta. E.g., in renormalizing a second order vertex contribution, the integral ($p_0^2 = m^2$)

$$\begin{aligned}
&\int dk[k^2-m^2]^{-1}\{[(k+\lambda p)^2-m^2]^{-1}-[(k+p_0)^2-m^2]^{-1}\} \\
&\quad = \int dk[k^2-m^2]^{-1}\{[(k+\lambda p)^2-m^2]^{-1}-[k^2-m^2]^{-1}\}+\text{const} \\
&\quad = \int_0^{\lambda} d\lambda' \int dk[k^2-m^2]^{-1} p(\partial/\partial k)\,[(k+\lambda' p)^2-m^2]^{-1}+\text{const} \\
&\quad = 2\int_0^{\lambda} d\lambda' \int dk[k^2-m^2]^{-2} kp[(k+\lambda' p)^2-m^2]^{-1}+\text{const}
\end{aligned}$$

arises. The last integral admits $m^2 \to 0$ for all $\lambda' > 0$ and then is const$'\,\lambda'^{-1}$ with const$'$ (for $p \neq 0$) p-independent, thus leading to behaviour const$'$ $\ln\lambda + 0(1)$ of the integral as $\lambda \to \infty$. This increase is, after $k \to \lambda k$, seen to be due to a UR divergence of the subtraction term. More complicated integrals behave similarly. – What we found in this section does not prove that a zero-mass theory does not exist, but that it must be constructed in a more sophisticated manner.

II. Mass Vertex Insertion

Coleman [5] showed that in the theory described by

$$\begin{aligned}
L = {} & \tfrac{1}{2}(\partial_\mu A\,\partial^\mu A + \partial_\mu B\,\partial^\mu B) - \tfrac{1}{2}m^2(A^2+B^2) \\
& - \tfrac{1}{24}g(A^2+B^2)^2 + (Z_3'-1)\,[\tfrac{1}{2}(\partial_\mu A\,\partial^\mu A + \partial_\mu B\,\partial^\mu B) - \tfrac{1}{2}m^2(A^2+B^2)] \\
& + Z_3'\tfrac{1}{2}\delta m'^2(A^2+B^2) - (Z_1'-1)\,\tfrac{1}{24}g(A^2+B^2)^2
\end{aligned}$$

the constants can be chosen such that L yields (in perturbation theory) a finite $O(2)$-invariant theory, and that the theory remains finite, though that symmetry is lost, under addition to L of a term cA^2 with c suitably chosen (logarithmically divergent). For the finiteness argument, however, the presence of the B field is immateriel, which means that instead of $L(\phi, \partial\phi)$ of (I.1) we may consider

$$L^s(\phi, \partial\phi) = L(\phi, \partial\phi) - s\tfrac{1}{2}\Delta m_u^2 Z_3 \phi^2 \tag{II.1}$$

and obtain, with Δm_u^2 suitably chosen (logarithmically divergent), a family of finite theories parametrized by s. The general member of this family describes, of course, again a hermitean scalar field in quartic self coupling, but with s-dependent parameters*

$$\Gamma^s(p_1 \ldots p_{2n})(m^2, g) = Z(s)^{-n}\Gamma(p_1 \ldots p_{2n})(m^2(s), g(s)) \tag{II.2}$$

whereby the multiplicative factor arises since $\Gamma^s(p(-p))$ will, at its zero at $p^2 = m^2(s)$, in general not have slope i (cp. (II.7b) below). From (II.2),

$$\begin{aligned}(\partial/\partial s)\,\Gamma^s(p_1 \ldots p_{2n})(m^2, g)|_{s=0} &\equiv \Delta\Gamma(p_1 \ldots p_{2n})(m^2, g)\\ &= [\Delta m^2(\partial/\partial m^2) + \Delta g(\partial/\partial g) - 2n\Delta z]\,\Gamma(p_1 \ldots p_{2n})(m^2, g)\end{aligned} \tag{II.3}$$

where

$$\Delta m^2 = (\partial/\partial s)\, m^2(s)|_{s=0}, \quad \Delta g = (\partial/\partial s)\, g(s)|_{s=0}, \quad \Delta z = \tfrac{1}{2}(\partial/\partial s)\, Z(s)|_{s=0},$$

while [1], on the other hand,

$$\begin{aligned}(\partial/\partial s)\,\Gamma^s(x_1 \ldots x_{2n})(m^2, g)|_{s=0} &= -i Z_3 \Delta m_u^2 \delta(x_1 - x_2)\,\delta_{n1}\\ &\quad -\tfrac{1}{2} i Z_3 \Delta m_u^2 \int dx\, dz\, du\, G(xz)\, G(xu)\,[\Gamma(zu\, x_1 \ldots x_{2n})\\ &\quad + \text{factorizing terms if } n > 1].\end{aligned} \tag{II.4}$$

This equation is illustrated in Fig. 2 for $n=1$ and $n=2$, where bubbles are vertex functions, dashed lines are propagators $G(\ldots)$, the BS-bubble is the Bethe-Salpeter kernel (which includes a logarithmically divergent constant in the present model) and the "2 irr" vertex function ** does not allow a two-particle cut at the broken line.

(II.4) means that $\Delta\Gamma$ is obtained from Γ by inserting one mass vertex into all Feynman diagrams in all possible ways. Thus, formula (I.3) gives $D_{\Delta\Gamma} = 2 - 2n$. Therefore, as *Coleman* [5] showed, only $\Delta\Gamma(p(-p))$ needs to be specified at one momentum, whereupon it can be calculated for

* That mass vertex insertion not merely changes the observed mass, but also dimensionless renormalized parameters, is due to the need of regularizing the theory before one can consider the Lagrangian density naively, and the regularized form of the mass vertex originally in (I.1) is rather more complicated, see, e.g., Ref. [11], than the one of the additional mass vertex in (II.1). This is intimately related to the fact that, while δm^2 in (I.1) is quadratically divergent, Δm_u^2 in (II.1) is only logarithmically divergent.

** For these concepts and relevant equations, see e.g. Refs. [12–14].

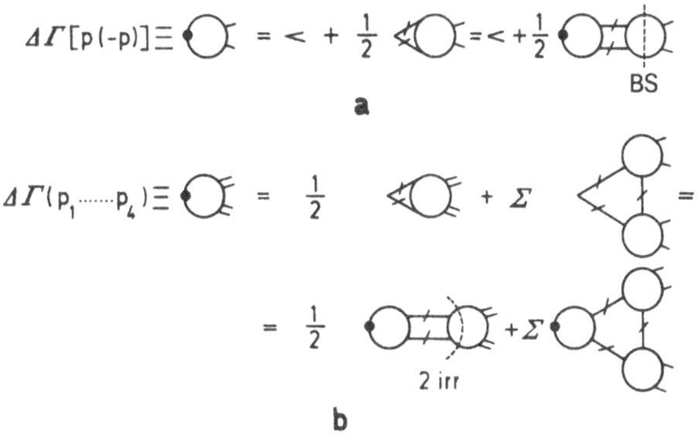

Fig. 2. Eq. (II.4) for a) $n=1$, b) $n=2$

all momenta from e.g. the once-subtracted Bethe-Salpeter equation obtained from Fig. 2a, while Fig. 2b shows how the other $\Delta\Gamma$ are then obtained by quadrature. If we choose

$$\Delta\Gamma(p(-p))|_{p^2=m^2} = -im^2, \qquad \text{(II.5)}$$

Δm_u^2 in (II.4) is hereby fixed and, since no extrinsic parameter is introduced, for dimensional reasons we must have

$$\Delta m^2 = m^2\alpha(g), \quad \Delta g = \beta(g), \quad \Delta z = \gamma(g). \qquad \text{(II.6)}$$

These coefficient functions are determined by applying (II.4) to the Γ with $D_\Gamma \geqq 0$ at those momenta at which the renormalization parameters are specified:

$$\Gamma(p(-p))|_{p^2=m^2} = 0, \qquad \text{(II.7a)}$$

$$(\partial/\partial p^2)\,\Gamma(p(-p))|_{p^2=m^2} = i, \qquad \text{(II.7b)}$$

$$\Gamma(p_1 \dots p_4)|_{\text{symmetry point}} = -ig, \qquad \text{(II.7c)}$$

where the symmetry point is defined by $p_i p_j = \frac{1}{3}(4\delta_{ij} - 1)\,m^2$. One finds [1]

$$\alpha(g) = 1, \qquad \text{(II.8a)}$$

$$\beta(g) = b_0 g^2 + b_1 g^3 + \cdots, \qquad \text{(II.8b)}$$

$$\gamma(g) = c_0 g^2 + c_1 g^3 + \cdots \qquad \text{(II.8c)}$$

with

$$b_0 = 3(32\pi^2)^{-1}, \quad c_0 = (2^{11} 3\pi^4)^{-1}.$$

That $\beta(g)$ and $\gamma(g)$ are $O(g^2)$ is of some significance later on, and comes from the fact that $\Delta\Gamma(p_1 \dots p_4)$, see Fig. 2b, is $O(g^2)$ while in $\Delta\Gamma(p(-p))$, see Fig. 2a, the first term that survives the subtraction that imposes (II.5) is also $O(g^2)$.

III. The Breaking of Scale Invariance*

With (II.6) and (II.8 a), (II.3) can be written

$$2im^2(\partial/\partial m^2)\,\Gamma(p_1 \ldots p_{2n})(m^2, g) = 2i\Delta\Gamma(p_1 \ldots p_{2n})(m^2, g) \\ + 2i[2n\gamma(g) - \beta(g)(\partial/\partial g)]\,\Gamma(p_1 \ldots p_{2n})(m^2, g). \tag{III.1}$$

This is the Ward identy, taken at zero momentum, for the scale-transformation generating current. More precisely, the left hand side is the negative of the equal-time-commutator contributions in the Ward identity such that the right hand side is $\Gamma(p_1 \ldots p_{2n}, 0)$ in *Coleman*'s notation [5] and essentially the matrix element of the divergence of the current. (III.1) shows how complicated an operator that divergence is. Had it been the "soft operator" const ϕ^2, only the first term on the right hand side of (III.1) should have been present, and we shall show in the next section that this term is, at large momenta, negligible relative to Γ itself. The other two terms indicate "nonsoft operator" contributions to the divergence arising, in view of (II.8 b–c), first in order g^2. For these operators, one can obtain more explicit forms but, as long as we are interested in the Ward identity only at zero momentum i.e. in global scale transformation, (III.1) is more useful since it embodies already the renormalization prescription. We merely note that the term proportional $\gamma(g)$ could be absorbed by "change of operator dimension" as proposed by *Wilson* [15], but the apparently equally important term proportional $\beta(g)$ cannot. Its treatment is implicit in the derivation of asymptotic forms to be discussed next.

IV. Asymptotic Form of Vertex Functions

We rewrite (III.1) as

$$\Delta\Gamma(p_1 \ldots p_{2n})(m^2\lambda^{-2}, g) \\ = [-\lambda^2(\partial/\partial\lambda^2) + \beta(g)(\partial/\partial g) - 2n\gamma(g)]\,\Gamma(p_1 \ldots p_{2n})(m^2\lambda^{-2}, g). \tag{IV.1}$$

This is a linear inhomogeneous partial differential equation for $\Gamma(\ldots)$ with, in principle, known coefficients $\beta(g)$ and $\gamma(g)$. We introduce new variables

$$\ln\lambda^2 = u, \quad \int_{g_0}^{g} dg'\,\beta(g')^{-1} = v \tag{IV.2}$$

* For details, see the appendix of Ref. [1].

and set

$$\Gamma(p_1 \ldots p_{2n})(m^2\lambda^{-2}, g) = f(u, v),$$
$$\Delta\Gamma(p_1 \ldots p_{2n})(m^2\lambda^{-2}, g) = \Delta f(u, v),$$
$$2n\gamma(g) = c(v).$$

The characteristics of (IV.1) are $u+v=\text{const}$, and the general solution is

$$f(u,v) = f(\tfrac{1}{2}[u+v+h(u+v)], \tfrac{1}{2}[u+v-h(u+v)]) \exp\left[\int_{\frac{1}{2}[u+v-h(u+v)]}^{v} dv' c(v')\right]$$
$$+ \int_{u}^{\frac{1}{2}[u+v+h(u+v)]} du' \Delta f(u', v+u-u') \exp\left[\int_{v+u-u'}^{v} dv' c(v')\right]. \quad \text{(IV.3)}$$

The first term on the right hand side is the solution of the homogeneous equation to (IV.1) with initial values $f(u', v')$ on the arbitrarily chosen curve $u'-v'=h(u'+v')$, while the second term is the contribution from the inhomogeneous term in the propagation away from that curve.

We are interested in the behaviour of $f(u,v)$ for $u\to\infty$. (IV.3) shows that a simple form is then obtained if $h(u+v)$ can be set ∞, since then as $u\to\infty$ $f(u,v)$ would approach the first term on the right hand side, a solution of the homogeneous equation. This leads us to define the asymptotic form of $f(u,v)$

$$f_{\text{as}}(u,v) = f(u,v) - \int_{u}^{\infty} du' \Delta f(u', v+u-u') \exp\left[\int_{v+u-u'}^{v} dv' c(v')\right] \quad \text{(IV.4)}$$

which, if it exists, indeed is easily checked to be a solution of the homogeneous equation, and as such can also be written

$$f_{\text{as}}(u,v) = \phi(u+v) \exp\left[\int_{-\infty}^{v} c(v')\, dv'\right] \quad \text{(IV.5)}$$

with, because of (IV.3),

$$\phi(u) = \lim_{h\to\infty} f(\tfrac{1}{2}[u+h], \tfrac{1}{2}[u-h]), \quad \text{(IV.6)}$$

provided also the integral in (IV.5) exists.

We first consider, using (II.8b), the function

$$\varrho(g) = \int_{g_0}^{g} dg' \beta(g')^{-1} = -b_0^{-1} g^{-1} - b_0^{-2} b_1 \ln g + \text{power series in } g. \quad \text{(IV.7)}$$

Since we are, of course, assuming that the expansions (II.8b–c) are asymptotic ones meaningful for sufficiently small g, and since $b_0 > 0$, $\varrho(g)$ increases from $-\infty$ monotonically as g runs from zero through positive values, till g reaches a possible zero of $\beta(g)$. We will, for definiteness, choose for g some value within this interval although we know of no reason why (the renormalized) g should e.g. take positive rather than negative values*. We learn from (IV.2) and (IV.7) that $v\to-\infty$ means $g\to+0$, such that in (IV.5)

$$\frac{1}{2n}\int_{-\infty}^{v} \mathrm{d}v' c(v') = \int_{0}^{g} \mathrm{d}g' \beta(g')^{-1}\gamma(g') = b_0^{-1}c_0 g + \tfrac{1}{2}(b_0^{-1}c_1 - b_0^{-2}c_0 b_1)\, g^2 + \cdots \tag{IV.8}$$

which, of course, we assume to exist for g chosen as described.

Actually, in unadulterated perturbation theory, (IV.8) can be misleading since the integral in (IV.5) must, in conjunction with (IV.6), be interpreted as

$$\lim_{h\to\infty} 2n \int_{\varrho^{-1}(-h+\varrho(g))}^{g} \mathrm{d}g' \beta(g')^{-1}\gamma(g'), \tag{IV.9}$$

and

$$\varrho^{-1}(-h+\varrho(g)) = g + \sum_{n=1}^{\infty} (n!)^{-1}(-h)^n [\beta(g)\,(\partial/\partial g)]^{n-1}\beta(g) \tag{IV.10}$$

gives, inserted in (IV.9), no limit in the expansion in powers of g. This means that, because we shall argue below that $f_{\mathrm{as}}(u, v)$ exists even in perturbation theory, the limit (IV.6) does not exist in perturbation theory.

We turn to the integral in (IV.4), which is, in the original notation,

$$\Gamma(p_1 \ldots p_{2n})(m^2\lambda^{-2}, g) - \Gamma_{\mathrm{as}}(p_1 \ldots p_{2n})(m^2\lambda^{-2}, g) \tag{IV.11a}$$

$$= \int_{\lambda^2}^{\infty} \lambda'^{-2}\mathrm{d}\lambda'^2 \Delta\Gamma(p_1 \ldots p_{2n})(m^2\lambda'^{-2}, g(\lambda'/\lambda)) \exp\left[2n \int_{g(\lambda'/\lambda)}^{g} \mathrm{d}g' \beta(g')^{-1}\gamma(g')\right]$$

with

$$g(\lambda'/\lambda) \equiv \varrho^{-1}(\varrho(g) - \ln\lambda'^2 + \ln\lambda^2)\,. \tag{IV.11b}$$

The exponential factor is finite and has a limit as $\lambda'\to\infty$, while in perturbation theory it gives, because of (IV.10), in any finite order in g a polynomial in $\ln(\lambda'/\lambda)$. $g(\lambda'/\lambda)$ in the argument of $\Delta\Gamma$ goes to zero as $\lambda'\to\infty$ such that, if we at all accept perturbation theoretical information about $\Delta\Gamma$ as indicative, it should apply throughout the integration range. In this sense,

$$\Delta\Gamma(\lambda p_1 \ldots \lambda p_{2n})(m^2, g) = \lambda^{4-2n}\Delta\Gamma(p_1 \ldots p_{2n})(m^2\lambda^{-2}, g) \sim \lambda^{2-2n}(\ln\lambda)^{\beta},$$

by power counting and (I.3), for nonexceptional momenta as discussed in Section I, such that for such momenta

$$\Delta\Gamma(p_1 \ldots p_{2n})(m^2\lambda'^{-2}, g(\lambda'/\lambda)) \sim \lambda'^{-2}(\ln\lambda')^{\beta'}$$

* In quantum electrodynamics, the analog of g in (IV.7) is e^2, the square of the charge, see (VII.3) below, such that its sign is known.

which secures convergence of the integral in (IV.11a), the increase in logarithmic exponent (due to the zero external momentum acting at the mass vertex, cp. Section I, esp. Fig. 1a) being harmless. This is so even in perturbation theory where additional factors $\ln(\lambda'/\lambda)$ arise from the argument $g(\lambda'/\lambda)$ of $\Delta\Gamma$ and from the exponential as discussed before.

Thus, we have shown that Γ_{as} exists in perturbation theory for nonexceptional momenta, and that for such momenta

$$\Gamma(p_1 \dots p_{2n})(m^2\lambda^{-2}, \bar{g}(\lambda)) - \Gamma_{\text{as}}(p_1 \dots p_{2n})(m^2\lambda^{-2}, \bar{g}(\lambda)) = O(\lambda^{-2}(\ln\lambda)^{\beta}) \tag{IV.12}$$

where $\bar{g}(\lambda)$ should not have a stronger than logarithmic λ-dependence. Furthermore, we have given arguments that Γ_{as} exist, and satisfy (IV.12), also in the exact theory, provided $\bar{g}(\lambda)$ does not increase with increasing λ. Nonexceptional momenta are hereby such that the leading integrals for $\Delta\Gamma$, in the sense of Section I, do not develop quadratic or stronger UR divergences.

Since Γ_{as} does satisfy the homogeneous equation to (IV.1), even in perturbation theory, it has the form

$$\Gamma_{\text{as}}(p_1 \dots p_{2n})(m^2\lambda^{-2}, g) = a(g)^n \phi_{p_1 \dots p_{2n-1}}(-\ln m^2 + \ln\lambda^2 + \varrho(g)) \tag{IV.13a}$$

with

$$a(g) = \exp\left[2\int_0^g \mathrm{d}g'\,\beta(g')^{-1}\gamma(g')\right] \tag{IV.13b}$$

such that also ϕ, in the definition (IV.13a) rather than (IV.6), exists in perturbation theory. The use of these asymptotic forms, and of their analoga in other theories, in conjunction with perturbation theory, e.g. to sum leading, next-to-leading etc. logarithmic terms, is well known (Refs. [3, 16], see also Ref. [17]) and need not concern us here.

V. The Zero-Mass Theory

We rewrite (IV.13a) as

$$\begin{aligned}\Gamma_{\text{as}}(\mu p_1 \dots \mu p_{2n})&(m^2\lambda^{-2}, g)\\ &= \mu^{4-2n} a(g)^n \phi_{p_1 \dots p_{2n-1}}(-\ln m^2 + \ln\mu^2 + \ln\lambda^2 + \varrho(g)).\end{aligned} \tag{V.1}$$

The right hand side cannot be expected to have a limit as $\lambda \to \infty$ and has none in perturbation theory. Thus, along with λ we change g so that

$$\ln\lambda^2 + \varrho(g) = \text{const} = \varrho(\bar{g})$$

such that (V.1) reads, using also (IV.12) which is applicable because of (IV.10),

$$\lim_{\lambda\to\infty} \{a(\bar{g}(\lambda))^{-n}\Gamma(\mu p_1 \dots \mu p_{2n})(m^2\lambda^{-2}, \bar{g}(\lambda))\}$$
$$= \mu^{4-2n}\phi_{p_1\dots p_{2n-1}}(-\ln m^2 + \ln\mu^2 + \varrho(\bar{g})) \qquad \text{(V.2a)}$$

where

$$\bar{g}(\lambda) = \varrho^{-1}(-\ln\lambda^2 + \varrho(\bar{g}))\,, \qquad \text{(V.2b)}$$

for nonexceptional momenta. (V.2) means that the zero-mass theory exists as the limit of the nonzero-mass theory provided we also change the conventional coupling constant during the limiting process and apply a certain amplitude renormalization factor, and that its vertex functions possess expansions* in powers of $\bar{g}$. As discussed in Section IV, in the limit the conventional coupling constant $\bar{g}(\lambda)$ actually goes to zero and the factor $a(\bar{g}(\lambda))$ goes to one, such that in the limit the Γ, i.e. the ϕ, are again conventionally normalized, but this normalization cannot be verified in the expansion in powers of $\bar{g}$ since the factor $a(\bar{g}(\lambda))$ itself has no limit in the expansion in powers of $\bar{g}$.

That the momenta in (V.2a) must be nonexceptional simply means that the ϕ have, as functions of the momenta, singularities where the nonzero-mass-theory Γ have none. The renormalization conditions for the zero-mass theory can be set also directly, but only at nonexceptional momenta. E.g., for $\phi_{p_1p_2p_3}(-\ln m^2 + \varrho(\bar{g}))$, $p_1 = p_2 = p_3 = 0$, the point of conventional definition of the coupling constant, is exceptional since, upon mass vertex insertion, Fig. 1b applies. Thus, in the zero-mass theory, a coupling constant could be defined as i times the value of the four-point vertex at the symmetry point to $p_i^2 = -a < 0, \forall i$. The amplitude renormalization can be fixed at any momentum $p^2 \leqq 0$, but in perturbation theory not at $p^2 = 0$, such that correctly normalized scattering amplitudes do not exist in perturbation theory due to UR-divergences which, as our arguments show, should be fictitious. The expansion so constructed is related to the one in powers of $\bar{g}$ obtained from (V.2) in that the new expansion parameter is a power series in $\bar{g}$.

Apart from the arbitrariness in amplitude renormalization (there is none, however, in (V.2)) the zero-mass theory has only one parameter, $\bar{g}$, which moreover can be absorbed by a rescaling of length, due to the non-scale invariance of the theory, i.e. the nontrivial μ^2-dependence of the right hand side of (V.2a). Thus, the zero-mass theory does not have an intrinsic dimensionless parameter.

* The existence of the limit in (V.2a) means that in every order of $\bar{g}$, all powers of $\ln\lambda$ not accompanied by a factor λ^{-2} and higher cancel.

If we go from the theory (I.1) of one hermitean scalar field to the one of an N-tuplet with $O(N)$ symmetry, all our conclusions again hold. The zero-mass theory in this case marks the transition between the Goldstone mode and the symmetric theory for the model discussed in Sections IV–VI of Ref. [6]. That it cannot be conventionally normalized in perturbation theory means that, in perturbation theory, it cannot be obtained as the naive common limit of the two modes mentioned, in each of which the "π-field" is conventionally normalized.

VI. Miscellaneous Remarks

The result of Section IV, i.e. (IV.12) and (IV.13) in conjunction with (I.4), valid for nonexceptional momenta, refers to overall scaling. Thus, the high-energy behaviour of scattering amplitudes is describable only insofar as the change of the vertex functions in the extrapolation from the mass shell to $p_i^2=0, i$, i.e. to zero "external masses", may then be neglected, an assumption commonly made and supported by perturbation theoretical examples (e.g., Ref. [17]). However, that exceptional momenta must be avoided means that e.g. in elastic two-particle scattering, $s=(p_1+p_2)^2$, $t=(p_1+p_3)^2$, and $u=(p_1+p_4)^2$ must all be different from zero since otherwise upon mass vertex insertion Fig. 1b applies, and $\frac{s}{t}=-1-\frac{u}{t}$ fixed, i.e. the scattering angle must be fixed and not* be zero or π since then, as $\lambda\to\infty$, $\Delta\Gamma$ is not negligible relative to Γ in (IV.1). Thus, also the high-energy behaviour of the total cross section is not analyzable by the present method. Similar restrictions hold for the application of the present method to other scattering amplitudes.

From (IV.12) and (IV.13) with (I.4) we can obtain the analog of the famous result of *Gell-Mann* and *Low* [2] on the unrenormalized coupling constant g_u. Namely, conventionally, cp. (I.1),

$$\begin{aligned}&-ig\,Z_1(g)\\&=\lim_{\lambda\to\infty}\Gamma(\lambda p_1\ldots\lambda p_4)(m^2,g)=a(g)^2\lim_{\lambda\to\infty}\phi_{p_1\ldots p_3}(-\ln m^2+\ln\lambda^2+\varrho(g))\end{aligned}$$

and

$$\begin{aligned}&ip^2Z_3(g)\\&\quad=\lim_{\lambda\to\infty}\lambda^{-2}\,\Gamma(\lambda p(-\lambda p))(m^2,g)=a(g)\lim_{\lambda\to\infty}\phi_p(-\ln m^2+\ln\lambda^2+\varrho(g))\end{aligned}$$

such that

$$Z_1(g)\,Z_3(g)^{-2}g\equiv g_u=\lim_{\lambda\to\infty}F(-\ln m^2+\ln\lambda^2+\varrho(g))$$

---

* That the renormalization group yields no result for scattering angles zero and π was observed by *Huang* and *Low* [18].

is g-independent. The solution of this "paradox" [19] appears to have been given by *Astaud* and *Jouvet* [20]: The asymptotic-behaviour analysis rests on the comparison with the zero-mass theory, for which the condition $m^2 = m_u^2 f(g_u) = 0$, with m_u the unrenormalized mass, fixes g_u since the solution $m_u^2 = 0$ would lead to a scale-invariant theory and thus is excluded. Less formally, the relevant observation is the one in Section V that the zero-mass theory does not have an intrinsic dimensionless parameter.

VII. Quantum Electrodynamics

Here the analysis [1] proceeds essentially in complete analogy to the scalar case, the insertion into the Lagrangian density being an electron mass vertex. If sufficiently carefully defined, this vertex is invariant under gauge transformations of the second kind, which implies that the $\Delta\Gamma$ for the negative inverse photon propagator* is transverse, whereas that inverse propagator itself in general is not transverse, which means that electron mass vertex insertion effects a gauge change that must be taken into account in the analog of (II.2), leading to additional terms in (II.3) and thus (IV.1).

Only the Landau gauge, whose photon propagator is transverse, remains unchanged under electron mass vertex insertion and thus leads to a simple formula for $\Delta\Gamma$. Let the Fourier transform of the vertex function to the Green's function

$$\langle(\psi(x_1)\ldots\psi(x_n)\,\bar{\psi}(y_1)\ldots\bar{\psi}(y_n)\,A(z_1)\ldots A(z_l))_+\rangle$$

in this gauge be denoted by $\Gamma(p_1\ldots p_n, q_1\ldots q_n, k_1\ldots k_l)$. Then (IV.1) becomes replaced** by (m is the electron mass, e the renormalized charge)

$$\begin{aligned} e^{-l}\Delta\Gamma(p_1\ldots p_n, q_1\ldots q_n, k_1\ldots k_l)\,(m\lambda^{-1}, e) \\ = [-\alpha(e^2)\,\lambda^2(\partial/\partial\lambda^2) + \beta(e^2)\,(\partial/\partial e^2) - 2n\gamma_2(e^2)] \\ \times e^{-l}\Gamma(p_1\ldots p_n, q_1\ldots q_n, k_1\ldots k_l)\,(m\lambda^{-1}, e) \end{aligned} \tag{VII.1}$$

where, upon suitable normalization [1],

$$\begin{aligned} \alpha(e^2) &= 1 + a_1 e^2 + \cdots, \\ \beta(e^2) &= b_0 e^4 + b_1 e^6 + \cdots, \\ \gamma_2(e^2) &= c_0 e^4 + c_1 e^6 + \cdots \end{aligned}$$

* We have in mind only covariant gauges. For noncovariant gauges such as the Coulomb gauge, renormalization seems not to have been studied to all orders.

** In Landau gauge only the longitudinal part of the two-photon vertex function does not exist. For that part, (VII.1) is empty.

with $b_0=(12\pi^2)^{-1}$. Thus, by the argument that led to (IV. 13), e.g. for the negative inverse of the (transverse part of the) photon propagator,

$$\Gamma_{as}(,,k(-k))(m\lambda^{-1},e)=e^2\phi_k(-\ln m^2+\ln\lambda^2+\varrho(e^2)) \quad \text{(VII.2)}$$

where

$$\varrho(e^2)=\int_{e_0^2}^{e^2} \mathrm{d}e'^2\beta(e'^2)^{-1}\alpha(e'^2)=-b_0^{-1}e^{-2}+(b_0^{-1}a_1-b_0^{-2}b_1)\ln e^2+\cdots. \quad \text{(VII.3)}$$

Zero-mass quantum electrodynamics exists* and is not scale-invariant. It can, similarly as in Section V, be obtained by a limiting process whereby $\ln\lambda^2+\varrho(e^2)=\varrho(\bar{e}^2)$ is kept fixed while $\lambda\to\infty$, such that the conventional charge goes to zero, though to infinity in perturbation theory, cp. (IV. 10). The photon propagator can only be normalized away from $k^2=0$ due to the factor e^2 in (VII.2).

VIII. Conclusion

We have studied the small-distance behaviour in quantum field theory, in the sense of overall scaling, with the help of the infinitesimal operation to mass vertex insertion into the Lagrangian density. While the result is closely related to the one obtained using the renormalization group, the present method, resting only on the familiar concept of power counting, is more direct and is based on the Eq. (IV. 1) rather than on merely asymptotic relations. As a consequence the coefficient functions $\beta(g)$, $\gamma(g)$ that also figure in the renormalization group result (IV. 13) are determined by relatively rigorous analysis rather than only by asymptotic considerations, such that non-perturbation theoretical information on these functions could perhaps be obtained. Moreover, the definition (IV. 11) of the asymptotic forms of vertex functions makes these forms accessible to "axiomatic" analysis concerning e.g. analytic properties in momenta, complementary to the information obtained from the identification of these forms as vertex functions of a certain zero-mass theory. This theory we here discussed in detail.

---

* This has been observed in Ref. [2]. Cp. also Refs. [9] and [16]. *Lee* and *Nauenberg* [10] argue, in effect, that this theory, obtained as the limit of vanishing electron mass, is not γ_5-invariant. The limit of vanishing photon mass, with strictly zero electron mass, cp. Ref. [21], would, if it existed naively, lead to a γ_5-invariant theory. It appears, however, that the γ_5-invariance of the zero-mass theory cannot be imposed through the renormalization conditions. The author thanks *T. T. Wu* for a discussion. [In perturbation theory, the two limites mentioned here lead to the same (up to length scale and normalization change unique) γ_5-invariant limiting theory.]

As a byproduct, the present method leads to as complete a picture of the breaking of (global) scale invariance as one can obtain using formal tools only.

Note added in proof:

Arguments for the Eqs. (III.1) have been given independently of Ref. [1] by *C. G. Callan Jr.*, Phys. Rev. D **2**, 1541 (1970).

References

1. *Symanzik, K.*: Commun. Math. Phys. **18**, 227 (1970).
2. *Gell-Mann, M., Low, F. E.*: Phys. Rev. **95**, 1300 (1954)
3. *Bogoliubov, N. N., Shirkov, D. V.*: Introduction to the Theory of Quantized Fields, Chapter VIII. New York: Interscience Publ. 1959.
4. *Landau, L. D., Abrikosov, A., Halatnikov, L.*: Nuovo Cimento Suppl. **3**, 80 (1956).
5. *Coleman, S.*: (to be published).
6. *Symanzik, K.*: Commun. Math. Phys. **16**, 48 (1970).
7. *Weinberg, S.*: Phys. Rev. **118**, 838 (1960).
8. *Fink, J. P.*: J. Math. Phys. **9**, 1389 (1968).
9. *Kinoshita, T.*: J. Math. Phys. **3**, 650 (1962).
10. *Lee, T. D., Nauenberg, M.*: Phys. Rev. **133**, B 1549 (1964).
11. *Callan, C. G., Jr., Coleman, S., Jackiw, R.*: Ann. Phys. (N.Y.) **59**, 42 (1970).
12. *Symanzik, K.*: In: Lectures in High Energy Physics, Ed. *B. Jakšić*, Zagreb: 1961. Reprinted: New York: Gordon and Breach 1966.
13. *Taylor, J. G.*: Nuovo Cimento Suppl. **1**, 857 (1963).
14. *Johnson, R. W.*: J. Math. Phys. **11**, 2161 (1970).
15. *Wilson, K.*: Phys. Rev. **179**, 1499 (1969), Phys. Rev. D **2**, 1478 (1970).
16. *Eriksson, K. E.*: Nuovo Cimento **27**, 178 (1963).
17. *Appelquist, T., Primack, J. R.*: Phys. Rev. **1** D, 1144 (1970).
18. *Huang, K., Low, F.*: JETP **46**, 845 (1964), Engl. transl. **19**, 579 (1964).
19. *Bjorken, J. D., Drell, S. D.*: Relativistic Quantum Fields, Sect. 19.15. New York: McGraw-Hill 1965.
20. *Astaud, M., Jouvet, B.*: Nouvo Cimento **63** A, 5 (1969); **66** A, 111 (1970).
21. *Adler, S. L.*: Phys. Rev. **177**, 2426 (1969).

Professor Dr. *K. Symanzik*
DESY
Notkestieg 1
D-2000 Hamburg 52

Physics on the Light Cone

R. A. BRANDT ★

In this lecture I shall describe some attempts [1–6] made by myself and others in the past two years to increase the domain of applicability of quantum field theory in particle physics. The introduction of current algebra some years ago led to a revival of field theoretic configuration space techniques, but the observable consequences of the equal-time commutation of current algebra (with PCAC) are limited to low energy theorems and indirect information obtained from knowledge of amplitudes in the unphysical Bjorken limit. We shall see here how knowledge of the behaviour of current products near the light cone $x^2 = 0$, rather than just at equal time $x_0 = 0$, leads to a description of a much richer class of physical phenomena.

The relevance of the light cone (LC) to these processes was pointed out in Ref. [1], and further studied in Refs. [2] and [3]. These investigations involved essentially only the LC behaviour of matrix elements of the current products. In the last year, *Preparata* and myself have derived operator expressions which describe this LC behaviour [6]. Most of this lecture will be concerned with the properties and applications of these operator expansions.

We begin by showing what physical quantities and in what region of momentum space the LC is relevant.

Let us first consider the weak or electromagnetic scattering of a lepton off of a hadronic system α to produce an arbitrary final hadronic state. Calling q the momentum transferred to the lepton and p the total momentum of the system α, the total cross-section of interest has the form

$$\frac{d^2\sigma^r}{dq^3\, d\nu} \propto \int d^4k\, \theta(k_0)\, \delta(k^2 - q^2)\, \delta(k\cdot p - \nu) \int d^4x\, e^{-ik\cdot x}\, {}_{\text{in}}\langle\alpha| J_\mu(x)\, J_\nu(0)|\alpha\rangle_{\text{in}}\, \varepsilon_r^\mu \varepsilon_r^\nu\,. \tag{1}$$

Here $\nu \equiv q\cdot p$ is the initial energy variable, r is the polarization of the leptons, and J_μ is the hadronic current to which the leptons couple. The matrix element in (1) is connected and spin averaged. Changing the orders of integration in (1) gives

$$\frac{d^2\sigma^r}{dq^3\, d\nu} \propto \int d^4x\, \Delta_+(x, p, \nu, q^2)\, \langle\alpha| J_\mu(x)\, J_\nu(0)|\alpha\rangle\, \varepsilon_r^\mu \varepsilon_r^\nu\,, \tag{2}$$

★ A.P. Sloan Foundation Fellow.

where we have defined

$$\Delta_+(x, p, \nu, q^2) = \int d^4k\, e^{-ik\cdot x} \theta(k_0)\, \delta(k^2 - q^2)\, \delta(k \cdot p - \nu)\,. \tag{3}$$

The form (2) is very useful for our configuration space purposes. The integral (3) can be simply evaluated in the frame $p = (1, \mathbf{0}) \equiv \zeta$ (we take $p^2 = 1$) to give

$$\Delta_+(x, \zeta, \nu, q^2) = \frac{\pi}{ir} \left[e^{i(\nu^2 + q^2)^{1/2} r} - \text{h.c.}\right] e^{-i\nu t}\,, \tag{4}$$

where we have written $r = |\boldsymbol{x}|$, $t = x_0$. We refer to the *Bjorken* [7] scaling limit as the A limit

$$\nu \to \infty,\ q^2 \to \infty,\ \omega \equiv \frac{q^2}{2\nu} \text{ fixed}\,, \tag{5}$$

and obtain

$$\Delta_+(x, \zeta, \nu, q^2) \underset{A}{\rightarrow} \frac{\pi}{ir} \left[e^{i\nu(r-t) + i\omega r} - e^{-i\nu(r+t) - i\omega r}\right]. \tag{6}$$

Only the regions $r = \pm t$ in the (first/second) term are important and so we can deduce the covariant result

$$\Delta_+(x, p, \nu, q^2) \underset{A}{\rightarrow} \frac{\pi}{i\tau} \sin\left(\frac{\nu x^2}{2\tau} + \omega\tau\right) \tag{7}$$

in terms of the "transverse" variable

$$\tau^2 \equiv (p \cdot x)^2 - x^2\,. \tag{8}$$

It follows from (7) that, in the A limit $\Delta_+(x, p, \nu, q^2)$ is highly oscillating outside of the region

$$\frac{x^2}{2\tau} \lesssim \frac{1}{\nu}, \quad \tau \lesssim \frac{1}{\omega} \quad \text{or} \quad x^2 \lesssim \frac{2}{q^2}$$

and, therefore, has effective support on the LC $x^2 \sim 0$. Referring back to (2), we see that the A limit of the cross-section can be obtained simply from the behaviour of $\langle \alpha | J_\mu(x) J_\nu(0) | \beta \rangle$ on the LC.

For reactions in which a hadronic system produces a lepton pair and an arbitrary final hadronic state, the analysis becomes more complicated because of kinematic restrictions on the k space integration region in the analogue of (1). In the frame $p = (\sqrt{s}, \mathbf{0})$, the physical k space region will have the form $\{|\boldsymbol{k}| < \varkappa(q^2, s)\}$ for some functions $\varkappa(q^2, s)$. Proceeding as above it can be shown that the relevant configuration space region is given by

$$|x^2 - \varkappa^{-2}| \lesssim \frac{1}{q^2}\,. \tag{9}$$

So, for large q^2 *and* $\varkappa$, the LC is again the dominant region. The extra condition that $\varkappa(q^2, s)$ be large can be satisfied in many cases of interest.

Another important use of the LC is to determine the behaviour of amplitudes for large values of a single mass variable. We illustrate this by consideration of a scalar vertex function

$$A(q^2, k^2) = \int d^4x e^{-iq\cdot x}\langle 0|T[A(x), B(0)]|p\rangle . \quad (10)$$

Here $A(x)$ and $B(x)$ are scalar currents and $|p\rangle$ is a state of one scalar particle of momentum $p = q - k$ and mass $p^2 = m^2$. We can write (10) as

$$A(q^2, k^2) = \frac{1}{4\pi}(v^2 - m^2 q^2)^{-\frac{1}{2}} \cdot \int d^4x \Delta_+(x, p, v, q^2)\langle 0|T[A(x), B(0)]|p\rangle , \quad (11)$$

where $v = q \cdot p = \frac{1}{2}(q^2 + k^2 - m^2)$. Thus, the behaviour of $A(q^2, k^2)$ for $q^2 \to \infty$ and fixed $q^2/2v$, i.e., for fixed q^2/k^2, is determined by the LC behaviour of $A(x)\,B(0)$. A special case of this limit is the old *Bjorken* [8] limit $q^2/k^2 \to 0$, in which case only the equal time behaviour of $A(x)\,B(0)$ is relevant. Another special case is the limit $q^2 \to \infty$ with k^2 fixed so that $q^2/2v \to 1$. This limit is important because it determines the number of subtraction needed in fixed k^2 dispersion relations.

Having established the relevance of the LC, we turn to a description of what happens there.

The behaviour of products $A(x)\,B(0)$ of (renormalized) local field operators at short distances $x^\mu \to 0$ has, in recent years, been understood in renormalized perturbation theory [9–11] and in soluble field theoretic models. One obtains operator expansions of the form

$$A(x)\,B(0) \underset{x\to 0}{\longrightarrow} \sum_{i=0}^{N} F_i(x)\,O_i(0) , \quad (12)$$

where $O_1, \ldots, O_N$ is a finite set of local field operators and the $F_i(x)$ are functions with singularities $(x)^{d_i - d_A - d_B}$ (apart from log's), where the d's are the dimensions of the fields [12]. The momentum space limit corresponding to (12) is, however, unphysical and so, although there are applications, the usefulness of expansions like (12) is somewhat limited. Of much greater physical interest is the behaviour of products like $A(x)\,B(0)$ near the light cone (LC) $x^2 \to 0$.

It can be shown that operator product expansions near the LC exist and have the form

$$A(x)\,B(0) \underset{x^2\to 0}{\longrightarrow} \sum_{i=0}^{M} \sum_{n=0}^{\infty} F_{in}(x)\, x^{\alpha_1} \ldots x^{\alpha_M} O^{(n)}_{i\alpha_1\ldots\alpha_M}(0) , \quad (13)$$

where $\{O^{(n)}_{i\alpha_1\ldots\alpha_n}\}$ in an infinite set of local field operators satisfying

$$\dim O^{(n)}_{i\alpha_1\ldots\alpha_n} = d_i + n$$

and $F_{in}(x) \sim F_i(x)$ for $x^2 \to 0$. Thus, rather than a finite number of fields as occurs in the short distance case (12), an infinite number of fields occur in the LC expansion (13). Each term

$$T_{in}(x) \equiv x^{\alpha_1} \ldots x^{\alpha_n} O^{(n)}_{i\alpha_1\ldots\alpha_n}(0)$$

has the same dimension d_i and carries the same LC singularity $\sim F_i(x) \sim x^{d_i - d_A - d_B}$. The short distance behaviour of $T_{in}(x)$ is, however, $x^{d_i - d_A - d_B + n}$ and hence it only contributes a short distance singularity if $n \leqq d_A + d_B - d_i$.

We shall not go into the derivation of expansions like (13), but, will exhibit a specific example in φ^4 theory. For notational simplicity, we shall here ignore all logarithmic factors even though they actually occur in perturbation theory. We want to discuss the renormalized scalar current operator $j(x) \equiv :\varphi(x)\,\varphi(x):$. Here and elsewhere we use the colon notation $:A(0)\,B(0):$ to denote a generalized Wick product of renormalized fields obtained from the ordinary product $A(x)\,B(0)$ by first subtracting off the singular expansion (12) (or a trivial modification of it) and then taking the limit $x^\mu \to 0$. The resulting quantity can be shown to be a finite local field operator having the same quantum numbers as the free field ordinary Wick product $:A(0)\,B(0):$ [9–11]. The relevant dimensions in the theory under consideration are $\dim I = 0$, $\dim \varphi = 1$, $\dim \partial_\alpha \varphi = 2$, $\dim :\varphi\varphi: = 2$, etc. Here I is the unit operator – the trivial local field.

We first exhibit the short distance behaviour:

$$j(x)\,j(0) \underset{x \to 0}{\longrightarrow} c_1 \left(\frac{1}{x^2}\right)^2 I + c_2 \left(\frac{1}{x^2}\right) j(0) + c_3 \left(\frac{1}{x^2}\right) x^\alpha :\varphi \partial_\alpha \varphi:(0)\,. \quad (14)$$

Here and elsewhere x^2 means $x^2 - i\varepsilon x_0$.

Note that, for $x \to 0$, $(1/x^2)$ is a power more singular than $(1/x^2)x^\alpha$. Near the LC, however, each function has the same singularity and, in fact, an infinite number of terms with this singularity occurs in the LC expansion. The result is

$$j(x)\,j(0) \underset{x^2 \to 0}{\longrightarrow} c_1 \left(\frac{1}{x^2}\right)^2 I + \frac{1}{x^2} \sum_{n=0}^{\infty} x^{\alpha_1} \ldots x^{\alpha_n} O^{(n)}_{\alpha_1\ldots\alpha_n}(0)\,, \quad (15)$$

where $\dim O^{(n)} = n + 2$. Thus, each term in the sum has dimension two and carries a LC singularity $1/x^2$. For consistency with (14), we must have

$$O^{(0)} = c_2 j \quad \text{and} \quad O^{(1)}_{\alpha_1} = c_3 :\varphi \partial_{\alpha_1} \varphi:\,.$$

The other terms in (15) do not contribute to the short distance limit (14).

We can now calculate the LC behaviour of, for example, the expectation value of $j(x)j(0)$ in the one-particle state of momentum p. We can write

$$\langle p|O^{(n)}_{\alpha_1\ldots\alpha_n}(0)|p\rangle = a_n p_{\alpha_1}\cdots p_{\alpha_n} + b_n g_{\alpha_1\alpha_2} p_{\alpha_3}\cdots p_{\alpha_n} + \cdots, \tag{16}$$

where the omitted terms each involve at least one $g_{\alpha\beta}$. Only the first term in (16) therefore contributes to the leading LC singularity of (15). Thus, defining

$$f(\lambda) = \sum_{n=0}^{\infty} a_n \lambda^n, \tag{17}$$

we obtain

$$\langle p|j(x)j(0)|p\rangle_c \xrightarrow[x^2\to 0]{} \frac{1}{x^2} f(x\cdot p) \tag{18}$$

as the leading LC singularity of the connected matrix element.

Expansions of the form (13) exist and describe the LC behaviour of the product of any local field operators in each order of renormalized perturbation theory and, more generally, in any theory in which expansions of the form (12) exist for all local field products at short distances. They might therefore be abstracted from these models and assumed to be true in the real world. The usefulness of such expansions stems from the fact that they describe the configuration space limit corresponding to the physical momentum space limit in which a very massive current interacts with a hadronic systems in a high energy inelastic collision. Knowledge of the LC expansion for the relevant current product determines the behaviour of the appropriate inelastic cross-sections in the specified limits. The relevance of the LC behaviour of the proton-proton *matrix element* of the product of electromagnetic currents to deep inelastic electron-proton scattering has already been noted [1–3]. The advantages provided by the *operator* expansions are, however, numerous. For example, they predict (by dimensional analysis) the *strength* of the LC singularity and thereby provide a means of measuring the dimensions of interacting fields, they determine, by their form, properties of inelastic form factors in *several* variables, and they provide relations between form factors describing *different* experiments corresponding to different matrix elements of the current products. Several such applications of LC operator expansions will be described below. The main conclusion we reach is the following: the present experimental results are in good agreement with the canonical singularity structure of renormalizable field theory – they provide no evidence for unrenormalizable theories and/or non-canonical dimensionality.

Expansions similar to (15) are valid for any products in any renormalizable field theory. As an example which will be applied below, we display the result for conserved vector currents:

$$\begin{aligned} j_\mu(x)\, j_\nu(0) \underset{x^2\to 0}{\longrightarrow} & (\partial_\mu\partial_\nu - g_{\mu\nu}\Box)\, x^{-2} \sum_n x^{\alpha_1}\ldots x^{\alpha_n} R_{0\alpha_1\ldots\alpha_n}(0) \\ & + i\varepsilon_{\mu\nu\alpha\beta}\,\partial^\alpha x^{-2} \sum_n x^{\alpha_1}\ldots x^{\alpha_n} R^\beta_{1\alpha_1\ldots\alpha_n}(0) \\ & + [g_{\mu\nu}\partial_\alpha\partial_\beta - g_{\alpha\nu}\partial_\beta\partial_\mu - g_{\alpha\mu}\partial_\beta\partial_\nu + g_{\alpha\mu}g_{\beta\nu}\Box] \\ & \cdot (\log - x^2) \sum_n x^{\alpha_1}\ldots x^{\alpha_n} R^{\alpha\beta}_{2\alpha_1\ldots\alpha_n}(0)\,. \end{aligned} \tag{19}$$

Here the $(\log - x^2)$ term does not violate our neglect of logs since the log goes away after it is differentiated.

As we have already mentioned, an immediate consequence of our derivations is that operator product expansions near the LC will be valid in any theory in which operator product expansions at short distances are valid. In addition to the renormalizable perturbation theories, this class of theories includes all known exactly soluble field theoretic models. More generally, or rather more phenomenologically, if, as seems to be indicated, short distance expansions are valid in the real world, then so are LC expansions.

Because of a phenomenon which occurs in the Thirring model, we are forced at this point to be more precise about the notion of dimensionality which we have been using. One says that a local field $\chi(x)$ has dimension d if there exists a one-parameter group $U(s)$ of unitary transformations such that

$$U(s)\,\chi(x)\,U^{-1}(s) = s^d \chi(sx)\,. \tag{20}$$

Examples are the free massless scalar field with $d=1$ and free spinor field with $d=3/2$. We shall refer to this notion of dimension as "dynamical" dimension. For the usual fields in free field theories, dynamical dimension coincides with naive dimension. We shall say a field has canonical dimension if it has a dynamical dimension equal to that of the corresponding free field. In a theory in which all local fields have dimensions and short distance expansions such as (12) are valid, application of (20) to (12) implies that the $F_i(x)$ behave as stated like $(x)^{d_i-d_A-d_B}$, with no logarithmic factor. This is what happens in free field theories.

In any finite order of a renormalizable perturbation theory, because of the *occurrence* of logarithmic factors, the renormalized fields do not have well-defined dynamical dimensions. Nevertheless, the short distance behaviour of any Wightman function is, apart from logarithmic factors, the same as it would be if the fields did have canonical dynamical dimensions. Put differently, the short distance behaviour is determined, apart from logs, by the naive dimensions of the fields. In particular, the nature of short distance expansions, and, by our analysis, of LC expansions are so determined. We shall describe this situation by saying that the fields have effective canonical dimensions.

In any theory with effective canonical dimensionality and with short distance expansions, LC expansions very similar to those given above will exist. Included in such theories are free field models, renormalizable perturbation theories in any finite order, and most of the known exactly soluble models. In theories with effective non-canonical dynamical dimensionality and with short distance expansions, our derivations show that LC expansions will also exist. In these theories, the singular functions will, of course, be somewhat different from those encountered above. The Thirring model is the only one we know of that exhibits non-canonical dynamical dimensionality [13].

A final point we should mention concerns the nature of the sum (if it exists) of the perturbative expansions of the renormalizable field theories. It is possible, and has been suggested, that the logarithmic factors occurring in each order sum up to a power and so change the dynamical dimensions of the fields. This is what happens in the Thirring model. We suspect that this Thirring model phenomenon arises because of the zero mass particle present and will not occur in realistic models with no massless particles. There is no evidence that the logs of renormalizable perturbation theories add up to a power [14]. In the following we shall point out the existence of empirical indications of the validity of effective canonical dimensionality.

Our purpose now is to use the expansion (19) to study the process $e+p\to e+$ anything. The relevance of the LC behaviour of the matrix element $\langle p|\,[J_\mu(x), J_\nu(0)]\,|p\rangle$ to the A limit this reaction has been known for some time [1–3]. Our use of the *operator* expansion will, however, enable us to deduce a number of new results from the observed scaling behaviour. We follow the notation of Ref. [3], where more details and references can be found. We shall first work with the result (19) of ignoring logarithmic factors and afterwards discuss the effect of these logs and of their possible role in changing the singularity structure.

The total cross-section (1) of interest can be written

$$\frac{d^2\sigma}{dq^2\,d\nu} = \frac{\pi\alpha^2}{E^2|q|^2\sin^2\frac{\theta}{2}}\left[W_2(q^2,\nu)\cos^2\frac{\theta}{2} + 2W_1(q^2,\nu)\sin^2\frac{\theta}{2}\right], \tag{21}$$

where E is the initial electron energy and θ the scattering angle and we have set the proton mass equal to unity: $p^2=1$. The structure functions are defined by

$$\frac{1}{2\pi}\int d^4x\, e^{iq\cdot x}\langle p|[J_\mu(x), J_\nu(0)]|p\rangle = (p_\mu - \varrho q_\mu)(p_\nu - \varrho q_\nu) W_2(q^2, \nu) - \left(g_{\mu\nu} - \frac{q_\mu q_\nu}{q^2}\right) W_1(q^2, \nu), \tag{22}$$

where $\varrho \equiv -\nu/q^2 = (2\omega)^{-1}$, and the A limits are

$$\lim_A \nu W_2(q^2, \nu) = F_2(\varrho), \tag{23}$$

$$\lim_A W_1(q^2, \nu) = F_1(\varrho). \tag{24}$$

The transverse and longitudinal structure functions are

$$F_T = F_1, \quad F_L = \varrho F_2 - F_1. \tag{25}$$

Experimentally [15], (23) is well satisfied in a non-trivial way [$F_2(\varrho) \sim$ const. for $\varrho \geqq 2$] and F_L/F_T is small, as suggested by the gluon model [16].

It is convenient to introduce new structure functions by writing

$$\frac{1}{2\pi}\int d^4x\, e^{iq\cdot x}\langle p|[J_\mu(x), J_\nu(0)]|p\rangle = [q^2 p_\mu p_\nu - \nu(p_\mu q_\nu + q_\mu p_\nu) + \nu^2 g_{\mu\nu}] V_2(q^2, \nu) - (q^2 g_{\mu\nu} - q_\mu q_\nu) V_1(q^2, \nu). \tag{26}$$

Eqs. (22)–(26) imply

$$\lim_A (-\nu^2) V_2(q^2, \nu) = \varrho F_2(\varrho), \tag{27}$$

$$\lim_A \nu V_1(q^2, \nu) = \varrho F_L(\varrho). \tag{28}$$

In configuration space (26) reads

$$\frac{1}{2\pi}\langle p|[J_\mu(x), J_\nu(0)]|p\rangle = -[\Box p_\mu p_\nu - (p\cdot\partial)(p_\mu\partial_\nu + p_\nu\partial_\mu) + (p\cdot\partial)^2 g_{\mu\nu}] \hat{V}_2(x^2, x\cdot p) - (\partial_\mu\partial_\nu - \Box g_{\mu\nu}) \hat{V}_1(x^2, x\cdot p), \tag{29}$$

in terms of the Fourier transforms

$$V_i(q^2, \nu) = \int d^4x\, e^{iq\cdot x} \hat{V}_i(x^2, x\cdot p). \tag{30}$$

The A limit of the V_i's is given by the LC behaviour of the $\hat{V}_i$'s and these can be determined from (19). We define as in (4)–(7) the matrix elements

$$\langle p|\sum_n x^{\alpha_1}\dots x^{\alpha_n}R_{0\alpha_1\dots\alpha_n}(0)|p\rangle = f_0(x\cdot p)+O(x^2)\,, \tag{31}$$

$$\begin{aligned}&\langle p|\sum_n x^{\alpha_1}\dots x^{\alpha_n}R^{\alpha\beta}_{2\alpha_1\dots\alpha_n}(0)|p\rangle\\ &\qquad = g^{\alpha\beta}f(x\cdot p)+p^\alpha p^\beta f_2(x\cdot p)+O(x^2)\,.\end{aligned} \tag{32}$$

Comparison with (29), using

$$\mathrm{Im}(x^2-i\varepsilon x_0)^{-1}=\pi\varepsilon(x_0)\,\delta(x^2),\quad \mathrm{Im}\log(-x^2+i\varepsilon x_0)=\pi\varepsilon(x_0)\,\theta(x^2), \tag{33}$$

gives the results

$$\hat{V}_2(x^2,x\cdot p)\xrightarrow[x^2\to 0]{}-\varepsilon(x_0)\,\theta(x^2)\,f_2(x\cdot p)\,, \tag{34}$$

$$\hat{V}_1(x^2,x\cdot p)\xrightarrow[x^2\to 0]{}-\varepsilon(x_0)\,\delta(x^2)\,f_0(x\cdot p)\,. \tag{35}$$

These are precisely the LC behaviours shown in Ref. [3] to be equivalent to the scaling laws (27) and (28) or (23) and (24). Indeed, direct substitution of (34) and (35) into (30) gives the results (27) and (28) with

$$\varrho F_2(p) = -2\pi\int d\lambda\, e^{-i\lambda/2\varrho}\lambda f_2(\lambda)\,, \tag{36}$$

$$\varrho F_L(p) = -\frac{\pi i}{2}\int d\lambda\, e^{-i\lambda/2\varrho} f_0(\lambda)\,. \tag{37}$$

We have thus derived the validity of the scaling laws (27) and (28) in the large class of theories in which (19) holds. Strictly speaking, because these theories really give extra logarithmic factors, we can only deduce that (27) and (28) are valid apart from powers of $\log q^2$. Indeed, it is a known fact that in low orders these theories give scaling apart from logs. This is satisfactory since the presence of such logarithmic factors could easily escape experimental detection at present. One might think that this result is trivial because we have built in scaling via the mass independence of (19). The point is, however, that, because of the possibility of non-canonical dimensions, mass independence is not equivalent to scaling. We can thus reach the strong conclusion that the observed scaling behaviour is consistent with canonical dimensionality but not with many types of non-canonical dimensionality. If a single operator with a non-vanishing proton-proton matrix element in (19) had a dimension significantly less than its canonical one, then the scaling limits (27) and/or (28) would be divergent by the corresponding power of q^2. If we further believe in the relevance of a perturbative model, then we can conclude that the logarithmic factors in the model do not add up to significantly change the singularity structure. It is therefore an interesting and non-trivial fact that *non-trivial scaling is equivalent to the presence of the LC singularity structure required by canonical dimensioality*.

We have thus seen how the existence of operator product expansions like (19) enables one to correlate experimental results with the nature of possible field theoretic models for the hadrons. In addition to providing evidence for essentially canonical dimensionality, the observed scaling strongly suggests the relevance of renormalizable field theories. Non-renormalizable models, made finite, say, by the introduction of infinitely many subtraction constants, possess much worse LC singularities. A further major advantage of our formalism is that, unlike the matrix element statements like (34) and (35), it enables one to compare and relate different processes since these processes simply involve different matrix elements of the same operators.

Before leaving electroproduction, we wish to comment on what happens if $F_L = 0$. It is clear from the above analysis that, neglecting the unlikely possibility that the proton-proton matrix element of each $R_{0\alpha_1\ldots\alpha_n}$ in (19) vanishes, $F_L = 0$ means that the leading allowed singularity $1/x^2$ in the first piece of (19) is not present. Assuming canonical dimensionality, this means that the $(1/x^2)$ must be replaced by $(\log x^2)$. This comes from both the non-leading contributions of the given operators satisfying $\dim R_{\alpha_1\ldots\alpha_n} = n+2$ and from the leading contributions of additional operators satisfying $\dim R_{\alpha_1\ldots\alpha_n} = n+4$. Calling the matrix element of the sum of these operators still $f_0(x\cdot p)$ as in (31), (35) becomes replaced by

$$\hat{V}_1(x^2, x\cdot p) \xrightarrow[x^2\to 0]{} -\varepsilon(x_0)\,\theta(x^2)\,[f_0(x\cdot p) + 2f(x\cdot p)]\,. \tag{38}$$

To conclude, we shall briefly mention some further applications of the formalism we have developed. Detailed treatments of these applications have been given elsewhere [4, 5].

An especially interesting application [5] is to study the recent measurement [17] of the cross-section $d\sigma/dq^2$ for the reaction proton + proton $\to \mu$ pair + anything. Here the double-proton matrix element $\langle pp'|J_\mu(x)\,J_\nu(0)|pp'\rangle$ is involved and the relevant region is given by (9) with $\varkappa^2 = [(s+q^2-4m^2)/4s] - q^2$ even though $s=(p+p')^2 \to \infty$ in the scaling limit. Most of the experimental points correspond to the LC $x^2 \sim 0$ and so the expansion (19) can be used. The matrix element of $R \equiv \sum x^{\alpha_1} \ldots x^{\alpha_n} R_{0\alpha_1\ldots\alpha_n}$ etc., is now more complicated, but Regge theory provides an enormous simplification in that it implies that, for example $\langle pp'|R|pp'\rangle \xrightarrow[s\to\infty]{} s^\alpha[h(p\cdot x) + h(p'\cdot x)]$ [5]. Further use of Regge theory for the dimensionless structure functions above the value $\varrho \sim 2$ suggested by SLAC enables us to express the cross-sections in terms of a two-parameter function which can be fit very nicely to the data which fall ~ 5 orders of magnitude for $2 \leqq q^2 \leqq 30\ \mathrm{GeV}^2$.

Another class of applications involves the determination of the asymptotic behaviour of vertex functions when a mass becomes large, as described above [4]. Knowledge of this asymptotic behaviour, plus the information, learned from the electroproduction and μ pair results, that asymptotia sets in at $q^2 \sim 2\,\mathrm{GeV}^2$, enables us to write the analogue of finite energy sum rules in the mass variable and thus approximately calculate both the "infinite" mass contributions and the continuum contributions. These contributions give corrections to the result of simply saturating the mass dispersion relation with a low lying meson. In this way we can understand, for example, pion pole dominance of matrix elements of the divergence of the axial vector current and we can estimate corrections to vector meson dominance which are in good agreement with experiment. Consider, for example, the $\pi \to 2\gamma$ off-shell amplitude. The asymptotic behaviour in the photon mass variable is determined from the second piece of Eq. (19). This information enables us to obtain an experimentally correct determination of the rate and also to relate the amplitude to, say, the electroproduction ones via (19).

References

1. *Brandt, R. A.:* Phys. Rev. Letters **22**, 1149 (1969).
2. — Phys. Rev. Letters **23**, 1260 (1969).
3. — Phys. Rev. D, **1**, 2808 (1970).
4. — *Preparata, G.:* Phys. Rev. Letters **25**, 1530 (1970).
5. *Altarelli, G., Brandt, R. A., Preparata, G.:* Phys. Rev. Letters **26**, 42 (1971).
6. *Brandt, R. A., Preparata, G.:* Nuclear Physics B **27** (1971).
7. *Bjorken, J. D.:* Phys. Rev. **179**, 1547 (1969).
8. — Phys. Rev. **148**, 1469 (1966).
9. *Wilson, K.:* Unpublished.
10. *Brandt, R. A.:* Ann. Phys. (N.Y.) **44**, 221 (1967).
11. *Zimmermann, W.:* Commun. Math. Phys. **6**, 161 (1967).
12. This dimensional concept is discussed in Refs. 9.–11. and below.
13. This property of the Thirring model was discovered by *K. Wilson*, Phys. Rev. D **2**, 1473 (1970).
14. The most recent and elegant account can be found in *K. Symanzik*, Cargèse Summer School Lectures 1970.
15. *Breidenbach, M., et al.:* Phys. Rev. Letters **23**, 935 (1969).
16. *Callan, C., Gross, D. J.:* Phys. Rev. Letters **22**, 156 (1969).
17. *Christenson, J. H., Hicks, G. S., Lederman, L. M., Limon, P. J., Pope, B. G., Zavattini, E.:* Phys. Rev. Letters **25**, 1523 (1970).

Professor Dr. *R. A. Brandt*
Department of Physics
New York University
New York, USA

Course on Padé Approximants

J. Zinn-Justin

Contents

Introduction

In Quantum Electrodynamics, Perturbation theory has proved to be an extremely powerful tool to get quantitative physical results. The reason for this is that, although the perturbative expansion is presumably divergent for all values of the expansion parameter, it has the property of an asymptotic series, which gives very good numerical results if the parameter is small enough which is the case.

In Strong Interactions, the perturbative expansion has been used for a long time only in order to understand the analytic properties of the scattering amplitudes or to estimate Born terms. The reason was that the coupling constants were found to be relatively large, and therefore in many cases the perturbation series was useless.

Moreover, it was not in general possible to find a corresponding approximate classical problem to compute mass spectra as in Electrodynamics, so that the problem of finding bound states or resonances starting from an interaction lagrangian was open.

A way to solve these two problems is to find a method of summation for divergent series.

Among these methods, the Pade approximation seems to have very pleasant mathematical and physical properties.

I. Elementary Mathematical Properties of Pade Approximants

A. Definition and Algebraic Properties

1. Let $f(z)$ be an analytic function defined by its Taylor series:

$$f(z) = \sum_{0}^{\infty} a_n z^n . \tag{1}$$

We call Pade approximant $f^{[N,M]}(z)$ of $f(z)$ the following rational fraction:

$$f^{[N,M]}(z) = \frac{P_N(z)}{Q_M(z)} = f(z) + O(z^{N+M+1}) \tag{2}$$

where $P_N(z)$ and $Q_M(z)$ are polynomials of degree N and M respectively.

If $N = M$ we call $f^{[N,N]}(z)$ a diagonal Pade approximant.

2. *Uniqueness of the Pade Approximants*

Let us assume that there exist two different $[N, M]$ Pade approximants satisfying

$$f(z) = \frac{P_N(z)}{Q_M(z)} = \frac{R_N(z)}{S_M(z)} O(z^{N+M+1}) . \tag{3}$$

We deduce from this equation:

$$P_N(z)\, S_M(z) - Q_M(z)\, R_N(z) = 0 \qquad O(z^{N+M+1}) . \tag{4}$$

The left hand side expression is a polynomial of degree $N+M$, therefore if it is of order $N+M+1$, it is identically zero.

$$\frac{P_N(z)}{Q_M(z)} = \frac{R_N(z)}{S_M(z)} .$$

Therefore the Pade approximant, if it exists, is unique. From this we deduce that if a function is a rational fraction $\frac{R_N(z)}{S_M(z)}$, then the Pade approximant $[N', M']$ is identical to the function if: $N' \geqq N$ and $M' \geqq M$.

3. *Groups of Transformation on the Pade Approximants*

The Taylor series expansion induces an homomorphism of algebras between functions and theirs expansions; the Pade approximation is a non linear operation, therefore the situation is completely different: we have groups of transformations.

We have to look for transformations $\mathcal{T}$ satisfying the following properties:

$$\text{a)}\ \mathcal{T}\left[\frac{P_N(z)}{Q_M(z)}\right]=\frac{R_{N'}(z)}{S_{M'}(z)} \quad \text{with} \quad N'+M'\leqq N+M \tag{5}$$

where $P_N, Q_M, R_{N'}, S_{M'}$ are polynomials.

$$\text{b)}\ f(z)=g(z)+O(z^{N+M+1}). \tag{6}$$

Then:

$$\mathcal{T}[f(z)]=\mathcal{T}[g(z)]+O(z^{N+M+1}). \tag{7}$$

Using the uniqueness property of Pade approximants we deduce that:

If $f^{[N,M]}(z)$ is the $[N, M]$ Pade approximant of $f(z)$ then $\mathcal{T}[f^{[N,M]}(z)]$ is the $[N', M']$ Pade approximant of $\mathcal{T}[f(z)]$. In what follows we give a certain number of such transformations. It is easy in all cases to verify properties a) and b). We remark that the product of two transformations satisfying a) and b) satisfies also a) and b).

4. *Groups of Transformations on the Functions*

$$\{f(z)^{[N,M]}\}^{-1}=\{f(z)^{-1}\}^{[M,N]}. \tag{8}$$

The $[M, N]$ Pade approximant of the inverse of a function is the inverse of the $[N, M]$ approximant of the function.

$$\{f^*(z^*)\}^{[N,M]}=\{f(z^*)^{[N,M]}\}^*. \tag{9}$$

This shows that the Pade approximant of a real function is a real function.

The product of the two first properties gives the following result: If $S(z)$ is a function satisfying:

$$S(z)\,S^*(z^*)=1, \tag{10}$$

then:

$$S^{[M,N]}(z)\,\{S^{[N,M]}(z^*)\}^*=1. \tag{11}$$

The diagonal Pade approximants satisfy the same functional identity as the function.

Let α be an arbitrary constant:

$$\{\alpha f(z)\}^{[N,M]}=\alpha\{f(z)\}^{[N,M]}. \tag{12}$$

Let N and j be two non negative integers.

Let $Rj(z)$ be a polynomial of degree not larger than j, then:

$$[f(z)+Rj(z)]^{[N+j,N]}=f(z)^{[N+j,N]}+Rj(z). \tag{13}$$

From this we deduce the following:

Let $Rj(z)$ such that:

$$f(z) - Rj(z) = z^j fj(z) \tag{14}$$

where $fj(z)$ is a regular function, then (13) gives

$$\{z^j fj(z)\}^{[N+j,N]} = f(z)^{[N+j,N]} - Rj(z). \tag{15}$$

Dividing both sides by z^j gives:

$$[fj(z)]^{[N,N]} = \frac{f(z)^{[N+j,N]} - Rj(z)}{z^j}. \tag{16}$$

This relates diagonal and non diagonal Pade approximants.

The consequence of a certain number of preceeding properties is in the case of diagonal Pade approximants that: the homographic transform of a function has for diagonal Pade approximant the homographic transform of the diagonal Pade approximant of the function.

5. *Groups of Transformations on the Variable*

Let α be a non zero constant. The change of z in αz, in the function, commutes with the Pade approximation. Moreover the change of z in $z/(1+z)$ commutes with the diagonal Pade approximation. We deduce for the diagonal Pade approximants, that if a homographic transformation leaving the origin invariant, is made on the variable, the transformed approximant is the approximant of the transformed function.

B. A Particular Case: Continued Fractions

Let $\{f_n(z)\}$ be a set of analytic functions, which have a Taylor series expansion near the origin, and $\{a_n\}$ a set of complex numbers. The functions $f_n(z)$ satisfy the following recurrence equation:

$$f_n(z) = 1 + a_n z[f_{n+1}(z)]^{-1}. \tag{17}$$

This so defined, formal expansion is called the continued fraction expansion near the origin of the function $f_1(z)$. The continued fraction, truncated at order n, is a rational fraction. It is very easy to see that it is a Pade approximant of type $[N+1, N]$ or $[N, N]$. We can linearize equation (17), putting:

$$f_n(z) = \frac{u_n(z)}{u_{n+1}(z)} \tag{18}$$

(17) becomes:

$$u_n(z) = u_{n+1}(z) + a_n z\, u_{n+2}(z) \quad \text{with} \quad u_n(0) = 1 \tag{19}$$

and

$$f_1(z)=\frac{u_1(z)}{u_2(z)}. \tag{20}$$

We can take for instance $u_2(z)=1$.

The Eq. (19) can be used to construct practically in a recursive way the continued fraction expansion, and therefore the $[N, N]$ and $[N+1, N]$ Pade approximants. For convergence theorems see [1]. Knowing the coefficients $\{a_n\}$ of the continued fraction, it is possible to calculate the numerators and the denominators of the $[N, N]$ and $[N+1, N]$ approximants by the equations:

$$\left.\begin{aligned} P_{n+1}(z)&=P_n(z)+a_n z\, P_{n-1}(z) \quad \text{with} \quad P_0=P_1=1 \\ Q_{n+1}(z)&=Q_n(z)+a_n z\, Q_{n-1}(z) \quad \text{with} \quad Q_0=0, Q_1=1 \end{aligned}\right\} \tag{21}$$

n is here not the degree of the polynomials because P_n and Q_n correspond successively to the $[N, N]$ and $[N+1, N]$ Pade approximant.

C. Some Results on Diagonal Pade Approximants

a) Let now consider the diagonal Pade approximants of a function $f(z)$ at infinity:

$$f(z)=\frac{P_N(z)}{Q_N(z)} \quad O\left(\frac{1}{z^{2N+1}}\right). \tag{22}$$

Let $f(z)$ be analytic near the point at infinity, and C a complex contour out of which $f(z)$ is analytic.

Then it is easy to prove [2, 3]:

$$\oint_C f(z)\, Q_N(z)\, Q_{N'}(z)\, \mathrm{d}z=\alpha_N \delta_{NN'}. \tag{23}$$

The denominators of the diagonal Pade approximants are orthogonal polynomials on a complex contour.

This result is very important for proving the convergence of Pade approximants in certain cases. If $f(z)$ is a Stieltjes function, this means an analytic function satisfying:

$$\operatorname{Im}[f(z)] \operatorname{Im} z>0, \tag{24}$$

and having only singularities on the real axis, then the discontinuity is imaginary with a positive imaginary part and the polynomials $Q_N(z)$ are orthogonal on the real axis with a positive measure.

From this it is easy to derive a number of properties for the zeros of the diagonal Pade approximant denominators, specially, that the zeros are real and located on the support of the measure. This is the first step in proving the uniform convergence, out of the support of the measure, of the diagonal Pade approximants in case of Stieltjes functions. For more details see [4].

b) An other interesting property has recently been proved by *Nutall* [5]. He proved for all meromorphic functions the following: Let $f(z)$ be a meromorphic function.

Let ε and δ be two arbitrary positive numbers. Then it is possible to find an integer N for each function $f(z)$ such that its $[n,n]$ Pade approximant satisfies

$$\text{for} \quad n \geqq N: \left| f(z) - \frac{P_n(z)}{Q_n(z)} \right| < \varepsilon \quad \text{for} \quad |z| < R \tag{25}$$

except for a domain of measure δ.

This is not a convergence theorem in the ordinary sense, it is often called convergence in measure.

Using the same method as Nutall one can prove the uniform convergence for entire functions whose Taylor coefficients decrease essentially more rapidly than $\frac{1}{\sqrt{n!}}$, except on a domain of measure zero.

c) Gaussian Integration

It is easy to prove the connexion between diagonal Pade approximants and the Gaussian integration method [3]. The points and the weights used in the Gaussian integration, are respectively the locations of the poles and the residues of the Pade approximants of the function associated to the measure.

d) Pade Approximants of Analytic Matrices

1. Definition. Let $f(z)$ be a matrix analytic function of a complex variable z, and having a Taylor series expansion:

$$f(z) = \sum_{0}^{\infty} a_n z^n , \tag{26}$$

where $\{a_n\}$ is a set of matrices. The matrices giving a non commutative algebra, we can now define two types of Pade approximants depending on the fact that the denominator is putted on the left hand side or the right hand side

$$f(z) = P_N(z)\,[Q_M(z)]^{-1} = [\tilde{Q}_M(z)]^{-1}\,\tilde{P}_N(z) \quad O(z^{N+M+1}) \tag{27}$$

where $P_N(z), \tilde{P}_N(z), Q_M(z), \tilde{Q}_M(z)$ are polynomials of degree N and M respectively in the variable z.

2. Uniqueness. If a function has a right hand side and a left hand side $[N,M]$ Pade approximant, then they are identical and therefore all the right hand side approximants are identical so as all the left hand side one; the approximant in this case is therefore unique.

The proof is:

$$f(z)=P_N(z)\,[Q_M(z)]^{-1}=[\tilde{Q}_M^{-1}(z)]\,\tilde{P}_N(z)\;\;O(z^{N+M+1}). \tag{28}$$

From this we derive

$$\tilde{Q}_M(z)\,P_N(z)-\tilde{P}_N(z)\,Q_M(z)=O(z^{N+M+1}). \tag{29}$$

As before, a polynomial of degree $N+M$ satisfying (29) is identically zero.

3. Groups of Transformation. We have essentially the same properties as in the scalar case.

Two more transformations exist; the transposition and the hermitian conjugation. This gives the following results, in the case where a unique approximant exists.

a) If $T(z)$ is a symmetric matrix:

$$T(z)=T^t(z) \tag{30}$$

then all Pade approximants are symmetric.

b) If $H(z)$ is a hermitian analytic matrix:

$$H^+(z^*)=H(z) \tag{31}$$

then all Pade approximants have the same property.

c) If $S(z)$ is a unitary analytic matrix:

$$S(z)\,S^+(z^*)=1. \tag{32}$$

Then Pade approximants satisfy:

$$S^{[N,M]}(z)\,\{S^{[M,N]}(z^*)\}^+=1. \tag{33}$$

The diagonal Pade approximants are unitary.

4. Other Properties. Essentially the same results as in the scalar case can be obtained, on continued fractions, orthogonality properties and also convergence properties.

e) Pade Approximants of type II

1. Definition. Let $z_1, z_2, \dots, z_p$ be p complex numbers, $f(z)$ an analytic function. We call Pade approximant of type II $[N,M]$ the following rational fraction:

$$f^{[N,M]}(z)=\frac{P_N(z)}{Q_M(z)}$$

with

$$f(z_i) = f^{[N,M]}(z_i) \quad \text{for} \quad i = 1, 2, \dots p \quad \text{and} \quad p = N + M + 1 .$$

2. *Properties.* The uniqueness is proved in the same way as before. Only the group of transformations on the function remains.

One can prove in this case the equivalent of all other properties of the Pade approximants of first kind. See [3].

f) Some Applications of Pade Approximants

Pade approximants of type I can be used to accelerate convergence of numerical series [6] $\sum_0^\infty a_n$, being considered as the value of the function $f(z)$

$$f(z) = \sum_0^\infty a_n z^n$$

for $z = 1$.

They can be used also to construct a function satisfying to the hypothesis of *Carlson*'s theorem and to extrapolate the first terms of a set $a_1, a_2, \dots a_{2N}$ [3].

New numerical integration methods of singular functions have also been found.

Pade approximants of type II can be used also to accelerate the convergence of series in the case where the previous method does not apply.

$$S_N = \sum_{n=1}^{N} u_n$$

is in this case considered as function of N. Pade approximants of type II can also be used for solving equations [3].

II. Application of the Pade Approximation to Physics

A. Potential Scattering, Anharmonic Oscillator

In two cases the Pade approximation was proved to converge: 1. The first case is potential scattering. *Chrisholm* [7] shows that Pade approximants give an exact solution for the spectrum of an integral equation with a finite rank kernel, and that they give a convergent set of eigenvalues in the case of compact kernels. Following the ideas of *Chisholm, Garibotti* and *Villani* [8] show the connexion of Pade approximants with the minimal iteration method for solving linear integral equations. It is the following: Consider the equation:

$$f = g + \lambda K f \tag{1}$$

where K is a certain kernel $K(x, y)$ which we assume to be hermitian for simplicity. Consider now the orthogonal family of vectors $\{g_n\}$

$$g_0 = g\,,$$

$$g_n = K^n g + \sum_{p=0}^{n-1} \alpha_p K^p g \quad \text{and} \quad (g_i, g_j) = \alpha_i \delta_{ij}\,. \tag{2}$$

It is easy to compute g_n in a recursive way. Consider now equation (1) projected in the subspace spanned by the basis $(g_0, g_1, \dots g_N)$.

Solving this new equation gives an approximation for the function f and for the eigenvalues of K. This method is called the minimal iteration method. It is clear that this method gives an exact solution for (1) if K is a kernel of finite rank N and if the dimension of the basis is larger than or equal to N. In a certain number of cases this method was proved to converge, in particular for compact kernels. What is now the connexion with the Pade approximation? Let us take the series expansion of f, formal solution of (1)

$$f = \sum_0^{\infty} \lambda^n K^n g \tag{3}$$

and the scalar product:

$$\langle g | f \rangle = \sum_0^{\infty} \lambda^n (g, K^n g)\,. \tag{4}$$

It can be shown that the $[N, N+1]$ Pade approximant of series (4) has $N+1$ poles which are the same that the eigenvalues given by the minimal iteration method in a space of dimension $N+1$.

Now this can be applied to the Lippman-Schwinger equation. Garibotti and Villani show that the procedure converges for regular potentials and gives the same result, on the energy shell, and for the forward direction amplitude or for the partial wave amplitude as the diagonal Pade approximation computed from the Born series.

2. The second case is the case of the anharmonic oscillator. *Loeffel et al.* [9] show that the Pade approximants on the energy levels converge because the energies are Stieltjes functions of the coupling constant. This result is very encouraging because it is perhaps the first step in the direction of field theoretical models.

B. Practical Application of Pade Approximation to a Simple Field Theoretical Model

In order to show how the Pade approximation allows to study practically two body S matrix elements in field theory, we want to examine in some detail the case of $\lambda\varphi^4$ interaction between pions. The model is not completely realistic for the pion interaction, but owing to its simplicity it was studied first (10, 11) and it was possible to compute five distinct Pade approximants in this case [1, 1] [2, 1] [1, 2], and [2, 2] on the S matrix and the [2, 1] on the K matrix. The very good agreement between all the results, especially in the s wave is very surprising and seems to indicate that we have here a very good method for summing approximately the perturbation series.

1. The Perturbation Series

In (10) the amplitude is calculated up to fourth order in the coupling constant, for each value of isospin.

$$A(s,t)=\sum_1^4 \lambda^n A_n(s,t) \tag{1}$$

λ is the coupling constant, s, t are the ordinary Mandelstam variables. Let l be the angular momentum, $A(s,t)$ is projected on the partial wave l with $\mathrm{Re}\, l>0$ using the Froissart-Gribov formula

$$a_l(s)=\frac{1}{\pi k^2}\int_4^\infty Q_l\left(1+\frac{2t}{s-4}\right)\mathrm{Im}\, A(s,t)\frac{2\mathrm{d}t}{s-4}\,. \tag{2}$$

Because of the asymptotic behaviour of $\mathrm{Im}\, A(s,t)$, the s wave does not belong to the carlsonian set and must be projected directly.

We now have the following series for the S matrix

$$S^{l=0}=1+S_1\lambda+S_2\lambda^2+S_3\lambda^3+S_4\lambda^4\,. \tag{3}$$

For l different from zero, S_1^l vanishes because $A_1(s,t)$ is a pure s wave.

$$S^{l\neq 0}=1+S_2^l\lambda^2+S_3^l\lambda^3+S_4^l\lambda^4\,. \tag{4}$$

Now on this perturbation series, the Pade solution is applied.

2. The Pade Solution

We compute the Pade approximants, for instance for $l=0$ we have:

$$\begin{cases} S^{l=0}_{[1,1]}=1+g\,S_1[1-g\,S_2/S_1]^{-1} \\ S^{l=0}_{[2,1]}=1+g\,S_1+g^2 S_2[1-g\,S_3/S_2]^{-1} \end{cases} \tag{5}$$

and for $l\neq 0$

$$\begin{cases} S^l_{[1,1]}\ \text{does not exist because}\ S_1=0 \\ S^l_{[2,1]}=1+g^2 S_2[1-g\,S_3/S_2]^{-1} \\ S^l_{[1,2]}=1+g^2 S_2[1-g\,S_3/S_2-S_2 g^2]^{-1}\,. \end{cases} \tag{6}$$

If we compute the S matrix starting from the [2, 1] Pade approximant of the K matrix we find:

$$S^l(K_{[2,1]}) = 1 + g^2 S_2 [1 - g S_3/S_2 - (S_2/2) g^2]^{-1} . \qquad (7)$$

We have only written the simplest approximants. We can now make the following remarks.

a) Analytic Properties

Because the Pade approximants are rational functions of the coefficients of the perturbation series, the analytic properties in s are the same for the Pade solution and for the series, except for possible poles which can be present in the Pade solution, and not in the series. These poles can represent mass spectra and even Regge trajectories.

The study of Pade approximants in the case of Stieltjes functions, shows that Pade approximants contain poles which converge to the poles of the function, and others which reconstruct the cuts of the function. The question is how to recognize the two kinds of poles in practice. Actually this is very easy, because the first kind of poles is very stable and the second kind very unstable under changing the order of approximation and even a little change in the external parameters.

Therefore in the physical case we can recognize easily in general if the poles have really something to do with resonances or bound states.

Moreover in the physical case, "cut type" poles have wrong physical properties.

b) Unitarity

As it is shown in the mathematical section, a real analytic function has real analytic Pade approximants, this applies to the Pade approximants of the K matrix.

It is also shown that a function which satisfies

$$S^*(g^*)\, S(g) = 1$$

has Pade approximants which satisfy

$$S^{*[N,M]}(g^*)\, S^{[M,N]}(g) = 1 . \qquad (8)$$

Therefore the diagonal Pade approximants of the S matrix are unitary [15].

Using the transformation properties by the homographical group it is also easy to see that

$$\begin{cases} \operatorname{Im} T^{[N,M]}(g) = |T^{[N,M]}(g)|^2 \\ \text{if} \quad N \leqq M . \end{cases} \qquad (9)$$

In fact all the unitary approximants can be deduced from those of the K matrix.

What is very interesting is this strong connexion between Pade approximation and unitarity, and specially in the case of the diagonal Pade approximants, which give the same result for the S matrix, the K matrix and the T matrix.

c) Variational Principle

It is interesting to know that the Pade solution can be obtained from the variational principle of Lippman-Schwinger with the Ansatz of Cini-Fubini [12]. This gives another justification for using the Pade approximation.

d) Crossing Properties

The perturbation series has by construction the good crossing properties. The Pade approximation which is a non linear operation gives an amplitude which satisfies crossing properties only in the perturbative sense. Therefore the numerical study of the crossing properties of the Pade approximants is the most important point, because it can give us an idea of the domain of validity of the approximation. This has been completely investigated in the $\lambda\varphi^4$ model.

3. *The Results* (10, 11, 13)

a) The s Wave

For the S matrix the [1, 1], [2, 1], and [1, 2] approximants were studied for $4 < s < 200$ (the pion mass $= 1$) and also for s complex in the first and second sheet near the physical region. The result was the following: the difference between Pade [1, 1] and [2, 1] was not higher than 12% and the difference between [2, 1] and [1, 2] of the order of 1 or 2%. It is easy to see that the Pade [2, 1] of the K matrix gives result which lies in between the result of the Pade [2, 1] and [1, 2] of the S matrix, therefore the non diagonal Pade approximants of the S matrix which are not automatically unitary, satisfy nevertheless very well unitarity, also on the form $\operatorname{Im} T = \varrho |T|^2$ which has sense even if the amplitude is small.

These very good results can be understood as follows: First let us write the amplitude up to the third order for the three isospins

$$
\left.\begin{array}{l}
(\pi\pi\to\pi\pi)\;\begin{array}{r|l}
I=0 & 10\lambda + 50\lambda^2 I(s) + 30\lambda^2[I(t)+I(u)] \\
I=1 & 10\lambda^2[I(t)-I(u)] \\
I=2 & 4\lambda + 8\lambda^2 I(s) + 18\lambda^2[I(t)+I(u)]
\end{array} \\
\begin{array}{r|l}
I=0 & 250\lambda^3 I^2(s) + 600\lambda^3 J(s) + 110\lambda^3[I^2(t)+I^2(u)] + 440\lambda^3[J(t)+J(u)] \\
+\,I=1 & 70\lambda^3[I^2(t)-I^2(u)] + 80\lambda^3[J(t)-J(u)] \\
I=2 & 16\lambda^3 I^3(s) + 144\lambda^3 J(s) + 86\lambda^3[I^2(t)+I^2(u)] + 224\lambda^3[J(t)+J(u)]
\end{array}
\end{array}\right\}
$$

where the unitary equation is

$$\operatorname{Im} A(s,\cos\theta)=\tfrac{1}{2}\,\varrho(s)\int \mathrm{d}\Omega\, A(s,\cos\theta_1)\, A^*(s,\cos\theta_2) \tag{11}$$

and $I(s)$ and $J(s)$ defined as follows

$$\begin{cases} \operatorname{Im} I(s)=4\pi\varrho(s) & I(4/3)=0 \\ \operatorname{Im} J(s)=\varrho(s)\int \mathrm{d}\Omega\, I(t) & J(4/3)=0\,. \end{cases} \tag{12}$$

We can now make the following remarks: Since all functions are subtracted at the symmetry point $s=t=u=4/3$, and as λ has to be chosen so that no resonances appear too near threshold, the perturbation series has decreasing terms near the Mandelstam triangle as a "good" asymptotic or convergent series does. Typically the first order is ten times the second order. Therefore Pade approximants and perturbation give almost the same thing, and unitarity and crossing are satisfied by both with a good accuracy.

Now for higher energies the perturbation series is no more good in any sense and is very different of Pade approximant, but a new phenomenon appears. It is easy to see the following curious thing:

$$\tfrac{1}{2}\int I(t)\,\mathrm{d}\cos\theta \simeq \operatorname{Re} I(s) \quad \text{for} \quad s>4\,. \tag{13}$$

This becomes asymptotically good for $s\to\infty$.

This gives the following:

$$\begin{aligned} \operatorname{Im} J(s) &\simeq \operatorname{Im} I(s)\operatorname{Re} I(s)\,, \\ J(s) &\simeq \tfrac{1}{2} I^2(s)\,. \end{aligned} \tag{14}$$

Now the third order becomes

$$\begin{array}{r|l} I=0 & 550\lambda^3 I^2(s)+330\lambda^3[I^2(t)+I^2(u)] \\ I=1 & 110\lambda^3[I^2(t)-I^2(u)] \\ I=2 & \;88\lambda^3 I^2(s)+198\lambda^3[I^2(t)+I^2(u)]\,. \end{array} \tag{15}$$

One can also see that:

$$\tfrac{1}{2}\int I^2(t)\,\mathrm{d}\cos\theta \simeq [\operatorname{Re} I(s)]^2 \quad \text{for} \quad s\to\infty\,. \tag{16}$$

So that, first the same kind of arguments applies to understand the fourth order, and secondly that the s wave projection becomes:

$$l=0\begin{cases} I=0\begin{cases} 10\lambda+\lambda^2[110\operatorname{Re} I(s)+50i\operatorname{Im} I(s)] \\ +\lambda^3[1210(\operatorname{Re} I(s))^2+1100i\operatorname{Im} I(s)\operatorname{Re} I(s)] \end{cases} \\ I=2\begin{cases} 4\lambda+\lambda^2[44\operatorname{Re} I(s)+8i\operatorname{Im} I(s)] \\ +\lambda^3[484(\operatorname{Re} I(s))^2+176i\operatorname{Im} I(s)\operatorname{Re} I(s)]\,. \end{cases} \end{cases} \tag{17}$$

In these expressions it appears clearly that the amplitudes are respectively given by the expansion of:

$$\begin{cases} I=0 & 10\lambda[1-11\lambda \operatorname{Re} I(s)-5i\lambda \operatorname{Im} I(s)]^{-1} \\ I=2 & 4\lambda[1-11\lambda \operatorname{Re} I(s)-2i\lambda \operatorname{Im} I(s)]^{-1}. \end{cases} \tag{18}$$

This being also approximately true for the fourth order one can thereby understand that the perturbation series has really a geometrical character, and that the Pade approximation is very well suited to study field theoretical models.

b) Other Waves

In the three isospin channels, nearly degenerate rising Regge trajectories were found with an sensible intercept violating a little the Froissart bound only for very high values of the coupling constant and only in isospin zero.

λ could be chosen to obtain a ϱ resonance at its physical mass and a f_0 resonance at 1600 MeV, with widths of the order of 20 MeV.

The important point is that the Pade [2, 1] [1, 2], and [2, 2] give the same result for the two masses of the resonances within a few percent.

Moreover in order to have an idea of the crossing symmetry violation the following tests were made:

The [2, 1] Pade approximant of the complete ($\pi^0\pi^0 \to \pi^0\pi^0$) amplitude, which is crossing symmetric, was computed. This so computed amplitude has a pole in s which depends in general of $\cos\theta$.

The trajectory of the pole was found to be relatively flat in the physical region of the s-t-u plane giving a mass for the resonance not very different from the result of the partial waves. Moreover Pade approximant [2, 1] and [2, 2] for the complete isospin one amplitude were computed; it was found that the flatness of the pole trajectory called "poloïd" by the authors, increases surprisingly from one approximant to the other.

c) Other Tests of Crossing Properties

Basdevant et al. [14] have studied how well Pade approximants satisfy Roskies sum rules and Martin inequalities. They found that Roskies sum rules involving s and p waves are satisfied within 2%. Moreover all Martin inequalities are satisfied also.

In the light of what was said in section (a) it is easy to understand this very good result.

First the perturbation series satisfies the crossing relations and therefore Roskies sum rules. Also the perturbative series truncated at second order satisfies positivity, because its imaginary part is the square of the Born term, and therefore all Martin inequalities hold.

Now, as was said in section (a), in the Mandelstam triangle the successive terms of the perturbation series decrease roughly by a factor ten for each order, and the largest part, the s wave, has an obvious geometrical character. So Pade approximants and the perturbation series are not very different. This explains the success of the Roskies sum rules and of the Martin inequalities. For the last point it must be said that in all cases Pade approximants satisfy positivity because they satisfy unitarity.

d) Concluding remarks

These results seem to show that the Pade approximation is one of the best ways to unitarize in a systematic way, a Born term and a Born series in the following sense: It keeps the good analytic and crossing properties of the perturbation series as much as possible, and the behaviour of the series in the region where no summation resonances or bound states appear and where the series makes sense. Furthermore it constructs the first summation resonances which are generated by the forces present in the model. The only limitation is that, since in practice we can only compute a small number of Pade approximants, it is difficult to obtain more than one nonelementary pole per channel and per partial wave.

III. Review of a Certain Number of Models Using Pade Approximants

It seems that the Pade approximation has been applied first in Statistical Mechanics [4], especially in the Ising Model, for instance on the expansion at low densities. It is amusing to know that the Van der Waals equation is a [1, 1] Pade approximant of the pressure expanded in terms of the density. Subsequently the Pade approximation was used in potential scattering [15]. We only want to discuss the most recent applications to potential scattering and field theoretical models.

A. Potential Scattering

Although it has been proved in this case that diagonal Pade approximants converge, it is interesting to study numerically different models, in order to know to what extent the first lowest order approximants are accurate in reproducing the exact amplitude.

For the one dimension δ function potential, the Pade [1, 1] is of course identical to the exact solution. For the square well problem it is very easy to find analytically that the Pade [1, 1] is very good as long as only one bound state exists.

Basdevant and *Lee* [16] have studied the exponential potential in the s wave where an analytic solution is also known. *Caser et al.* [17] have studied numerically a single Yukawa potential term. What comes out of the analysis is that, as long as no more than one bound state exists in the exact solution, both the position of the bound state, and the physical phase shifts are obtained with a good accuracy with the first approximants, in the worst case with the [2, 2]. Of course, one can find more complicated potentials, those which actually cannot be well approximated by a few separable terms, insofar as one is concerned with physical region on the energy-shell amplitudes, for which it is necessary to compute Pade approximants of higher order.

These results are very encouraging for field theoretical models where the Born term is a pole and which in some aspects resemble potential models.

B. Meson-Meson Scattering Models From Field Theory

Because of its simplicity, pseudoscalar meson-pseudoscalar meson scattering has been studied in great detail.

1. The $\gamma \varphi^4$ Theory

Alexanian and *Wellner* [18] first studied pion pion scattering in the $\lambda \varphi^4$ theory, but they applied the Pade approximation to the phase shift, and this procedure generates essential singularities in the amplitude. After that, in the same theory *Masson* and *Copley* [11] computed the perturbation series up to third order and applied Pade to the K matrix, *Bessis* and *Pusterla* [10] computed the fourth order and applied Pade to the S matrix.

As was said in Chapter II, they found that Pade approximants were very well suited to the problem, but that the physical results were not completely satisfactory. The good features were:

The possibility to obtain a ϱ at its physical mass and a f_0 at 1600 MeV. Moreover the Regge trajectories of the ϱ and the f_0 had reasonable intercepts and slopes.

The unpleasant results were that:

a) The s waves had in both isospin channels a negative phase shift with a ratio of the scattering lengths

$$\frac{a_0}{a_2} \simeq 2$$

very near from the ratio 5/2 of the corresponding total amplitudes at the center of Mandelstam triangle.

b) The three isospin channels had nearly degenerated Regge trajectories, which is good for isospin zero and one but gives also an exotic resonance at the same mass as the f_0.

c) The widths of the resonances were too small by a factor of four or five.

2. The $\lambda\varphi^4$ Theory with Pions and Kaons

The features of this theory studied by *Basdevant et al.* [13] are of the same nature as those of the preceeding one.

It was possible in this model to fit the 7 well established resonances ϱ, f_0, f', φ, A_2, $K^*(890)$, $K^*(1420)$, with only three parameters, within an error of a few percent for the masses.

But, because in each J^P channel all the isospin amplitudes were nearly degenerate and because the lagrangian had a wrong $O(4)$ symmetry relating all the KK and $K\bar{K}$ states, a certain number of exotic resonances were found. Furthermore, the widths were all too small by a large factor.

3. The σ Model

Basdevant and *Lee* [19] had the feeling that, first the good results of the $\lambda\varphi^4$ theory could remain in the σ model of *Gell-Mann* and *Levy*, secondly the bad things would be corrected, especially the s waves and the isospin zero and two near degeneracies.

Starting from a lagrangian satisfying current algebra requirements and PCAC:

$$\mathcal{L} = \tfrac{1}{2}[(\partial\mu\sigma)^2 + (\partial\mu\boldsymbol{\pi})^2] - \frac{\mu^2}{2}(\boldsymbol{\pi}^2+\sigma^2) - \frac{g}{4}(\boldsymbol{\pi}^2+\sigma^2)^2 + c\sigma$$

where $\boldsymbol{\pi}$ is the pion field and σ an isospin zero field, they compute $(\pi\pi\to\pi\pi)$ phase shifts using the [1, 1] Pade approximant. They had two parameters g and c. The second parameter c is related to f_π and the decay rate of the pion. It was chosen within 20% of the experimental value.

The results were the following.

a) The s Waves

The isospin two phase shift was found small and in agreement with experimental results, as in $\lambda(\boldsymbol{\pi}^2)^2$.

The isospin zero phase shift was found in agreement with the so called "up down" experimental results, showing a very broad σ resonance.

The scattering length ratio was:

$$\frac{a_0}{a_2} \simeq -5$$

to be compared to -3.5 obtained by Weinberg.

b) p and d Waves

They found resonances in isospin zero and one which can be identified with the ϱ and the f_0 although the masses obtained are a little too small. In the isospin two channel they found also an exotic resonance, but now splitted from the f_0 of about 250 MeV. The width of the ϱ remains too small, the width of the f_0 is better but with an inelastic $\sigma\sigma$ contribution which is too large.

What can we conclude from the results of the model? The great defect of the $\lambda\varphi^4$ model was the dominance of s wave forces, giving a degeneracy beetween different isospin amplitudes, small widths, and a wrong isospin zero s wave. Due to current algebra properties, two forces are introduced in the σ model, a $\lambda\varphi^4$ force and a σ pole force, which are both dominant s wave forces, but with weights such that they tend to each other at low energies, leaving a strong p wave force.

This is the main modification to the $\lambda\varphi^4$ model. The things which are not completely satisfactory in this model, resonances too strongly bound, small widths, an isospin two exotic resonance which remains, can perhaps be due to a defect in the convergence of the [1, 1] Pade approximant; it would be very interesting to compute the next order.

4. Pion-Pion Interaction Via Yang Mills Fields

In order to use the good features of the σ model due to the introduction of a p wave force, and also to be more efficient in the sense that a dominant feature of the pion-pion interaction, the ϱ is automatically taken in account, *Basdevant* and *Zinn-Justin* [20] have studied a model of Yang Mills field interaction between pions. The basic idea is to construct a lagrangian with pions, which is gauge invariant through local gauge transformations of a symmetry group $SU(2)$, the isospin group. This can only be obtained if there exists in the lagrangian a vector field, transforming as the regular representation of the group, and universally coupled to the conserved isospin current.

The lagrangian is as follows:

$$\begin{cases} \mathscr{L} = \frac{1}{2}(\partial\mu\boldsymbol{\pi} - g\boldsymbol{\varrho}\mu X\boldsymbol{\pi})^2 - \frac{1}{2}\mu^2\boldsymbol{\pi}^2 \\ \quad -\frac{1}{4}(\partial\mu\boldsymbol{\varrho}_\nu - \partial\nu\boldsymbol{\varrho}_\mu - g\boldsymbol{\varrho}_\mu X\boldsymbol{\varrho}_\nu)^2 + \frac{1}{2}m^2\varrho_\mu\varrho^\mu \end{cases} \tag{1}$$

where $\boldsymbol{\varrho}_\mu$ is an isospin one vector field which can be related to the physical ϱ resonance. The two first orders of the perturbation series were calculated. Although the question of the renormalisability of such a lagrangian is still open, at second order, the behavior of the amplitude is the same as for a renormalizable theory, only one substraction is necessary.

Due to the fact that only the ϱ pole appears in the $(\pi\pi\to\pi\pi)$ Born amplitude, the Born term satisfies *Adler*'s consistency condition, therefore the substraction constant was fixed by asking the second order to vanish also at Adler point $s=t=u=1$.

Therefore the model depends on two parameters g and m, which are related to the ϱ mass and width. The Pade approximant [1, 1] was computed.

The results are:

a) s Waves

The isospin two phase shift is small and negative in satisfactory agreement with experiment.

In the isospin zero, a very broad resonance was found as in the σ model. In this sense it is possible to talk of a reciprocal bootstrap of the ϱ and the σ, the existence of one as force in the lagrangien implies the existence of the second.

The scattering lengths are very close to those of *Weinberg* with a ratio:

$$\frac{a_0}{a_3} = -3. \tag{2}$$

This can be easily understood; if one computes the Born term near $s=1, u=1, t=0$ where one applies current algebra constraints. The compatibility can be written as:

$$\frac{g^2 f_\pi^2}{m^2} = \frac{1}{3}, \tag{3}$$

which can be compared with the celebrated KSFR relation which reads

$$\frac{g^2 f_\pi^2}{m^2} = \frac{1}{2}. \tag{4}$$

The relation (3) is in very good agreement with an f_π deduced from G_A/G_V. The difference between (3) and (4) is due to the fact that the first

g is really related to the ϱ form factor at $s=m^2$, and the second to ϱ form factor at $s=0$.

b) *p Wave*

This wave is elementary, and the ϱ phase shift can be fitted without difficulties.

d) *d* Wave

Simply looking at the isospin crossing matrix shows that the ϱ force is attractive in isospin zero and one and repulsive in isospin two. Therefore no exotic states can be found. For the isospin zero a f_0 resonance is found very near the physical mass 1250 MeV, but with a little too broad width 220 to 260 MeV.

5. *Concluding Remarks on Meson-Meson Scattering*

It seems that the physics of meson-meson scattering can be well studied and understood with the use of the Pade approximation. We want to outline two points:

a) s and p consistent phase shifts have been obtained and the relation between them has been shown.

b) the 2^+ resonance has been found with no arbitrary parameters with a good physical mass and reasonable width. This was obtained in no other model, as far as I know.

What remains to do is: In the two last models to compute higher orders, in order to obtain a better approximation. To extend both models to the (π, K, η) system, for the Yang-Mills model, calculation by *D. Iagolnitzer et al.* is underway [21].

Also the application of the Yang-Mills interaction to the $\pi\varrho$ system can be of interest.

C. Pion Nucleon System

1. *Mignaco et al.* [22] and *Gammel* and *Kubis* [23] have studied pion nucleon in the γ_5 interaction:

$$\mathscr{L}_{\text{int}} = -ig\bar{N}\gamma_5\tau^\alpha N\pi^\alpha - \frac{\lambda}{4}(\pi^\alpha\pi^\alpha)^2 .$$

They compute the perturbation series in the one loop approximation and have constructed the [1, 1] Pade approximant.

It must be said that when the lagrangian contains many coupling constants, the good classification of the graphs in order to compute Pade approximants, seems to be a classification according to the number of loops. In the case where symmetries play a role, this is necessary in order to conserve the symmetry properties.

The results of the γ_5 interaction are: with the two parameters, a certain number of isospin 3/2 phase shifts can be obtained in reasonable qualitative agreement with experiment.

All the isospin 1/2 phase shifts are completely wrong.

The results show that the interaction which has been considered, cannot really explain the pion nucleon interaction. This was well known for the s waves scattering lengths, which are in the contrary very well given by current algebra.

2. The σ Model

Therefore it seems to be necessary to study a model including current algebra constraints. This has been recently done by *Mignaco* and *Remiddi* [24] using also the lagrangian of *Gell-Mann* and *Levy* with pions and nucleons.

The lagrangian is:

$$\mathscr{L} = \bar{N}[i\gamma_\mu \partial^\mu - g(\sigma + i\boldsymbol{\pi}\boldsymbol{\tau}\gamma_5)]N + \tfrac{1}{2}[(\partial\mu\boldsymbol{\pi})^2 + (\partial\mu\sigma)^2]$$
$$-\tfrac{1}{2}\mu[\boldsymbol{\pi}^2 + \sigma^2] - \tfrac{1}{4}\lambda[\boldsymbol{\pi}^2 + \sigma^2]^2 + c\sigma .$$

The results are better than in the γ_5 interaction. S_{31} and P_{33} phase shifts are remarkably good, P_{11} is resonating.

But what can seem surprising is that the scattering length in the S_{11} state cannot be reproduced for values of the parameters which agree with the Basdevant – Lee calculation and also which gives a good P_{33}, although the lagrangian satisfies current algebra constraints.

The reason seems to be a very bad convergence of the [1, 1] Pade approximant, due to a very large coupling constant, and also to the large renormalisation of the σ mass which from the real axis in the Born term goes very far in the second sheet in the solution.

In this model it is particularly crucial to compute higher order.

D. Nucleon Interaction

The first work has been done by *Bessis et al.* [25] in the simple γ_5 interaction. They compute the two first orders of the perturbation series and construct the Pade [1, 1]. In this model no free parameters appear, the coupling constant being known.

A complete description of low energy NN scattering is obtained which is in qualitative and sometimes even in quantitative agreement with experimental data. A deuteron pole is found with 5 MeV binding energy. The most important defect of the model is the $^1S_0$ wave, which has a completly wrong behaviour, due to the fact that the Born term in the s wave has the behaviour of a p wave. Even for $^3S_1$ the agreement with experimental data is not very good.

Therefore *Bessis* and *Turchetti* [26] have very recently begun to study the σ model, with also a new method of renormalization due to *Symanzik*.

They have computed the nucleon nucleon s wave scattering lengths and have found that they could fit the experimal results with one free parameter. The value of this parameter, the bare σ mass, was found to agree with that used by *Basdevant* and *Lee* in pion-pion interaction

This result is very encouraging and seems to indicate that it is possible to obtain a good quantitative description of nucleon-nucleon scattering at low energy from field theory and with only one parameter.

References

1. *Wall, H. S.*: Continued Fractions. New York: Van Nostrand 1948.
2. *Szegö*: Orthogonal Polynomials.
3. *Basdevant, J. L., Bessis, D., Zinn-Justin, J.*: Nuovo Cimento **60** A, 185 (1969) Mathematical Appendices.
4. *Baker, G. A.*: J. Adv. Theoret. Phys. **1**, 1 (1965).
5. *Nuttall, J.*: Preprint – Texas A & M University.
6. *Zinn-Justin, J.*: Thesis Paris (1968).
7. *Chisholm, R.*: J. Math. Phys. **4**, 1506 (1963).
8. *Garibotti, C. R., Villani, M.*: Nuovo Cimento **59** A, 107 (1969) and **63** A, 1267 (1969).
9. *Loeffel, J. J., Martin, A., Simon, B., Wightman, A. S.*: Phys. Letters **30** B, 656 (1969).
10. *Bessis, D., Pusterla, M.*: Nuovo Cimento **54** A, 243 (1968).
11. *Copley, L. A., Masson, D.*: Phys. Rev. **164**, 2059 (1967).
12. *Cini, M., Fubini, S.*: Nuovo Cimento **11**, 142 (1954).
13. *Basdevant, J. L., Bessis, D., Zinn-Justin, J.*: Nuovo Cimento **60** A, 185 (1969).
14. — *Cohen-Tannoudji, G., Morel, A.*: Nuovo Cimento **64** A, 585 (1969).
15. *Gammel, J. L., Mac Donald, F. A.*: Phys. Rev. **142**, 1145 (1966).
16. *Basdevant, J. L., Lee, B. W.*: Nucl. Phys. B **13**, 182 (1969).
17. *Caser, S., Piquet, C., Vermeulen, J. L.*: Nucl. Phys. B **14**, 119 (1969).
18. *Alexanian, M., Wellner, M.*: Phys. Rev. **137** B, 155 (1965).
19. *Basdevant, J. L., Lee, B. W.*: To be published in Phys. Rev.
20. — *Zinn-Justin, J.*: To be published in Phys. Rev.
21. *Iagolnitzer, D., Zinn-Justin, J., Zuber, J.*: in preparation.
22. *Mignaco, J. A., Pusterla, M., Remiddi, E.*: Nuovo Cimento **64** A, 733 (1969).
23. *Gammel, J. L., Mentzel, M. T., Kubis, J. J.*: Los Angeles preprint.
24. *Mignaco, J. A., Remiddi, E.*: Preprint
25. *Bessis, D., Graffi, S., Grecchi, V., Turchetti, G.*: Phys. Letters **28** B, 567 (1969).
26. — *Turchetti, G.*: To be submitted to Phys. Rev. Letters.
27. Other references on the subject.
    Badevant, J. L.: Pade Approximants. Published in: Methods in Subnuclear Physics, Vol. IV (M. Nikolic, Ed.). London: Gordon Breach 1969.

Masson, D.: J. Math. Phys. **8**, 511 (1967).
Nieland, H. M., Tjon, J. A.: Phys. Letters **27** B, 5 (1968).
Baker, G., Chrisholm, R.: J. Math. Phys. **7**, 1900 (1966).
Copley, L. A., Elias, D. K., Masson, D.: Broken Symmetry model of meson interaction (preprint).
Wortman, N. R.: Phys. Rev. **176**, 1762 (1968).

Professor Dr. *J. Zinn-Justin*
Centre d' Etudes Nucléaires de Saclay
F-91, Gif-sur-Yvette

SPRINGER TRACTS IN MODERN PHYSICS

Ergebnisse
der exakten Naturwissenschaften

Volume 57

Editor: G. Höhler

Reprint:

D. Atkinson

Some Consequences of Unitarity and Crossing
Existence and Asymptotic Theorems

Springer-Verlag Berlin Heidelberg GmbH 1971

SPRINGER TRACTS IN MODERN PHYSICS

Ergebnisse
der exakten Natur-
wissenschaften

Volume 57

Reprint:

G. Wanders
Analyticity, Unitarity and Crossing Symmetry Constraints for Pion-Pion Partial Wave Amplitudes

Springer-Verlag Berlin Heidelberg GmbH 1971

SPRINGER TRACTS IN MODERN PHYSICS

Ergebnisse der exakten Naturwissenschaften

Volume 57

Editor: G. Höhler

Reprint:

J. Hamilton
New Methods in the Analysis of π-N Scattering

Springer-Verlag Berlin Heidelberg GmbH 1971

SPRINGER TRACTS IN MODERN PHYSICS

Ergebnisse der exakten Naturwissenschaften

Volume 57

Reprint:

E. J. Squires
Regge-Pole Phenomenology

Springer-Verlag Berlin Heidelberg GmbH 1971

SPRINGER TRACTS IN MODERN PHYSICS

Ergebnisse der exakten Naturwissenschaften

Volume 57

Reprint:

A. P. Contogouris

Certain Problems of Two-Body Reactions with Spin

Springer-Verlag Berlin Heidelberg GmbH 1971

SPRINGER TRACTS IN MODERN PHYSICS

Ergebnisse
der exakten Natur-
wissenschaften

Volume 57

Reprint:

R. Oehme
Duality and Regge Theory

Springer-Verlag Berlin Heidelberg GmbH 1971

SPRINGER TRACTS IN MODERN PHYSICS

Ergebnisse
der exakten Natur-
wissenschaften

Volume 57

Reprint:

R. Oehme
Complex Angular Momentum

Springer-Verlag Berlin Heidelberg GmbH 1971

SPRINGER TRACTS IN MODERN PHYSICS

Ergebnisse der exakten Naturwissenschaften

Volume 57

Reprint:

H. Satz
An Introduction to Dual Resonance Models in Multiparticle Physics

Springer-Verlag Berlin Heidelberg GmbH 1971

SPRINGER TRACTS IN MODERN PHYSICS

Ergebnisse der exakten Naturwissenschaften

Volume 57

Reprint:

H. R. Rubinstein
Physical *N*-Pion Functions

Springer-Verlag Berlin Heidelberg GmbH 1971

SPRINGER TRACTS IN MODERN PHYSICS

Ergebnisse
der exakten Natur-
wissenschaften

Volume 57

Reprint:

W. Rühl
Application of Harmonic Analysis to Inelastic Electron-Proton Scattering

Springer-Verlag Berlin Heidelberg GmbH 1971

SPRINGER TRACTS IN MODERN PHYSICS

Ergebnisse der exakten Naturwissenschaften

Volume 57

Reprint:

K. Symanzik
Small-Distance Behaviour in Field Theory

Springer-Verlag Berlin Heidelberg GmbH 1971

SPRINGER TRACTS IN MODERN PHYSICS

Ergebnisse
der exakten Natur-
wissenschaften

Volume 57

Reprint:

R. A. Brandt
Physics on the Light Cone

Springer-Verlag Berlin Heidelberg GmbH 1971

SPRINGER TRACTS IN MODERN PHYSICS

Ergebnisse
der exakten Natur-
wissenschaften

Volume 57

Reprint:

J. Zinn-Justin
Course on Padé Approximants

Springer-Verlag Berlin Heidelberg GmbH 1971